# GEOLOGY OF THE ICE AGE NATIONAL SCENIC TRAIL

Publication of this volume has been made possible, in part, through support from alumni, friends, and the Board of Visitors of the Department of Geoscience at the University of Wisconsin–Madison; and from the Ice Age Trail Alliance, through federal financial assistance from the National Park Service. Unless otherwise indicated in the text of this document, the views and conclusions in this document are those of the authors and should not be interpreted as representing the opinions or policies of the National Park Service or U.S. Government.

# Geology of the
# Ice Age National Scenic Trail

David M. Mickelson,
Louis J. Maher Jr.,
*and*
Susan L. Simpson

THE UNIVERSITY OF WISCONSIN PRESS

The University of Wisconsin Press
1930 Monroe Street, 3rd Floor
Madison, Wisconsin 53711-2059
uwpress.wisc.edu

3 Henrietta Street
London WC2E 8LU, United Kingdom
eurospanbookstore.com

Printed in the United States of America

Library of Congress Cataloging-in-Publication Data
Mickelson, David M.
Geology of the Ice Age National Scenic Trail / David M. Mickelson, Louis J. Maher Jr.,
and Susan L. Simpson.
p.   cm.
Includes bibliographical references and index.
ISBN 978-0-299-28484-8 (pbk.: alk. paper) — ISBN 978-0-299-28483-1 (e-book)
1. Geology—Wisconsin—Ice Age National Scenic Trail. 2. Geology—Wisconsin—Ice Age
National Scientific Reserve. 3. Glacial landforms—Wisconsin—Ice Age National Scenic Trail.
4. Glacial landforms—Wisconsin—Ice Age National Scientific Reserve. 5. Ice Age National Scenic
Trail (Wis.)—Guidebooks. 6. Ice Age National Scientific Reserve (Wis.)—Guidebooks. I. Maher,
Louis J. (Louis James), 1933– II. Simpson, Susan L. III. Title.
QE697.M565   2011
557.75—dc22
2011013638

# CONTENTS

# ACKNOWLEDGMENTS

THIS BOOK WOULD NOT HAVE BEEN POSSIBLE without the help of many, many people, all of whom offered encouragement! Below we try to recognize that assistance, but there may be others who contributed whom we failed to list. To those we apologize. Full-color printing of the book was made possible in part by federal financial assistance from the National Park Service (NPS) and by the Board of Visitors and other alumni and friends of the Department of Geoscience (formerly Geology and Geophysics) at the University of Wisconsin–Madison. Our thanks go to many former students who contributed!

Contributions were made by many who are directly involved with the Ice Age Trail and its development across the state. Drew Hanson, formerly of the Ice Age Trail Alliance, and Eric Sherman of the Ice Age Trail Alliance provided a detailed review of the many maps and pieces of the manuscript, and reviews of individual segments and sections were provided by Bernie Brouchard, Marlin Johnson, David Lonsdorf, Russ Helwig, Gary Klatt, Fred Nash, Jean Clark, Joe Jopek, Don Erickson, Russ Werner, Herb Schotz, Ken Lange, Gerald "Buzz" Meyer, Patti Herman, Tammy Malchow, Tim McRaith, and Bob and Sally Freckmann.

Many people contributed photos to enhance the text. They are credited in the photo captions. They include Alan Mercer, Brad Singer, Bruce Jaecks, Diann Kiesel, Donna Harris, Gail Ashley, John Chapman, Vin Mickelson, Kent Syverson, Gene LaBerge, Magnus T. Gudmundsson, Richard Goldthwait (deceased), Richard Becker, and Thomas Gustavson. If the photos are not credited, they were taken by the authors. The low-level oblique photos were enabled by pilots Alan Carroll and Ben Abernathy.

Professor William Mode at the University of Wisconsin–Oshkosh and Professor Kent Syverson at the University of Wisconsin–Eau Claire provided detailed reviews of the entire manuscript. We appreciate all of their helpful comments and we exempt them from any responsibility for misstatements that we have made. Randy Maas, Bruce Brown, and Gene LaBerge provided information on Precambrian geology. Other valued colleagues who provided advice are Professors John Attig and Lee Clayton at the Wisconsin Geological and Natural History Survey, James Knox at the University of Wisconsin–Madison, and Tom Hooyer at the University of Wisconsin–Milwaukee. Professor Adam Cahow (retired) from the University of Wisconsin–Eau Claire provided Dave Mickelson with early inspiration to study the geology of the Ice Age Trail. Gary Werner and Mike Wollmer, along with the staff at the Ice Age Trail Alliance, continue to be important advocates for the Ice Age Trail and were enthusiastic supporters of this book. We appreciate the enthusiasm, guidance, and attention to detail of the staff at the University of Wisconsin Press including Sheila Leary, Gwen Walker, Terry Emmrich, Carla Marolt, and Sheila McMahon. We could not have reached publication without them!

The book contains many maps and diagrams that would not have been possible without help. Mike Czechanski and Pete Schoephoester created nearly all of the shaded relief maps. Deb Patterson assisted with many figures, and Carol McCartney and Linda Deith also helped with figures. All five are with the Wisconsin Geological and Natural History Survey. Michael Bricknell of the Wisconsin State Cartographer's Office also assisted with maps. Mary Diman of the Department of Geoscience at the University of Wisconsin–Madison was a major help with much of the artwork. All of the topographic maps were downloaded from the National Geographic Society's TOPO! program. The Ice Age Trail was then plotted on the maps using GIS files provided by Tiffany Stram and Drew Hanson of the Ice Age Trail Alliance. Tiffany Stram and Sharon Dziengel were especially helpful during revision of the manuscript, providing information about updated trail segments. We are also very grateful for helpful editing by Gail Schmitt and the staff at the University of Wisconsin Press. The authors also wish to thank the NPS Ice Age National Scenic Trail staff, and the NPS Harpers Ferry Center staff for permission to use several illustrations in this book. Moral, financial, and GIS support were provided by Tom Gilbert, Pam Schuler, and Doug Wilder of the NPS. Helpful suggestions were also provided by Juliann Fox, Lois Hanson, Brenda Rederer, and Jackie Scharfenberg—all with the Wisconsin Department of Natural Resources. Helpful suggestions and the patience and understanding of William Ehrenclou and Jane Maher are also greatly appreciated.

On most travels around Wisconsin and hikes on the Ice Age Trail, Dave was assisted by his wife, Vin Mickelson. Many thanks for being a great help, for providing encouragement and suggestions, and for putting up with his long hours spent putting this book together!

Finally, this book is dedicated to all the volunteers who have helped, and continue to help, make the Ice Age Trail a reality.

# GEOLOGY OF THE ICE AGE NATIONAL SCENIC TRAIL

# Introduction

A HIKE ALONG THE ICE AGE TRAIL (IAT) can be a leaf-dappled, peaceful walk in the summer or a snow-shrouded, silent reverie in winter. The landscape you walk eases you into a steady rhythm as Wisconsin's rich cropland rises to ridges, ridges circle deep lakes, and lakes feed wide wetlands bobbing with migratory birds.

But it was not always this serene. Whether you choose to hike a single segment or the entire length of the IAT, each footfall takes you over layers of often violent geologic history. If you know how to read the landscape, the Ice Age Trail can tell the story of how Wisconsin came to be. It is a story of mountain ranges and erupting volcanoes, massive glaciers, and torrents of rushing water bursting through ice and rock dams to create huge floods. Even today the story continues: your hike is but a moment in an ongoing geologic sculpting of the landscape. This book explains the geologic forces that shaped this land and will allow you to better read history in the ridges, rivers, and rocks you find along your way.

The story of the Ice Age Trail truly begins with the formation of the earth billions of years ago, but for the majority of this book we will talk about the land's history since the Laurentide Ice Sheet was here thousands of years ago.

Ice sheets are glaciers: often more than a mile thick and thousands of miles wide, they move constantly, but very, very slowly, across the land. The Laurentide Ice Sheet advanced into Wisconsin from Canada between 25,000 and 30,000 years ago. The glacier ice flowed down lowlands, forming lobes of ice; you can see these lobes as well as their names on figure 1. For thousands of years the ice sat over about two-thirds of the state without advancing or retreating great distances across the landscape. However, change was about to come: by 20,000 years ago the climate warmed and the massive ice sheet began to retreat. As it did, it left behind a dramatic set of landforms—drumlins, eskers, moulin kames, kettle lakes, tunnel channels, braided river channels, and plains of glacial till—that now define about two-thirds of Wisconsin's landscape.

As people moved into what is now Wisconsin, they shaped their lives around these glaciers and the landforms they left behind. Mammoths and mastodons roamed the slowly vegetating landscape, and people explored the new rivers and plains to fish and hunt. Where once the face of a melting glacier sent rivulets of sediment-laden water through a valley or

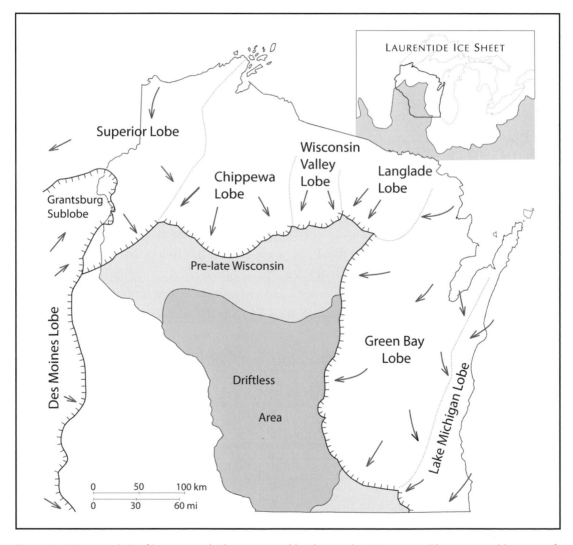

Figure 1. Wisconsin's Driftless Area, which was covered by the pre–late Wisconsin Glaciation, and location of ice lobes of the late Wisconsin Glaciation. Arrows indicate flow direction of the lobes of the glacier. Inset map: location of Wisconsin at the southern edge of the Laurentide Ice Sheet during the late Wisconsin Glaciation. (From Wisconsin Geological and Natural History Survey [WGNHS].)

across broad plains, farmers eventually recognized rich soils and planted their crops. Where glaciers had dropped rocks onto ridges, settlers saw good perches overlooking the thousands of glacier-made kettles.

As use of the land intensified, some people saw changes at the landscape level that threatened the very glacial formations that had shaped Wisconsin's natural and cultural identity. Gravel mining, highways, and especially public and private development were erasing signs of the glacier's action. To protect these landforms, concerned citizens in the late 1950s sought legislation that would establish nine Ice Age National Scientific Reserve areas. The purpose was to "assure protection, preservation, and interpretation of the nationally significant values of Wisconsin continental glaciation" (Reuss 1990). The bill was passed into law in 1964. At the same time, a nearly 1,200-mile IAT was planned to allow hikers to enjoy and appreciate Wisconsin's beautiful Ice Age landscape.

Today the Wisconsin DNR and the National Park Service jointly administer the reserve areas and the Ice Age National Scenic Trail, which was designated part of the National Trails System in 1980. At the time of writing, over half of the proposed trail has been built, and there are active chapters of the Ice Age Trail Alliance in all the counties through which the trail passes (fig. 2). You will find the nine Ice Age Reserve sites (in black) and the more than 120 IAT segments grouped geographically in figure 3.

The next section of this introduction provides an overview of the Ice Age Trail, followed by a short summary of the geology of Wisconsin. The following part of the book comprises twenty-two Science Briefs (referenced as SB #), which provide more detail about geologic processes, materials, and landforms that you'll encounter in the trail-segment descriptions. Because most hikers will not hike the segments in any particular order, we refer to the Science Briefs the first time a term is used in each segment description. For the through hiker, this may lead to some repetition; but for the occasional hiker this approach should enable a quick understanding of the formation of the feature on any given segment.

## Notes on the Ice Age Trail Segments

Much of the Ice Age Trail (IAT) follows the outermost end moraine (SB 6) of the last major glacial advance, the late Wisconsin Glaciation (figs. 2, 3). It occurred between about 30,000 and 10,000 cal. years ago (see SB 18 for a complete description of calendar years [cal. years] versus radiocarbon years), and the outermost moraine was mostly formed between about 26,000 and 20,000 cal. years ago. When the glacier came into Wisconsin it formed lobes that behaved somewhat independently of each other (SB 2). The curvature of the outermost moraine reflects these lobes, and this book is organized into chapters based on these lobes and their end moraines. An exception to this is in eastern Wisconsin, where the Green Bay Lobe abutted the Lake Michigan Lobe. Here an interlobate zone called the Kettle Moraine

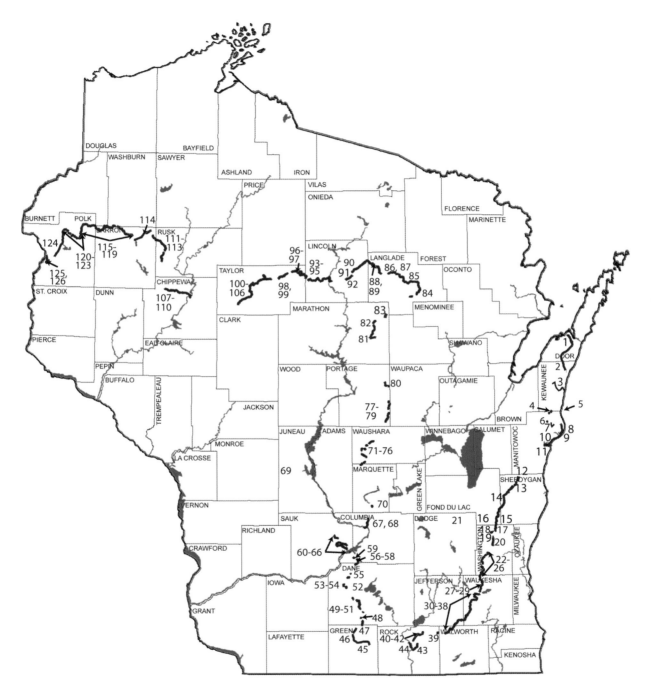

Figure 2. Individual trail segments of the Ice Age Trail (red). Numbers on the map correspond to the segment numbers used in the table of contents. Wisconsin counties are labeled. The Ice Age Trail Alliance has a chapter in each county through which the trail passes.

Figure 3. Moraines (bright green) in Wisconsin relative to the location of the Ice Age Trail (red), the groupings of IAT segments used in this book (black labels), and the Ice Age National Scientific Reserve units (black numbers): (1) Two Creeks Buried Forest, (2) Kettle Moraine State Forest–Northern Unit, (3) Campbellsport Drumlins, (4) Horicon Marsh, (5) Cross Plains, (6) Devil's Lake, (7) Mill Bluff, (8) Chippewa Moraine, (9) Interstate State Park. (Map based on Farrand et al. 1983, Goebel et al. 1984, Lineback et al. 1984, Hallberg et al. 1991.)

forms a ridge complex that is one of Wisconsin's largest and most striking glacial features. It is actually a collection of landforms that vary considerably along its 130-mile length between Walworth County in the south and Manitowoc County in the north. Finally, short segments of the IAT in Manitowoc, Kewaunee, and Door counties do not follow a moraine, but instead run along part of the Niagara Escarpment (SB 19) and a variety of other glacial landscapes.

Each of the trail chapters begins with a discussion of the regional geology and glacial history of that area's segments, and a list of additional readings for each chapter is included at the back of the book. We have intentionally avoided discussing plants, scenic views, hiking experiences, etc. because space is limited and these topics are covered in other guides. Included are regional maps based on a digital elevation model with shaded relief. Major highways are shown, as are symbols for enough villages and cities to allow orientation. Cultural features are kept to a minimum to provide the best possible view of the landscape. The best way to orient yourself is to compare these regional maps to a state highway map. Introductions to the trail chapters are followed by individual IAT segment descriptions. Descriptions of the Ice Age National Scientific Reserve units (fig. 3) are interspersed with the trail segment descriptions where appropriate.

The trail segments used here are the same, and presented in the same east to west order, as those in the *Ice Age Trail Companion Guide 2011*. Our book concentrates on the geology along the trail and typically does not give information about parking at the trailheads, the condition of trails, or the services available in the vicinity. Thus, this geology guide should not be used by itself as a trail guide. Instead, it should be used in conjunction with the *Ice Age Trail Companion Guide 2011* and the *Ice Age Trail Atlas*, both of which are available from the Ice Age Trail Alliance (http://www.iceagetrail.org/) and many bookstores. The trail routes change over the years, so be sure to obtain the most recent information from the website about the segments you plan to hike.

A few other points will help with the use of this guidebook.

Even though the trail changes orientation over short distances, we mostly use east and west to describe the direction along the trail segments. This direction refers either toward the eastern terminus in Sturgeon Bay or toward the western terminus, which is at Interstate State Park, in St. Croix Falls. We use actual compass directions to locate geological features. Be sure to bring a compass or GPS unit with you!

On the maps, the trail and related features are always shown in red. There are small arrows at the ends of the mapped trail segments on the topographic maps. If they point into the segment, it means that the trail does not directly connect to another segment. If an arrow points away from the segment being described, then that end connects directly to another segment. The geologic features along each segment of the IAT and in each unit of the Ice

Age Reserve are labeled on topographic maps. In general, we did not map wetlands, but the symbol on the topographic map indicates their presence. We use the abbreviations USH for U.S. highways, STH for state highways, and CTH for county roads.

If you need a refresher on how to interpret topographic maps, take some time to read SB 22. Topographic, or topo, maps are an invaluable tool for navigating on your hike. All topographic maps used in the segment descriptions are downloaded from TOPO! software by National Geographic. Unless otherwise noted, they are all from 1:24,000 scale quadrangles. Because of page-size limitations in this book, you will not be able to read all of the detail that is on the original, nor will you see surrounding areas that help in navigation. You'll see the name of the topographic map from which each segment map was made so you can purchase the full-size version if necessary.

## Notes on Safety and Ice Age Trail Access

No matter the length of your hike, you should be trail-smart about the things you bring along. This includes wearing comfortable shoes and having a daypack with a flask of water, high-energy foods, a first-aid kit, a raincoat, a pocketknife, a pencil and notebook, and a compass or GPS unit, among other things. Keep in mind: this book is meant to be a field guide to the geology of the trail, not a navigation tool. In many cases, reprinting the segment maps reduces the level of detail needed for driving and hiking directions. Finally, respect private land. Some of the features we describe are close to the trail, but on private land. View them from the trail or get permission if you want to get a closer look. Be especially careful of gravel pits and quarries even if you have permission to enter. The walls are usually unstable and can collapse suddenly! Beside using this book's maps to look for the highlighted geological features, always consult an area map. You will also want to look at regional maps for location of restrooms or legal parking areas. Parking is not available at all of the trailheads; some require that you be dropped off or use a car shuttle system.

As mentioned earlier, two publications from the Ice Age Trail Foundation provide information on access and services available near the IAT and are excellent companions to this volume: *Ice Age Trail Companion Guide 2011* and *Ice Age Trail Atlas*.

For many more photographic views along the trail, consider the book by Smith et al. (2008) listed in the suggestions for further reading at the back of the book. For more details about the geology of Wisconsin that is not on the IAT, consider Dott and Attig's book (2004). Other general books available for purchase are listed at the end of the book.

The Wisconsin Geological and Natural History Survey (WGNHS) is the main repository of geologic information and topographic maps in the state (http://www.uwex.edu/wgnhs/index.html).

Happy hiking!

## An Overview of Wisconsin Geology

Geology, the study of the earth, is both useful and fun. People have used the materials of the earth for many purposes: as tools, building materials, and fuel, and even for their pure beauty. For as long as we know, people have used pieces of flint and steel to start fires. The early Egyptians and Mayans knew where to find the proper stone to construct their giant pyramids, and the Romans used stone to build their extensive roads. Early archaeological records show that for thousands of years people have sought and guarded gold, silver, and gems not only for trading, but sometimes for purely aesthetic purposes that had little to do with their practical or monetary value.

The modern study of geology came to the United States from Europe. In both places, it first developed for very practical purposes: geologists often concentrated on materials used for industry and profit. Coal, for example, was used to smelt iron and other metal ores and fueled the Industrial Revolution and the railroads that soon stretched from the Pacific to the Atlantic. As cities grew, stone and metals formed the skeletons of the roads, walls, bridges, and buildings that in many cases still stand today. As the discipline of geology matured, geologists expanded their scope, seeking answers not only to how we could use the earth's resources, but also to how geologic processes had formed the land around us.

Wisconsin lies in the midst of the North American continent. It has no tall mountains, volcanoes, ocean shores, or glaciers today, but all of these existed at various times in the state's past. We can read the history of Wisconsin in its rocks and soils, both of which bear witness to the eons of rifts, lava flows, flooding, erosion, glaciation, sedimentation, and other events that shaped the land.

There are three major types of rocks, which are described in more detail in SB 21: igneous, sedimentary, and metamorphic. Igneous (fire-formed) rocks originate when melted silicate-rich magma cools and mineral crystals start to form. Igneous rocks can form deep in the earth in structures called batholiths. It takes a long time for large masses of magma to cool, and while they do, large mineral crystals can grow. Igneous rocks can also occur as planar (flat) structures that form when molten rock forces its way into cracks in existing rocks. There the liquid cools quickly, forming small crystals in sheetlike structures. Sheets that cut across the structure of the rocks are called dikes; sheets that parallel the structure are called sills. And of course, magma can also erupt at the surface and cool quickly to produce very small crystals that make up the igneous rock that most people envision when they hear about these fire-formed rocks.

One of the useful things about igneous rocks is that they can tell us when they formed. Magmas contain radioactive elements that tend to concentrate in specific types of minerals as the magmas cool and crystallize. These radioactive elements can be thought of as "parent" elements that are unstable and decay over time into stable "daughter" elements. The rate of decay of a radioactive element is constant; it is not subject to external conditions, such as heat or pressure. If we know the rate of decay for the elements in a rock, we can determine when it formed based on the ratio of daughter to parent elements left in the rock. Once we know the age of an igneous rock, we also know that any rock it penetrates must be older, because it had to have existed when the igneous rock formed and pushed its way through. By using these and other techniques, we can deduce the ages of most of Wisconsin's igneous rocks.

Sedimentary rock is another common type of rock in the state. These rocks are formed of layers of debris—sediment—that settled onto a surface as preexisting rock weathered. Sedimentary rocks may consist of large chunks of rock or of tiny grains of sand, silt, and clay. Others are made up almost entirely of precipitates, either chemical (which came directly from water) or biochemical (which were secreted by organisms). Sedimentary rocks may contain fossils of living organisms that became part of the rock layer and hence provide us with information about the history of life on earth.

The third type of rock is metamorphic, which means "change of form." Such rocks result when igneous, sedimentary, or even other types of metamorphic rocks are subjected to increased heat and pressure and undergo recrystallization. During this process, some minerals in the rock disappear, while others continue to grow into larger and larger crystals.

As you drive and hike around the state, you will see some rock outcrops, but more often the landscape changes from thick forests to waving prairies or rich farmland with little sign of exposed rocks. They are inconspicuous because they are covered by a layer of sediment or soil formed by the weathering and erosion of the rocks directly beneath the surface or carried in from afar by streams, glaciers, and wind.

Rocks in Wisconsin differ substantially in age. Those of the Precambrian Era (table 1), which lasted from the beginning of the rock record to about 550 million years ago, are almost all either metamorphic or igneous. All of the younger rocks are sedimentary. Note that the record is very incomplete. Hundreds of millions of years have no record remaining at all! Gaps in the rock record are not that rare: they result from long periods of erosion, when rock wears away, or of nondeposition, when no rock forms at all, and are called unconformities.

## PRECAMBRIAN WISCONSIN

The geologic history of Wisconsin is closely related to that of its neighboring states, Michigan and Minnesota, as well as to that of the Canadian province of Ontario. Several books listed as companion books to this introduction provide more details about the geologic history of the region. Precambrian rocks are the earliest rock record in Wisconsin (table 1), and they underlie much of the northern part of the state (fig. 4). The earliest rocks resulted from volcanic activity about 3,500 million (3.5 billion) years ago in what geologists call the

| ERA | Period | | Age at base | Events |
|---|---|---|---|---|
| CENOZOIC | Quaternary | Holocene | 11,700 yr | Glaciers continue to retreat, sea level rises, human activity increases, mammoths and mastodons become extinct. |
| | | Pleistocene | 2.6 my | Multiple glacial and interglacial periods culminating in the Late Wisconsin Glaciation, fluctuating sea level, large mammals. |
| | Tertiary | Pliocene | 5.3 my | |
| | | Miocene | 23 my | |
| | | Oligocene | 33.9 my | All evidence eroded from what is now Wisconsin. Cooling climate, age of mammals. |
| | | Eocene | 55.8 my | |
| | | Paleocene | 65.5 my | |
| MESOZOIC | Cretaceous | | 145.5 my | Rivers flowed across Wisconsin to a sea in what is now the Great Plains. Most evidence has been eroded. |
| | Jurassic | | 199.6 my | All evidence eroded from what is now Wisconsin. Age of dinosaurs. |
| | Triassic | | 251 my | |
| PALEOZOIC | Permian | | 299 my | |
| | Pennsylvanian | | 318.1 my | All evidence eroded from what is now Wisconsin. |
| | Mississippian | | 359.2 my | |
| | Devonian | | 416 my | Most of Wisconsin was beneath the sea. Most rocks of this age have now been eroded away. |
| | Silurian | | 443.7 my | Erosion Deposition of dolomite in shallow sea over much of area. Niagara dolomite is of this age. First true coral reefs. |
| | Ordovician | | 488.3 my | Erosion Deposition of dolomite, sandstone, then dolomite in shallow sea separated by periods of erosion. Brachiopod fossils common. |
| | Cambrian | | 542 my | Erosion Deposition of sandstone with some dolomite in shallow sea separated by periods of erosion. Trilobite fossils common. |
| PRECAMBRIAN | Proterozoic | | 2,500 my | Deposition of banded iron formations, Penokean Mountains formed, intrusion of granites and volcanic activity. Baraboo and Blue Hills Quarzites deposited, buried, folded. No fossils in state. |
| | Archean | | 4,000 my | Early continental plates collide, mountain building, then erosion. |
| | Hadean | | 4,600 my | Formation of the earth. |

Table 1. Geologic column showing eras, periods, and age of the beginning of each in millions of years (my) or years (yr). (Based on the International Stratigraphic Chart, 2009, International Commission on Stratigraphy.)

Archean Period (table 1), during which our continents first developed. Many of these rocks formed at the earth's surface then were deeply buried and intensively deformed and metamorphosed to create metamorphic rocks called gneiss (pronounced "nice") and greenstone. These very contorted rocks probably were metamorphosed at depths greater than 15 miles, where heat and pressure made the rocks soft and pliable like putty. Some of the rock partly melted, resulting in light-colored bands of pale minerals layered with bands of darker minerals, forming gneiss. These Archean rocks became the core of what is called the Superior Continent by about 2,500 million years ago, at the end of the Archean. There are no Archean rocks exposed along the IAT.

Throughout much of the late Archean and early Proterozoic, Wisconsin alternated between being above sea level, when winds whipped the landscape and caused intense erosion, and being below sea level, when sediment accumulated in thick layers. Proterozoic means "primitive life," and it is in the early Proterozoic marine sediments that geologists find the first evidence of life on earth.

About 2,000 million years ago, early in the Proterozoic, atmospheric oxygen levels had risen enough that exposed iron oxidized (turned rusty in the air), and for the first time, large amounts of iron-rich sediment interlayered with silica began to settle to the bottoms of seas around the world. The rock that formed has distinctive orange and gray banding (fig. 5). Miners have tapped this banded iron formation for iron ore in Minnesota, Michigan, and Wisconsin. In Wisconsin, these rocks extend from northeastern Iron County southwest to northwestern Sawyer County (fig. 2), so they are north of the IAT. Although you won't see this bedrock exposed along the IAT, you may find pieces of banded iron formation along the IAT that glaciers carried and deposited across the landscape. In addition to the distinctive banding, you can identify these rocks by their weight: they are heftier than other rocks of a similar size because of their iron content.

Shortly after the banded iron formation accumulated on the ocean bottoms, volcanoes again developed in central Wisconsin as the Superior Plate collided with another continental plate. The result: the Penokean Mountains. This high mountain range was oriented more or less east-west across what are now Minnesota, Wisconsin, and Michigan. Granites that intruded during this continental collision suggest these mountains formed anywhere between 1,900 to 1,800 million years ago. The roots of these ancient mountains cover a broad area from Michigan across Wisconsin at least as far west as Minnesota and as far east and north as Ontario, Canada. Much of the IAT in northern Wisconsin runs across rocks formed during the Penokean Mountain building. Rocks exposed along the Wisconsin River at the Grandfather Falls Segment of the IAT in the Wisconsin Valley Lobe (figs. 2, 3) are of this age. The granitic rocks at the Eau Claire Dells Segment in the northern Green Bay Lobe are also of this age.

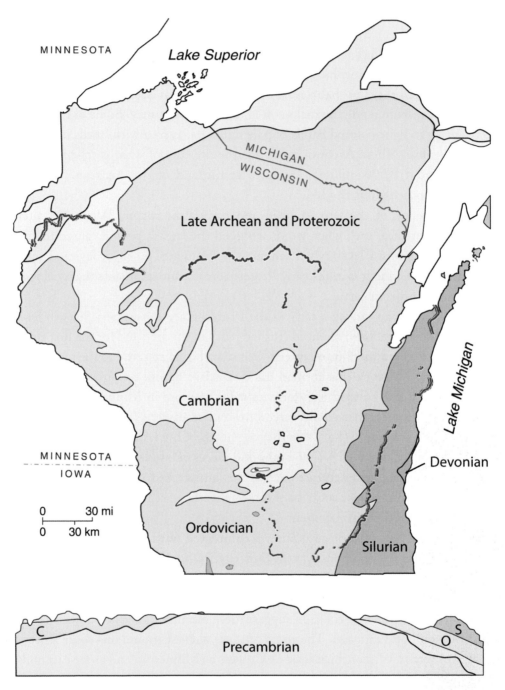

Figure 4. Generalized bedrock geology of Wisconsin and part of the Upper Peninsula of Michigan. The approximate location of the Ice Age Trail is shown in red. Cross section in lower part of figure is an east-west profile across central Wisconsin. More detailed bedrock geology maps appear with the segment descriptions. (Reproduced and modified by permission of Mountain Press from Dott and Attig 2004, 7; drafted by Mary Diman.)

By about 1,750 million years ago, violent volcanic activity began again in southern Wisconsin. Granite intruded and explosive volcanoes deposited rhyolite (SB 21). Observatory Hill, which is very close to the John Muir Memorial Park Segment in the southern Green Bay Lobe, is a remnant of this old volcanic rock. A period of quiescence followed: starting around 1,700 million years ago, a shallow sea lapped the rocks that had earlier burst from volcanoes, and quartz-rich sand settled on the sea bottom. The sandstone that formed was buried and underwent folding and metamorphism to produce quartzite about 1,650 million years ago. While erosion in later years removed much of this quartzite, you'll encounter it in a few places as you hike. You'll cross quartzite in the Southern Blue Hills, Northern Blue Hills, and Hemlock Creek segments of the IAT in the Chippewa Lobe (figs. 2, 3), as well as the Merrimac, Devil's Lake, and Sauk Point segments in the Southern Green Bay Lobe.

The next geologic activity still preserved in the rock record took place about 200 million years later, when granite intruded beneath present-day east central Wisconsin. This huge

Figure 5. Sample of banded iron formation of Proterozoic age from glacial deposits in the Chippewa Lobe. Quarter shows scale. (Photo by Alan Mercer.)

event, which dates to about 1,450 million years ago, created the Wolf River batholith, a huge mass of intruded granite. The granite in this batholith characteristically has large grains of feldspar (SB 21) and is pretty easy to identify. Mica grains, which are also present in the granite, are mostly black in fresh rock, but they turn a cream color when weathered. Batholiths usually have joints, or cracks, into which water can penetrate. As that happens, the mica and feldspar grains weather and expand in volume, breaking the rock preferentially along these joints. This process results in blocks of unweathered solid granite separated by very weak surfaces. Because of the way it weathered, large boulders of this granite were picked up, carried a short distance, and then left behind by the glacier. You'll find these boulders dotting the trailside from the Waupaca River Segment of the IAT in the western Green Bay Lobe to Langlade County in the northern Green Bay Lobe and Langlade Lobe (figs. 2, 3).

The final major geologic event that Precambrian rocks record for us is the development of a deep rift, or series of wide cracks, in the earth's crust across northwestern Wisconsin. About 1,100 million years ago, plate tectonic forces ripped apart the earth's crust, forming these deep trenches. Volcanoes erupted all along the rift, and molten liquid lava flowed over broad areas, producing sheets of a rock called basalt (SB 21). Streams flowing into the rift valleys eroded some of this Keweenawan basalt, creating sand and gravel layers between some of the basalt flows. Instead of widening the rifts further, which could have caused a new ocean to form, tectonic forces eventually changed and pushed the rifts together, uplifting the basalt, sand, and gravel layers. Overall, the structure of the rock that the rift system created appears to have been responsible for the southwest-northeast orientation of the Lake Superior Basin today. We occasionally find native copper from the Keweenawan basalt in glacial deposits throughout Wisconsin, and, particularly in northwestern Wisconsin, weathering of the basalt has freed Lake Superior agates that are found in glacial deposits and on beaches. IAT segments in the area of Keweenawan basalt are the Trade River, Gandy Dancer State Trail, and St. Croix Falls segments in the Superior Lobe (figs. 2, 3). In Interstate State Park, at the western terminus of the IAT, you'll find excellent outcrops of this basalt.

## PALEOZOIC WISCONSIN

Since the Keweenawan rifting and volcanic eruptions in the late Proterozoic, Wisconsin and the rest of central North America have been stable. The Appalachian Mountain range on the east coast of the United States developed during the Paleozoic, and the mountains of western North America formed even later. By early Paleozoic time, an upwarping of the Precambrian crust—a slight doming over a large area—created the Wisconsin arch. There was no mountain range at this time, but Wisconsin was at a somewhat higher elevation than the surrounding areas. Throughout the Paleozoic, the level of the land in Wisconsin remained slightly higher than the subsiding basins beneath what are now Iowa, Illinois, and Michigan. In general, when the sea level was high, sandstone and limestone (SB 21) formed in shallow water in Wisconsin, and deep-water shale deposits formed in the basins. When the sea level was lower, Wisconsin was an eroding landmass ringed by oceans to the west, south, and east.

During the Cambrian, the earliest period of the Paleozoic (table 1), North America was situated very differently on earth than it is today, and even the earth's magnetic field was different. We can trace this history by looking at igneous rocks. Magnetic fragments of certain iron minerals orient themselves parallel to the earth's magnetic field, which varied in a known way over time and across the face of the earth. As iron minerals were deposited, they oriented themselves relative to the magnetic field at the time and then became frozen as crystals within igneous rock. Over time, these iron minerals stayed in their original orientation, even if the rock changed orientation and moved thousands of miles with the migrating continental plate. Today these fossil magnets tell us that North America was in the tropics during Cambrian time; it lay parallel to the equator, and Wisconsin was in the southern hemisphere about 10 degrees south latitude.

Figure 6 shows the changing orientation and position of Wisconsin during the past 600 million years. The letters are abbreviations of the names of the geological periods. Note that during the Cambrian Period, Wisconsin was south of the equator and it "sat on its side": its present north direction faced east! It is fascinating to think of all the changes in the prevailing wind direction and temperature that this land experienced as plate tectonics moved

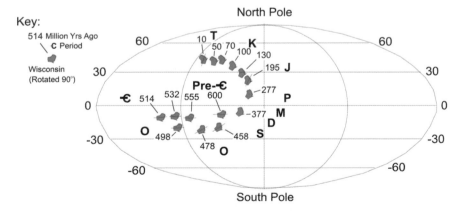

Figure 6. Location and orientation of Wisconsin (green) on the earth relative to the equator and the poles for the last 600 million years. Numbers are millions of years ago. Letters stand for geologic periods shown in table 1: (Ꞓ) Cambrian, (O) Ordovician, (S) Silurian, (D) Devonian, (M) Mississippian, (P) Permain, (J) Jurassic, (K) Cretaceous, (T) Tertiary. Latitude is present-day latitude. (Data from Scotese 1987.)

it north across the equator and well up into the northern hemisphere. If you should get tired while walking on Wisconsin's IAT, think about the tough job Wisconsin had getting to where it is today!

During the latest Precambrian and early Cambrian time, a few hundred million years of tropical weathering and erosion had reduced much of North America to a low, nearly flat landscape with hills of resistant rock rising above a vast plain. The Baraboo Hills and the Blue Hills looked much as they do today. A veneer of sand that wind and rivers deposited probably covered much of the land surface. Remember, there was no vegetation to block blowing sand, because land plants had not yet evolved.

Seas gradually submerged the sandy landscape, but only after several hundred million years of erosion on the continent, creating an unconformity in the rock record. The sea encroached on the North American continent some 600 million years ago and reached Wisconsin during the Late Cambrian, about 500 million years ago, when the oldest marine sediments preserved as sedimentary rock in Wisconsin were deposited (Dott and Attig 2004). Waves and currents redeposited the loose sand that winds had abraded for millions of years. In places where hills stood as islands in the shallow sea, storm waves eroded sea cliffs, producing coarse-grained conglomerate (SB 21) that was eventually buried by sandstone as the sea level rose. You can see examples of the unconformity and overlying conglomerate at Devil's Lake and Parfrey's Glen in the southern Green Bay Lobe, where sandstone overlies Precambrian quartzite, and at Interstate State Park in the Lake Superior Lobe, where Cambrian conglomerate overlies Precambrian basalt.

The first easily recognizable fossils are preserved in Cambrian sandstone, although finding these sparse fossils isn't an easy task. They are characterized by tracks of marine animals, burrows made by worms and other sea-bottom organisms, and the trilobite—the state fossil of Wisconsin! If you're lucky enough to spot one of these uncommon fossils, you're likely to find only fragments of body parts (fig. 7). Another feature common in the Cambrian sandstone of Wisconsin is ripple marks. These form on a sandy stream, lake, or ocean bottom beneath slowly lapping water. You can see modern ripples throughout the state on sandy lake and stream bottoms, but a particularly good place to see them is in the shallow water adjacent to the Point Beach Segment of the IAT.

Most Cambrian rocks in Wisconsin are sandstone, but during times when little sand was carried by rivers into the coastal waters, limestone was deposited (SB 21). Limestone is a calcium-rich rock (calcium carbonate) that accumulated on the sea bottom from the shells of tiny plants and animals. This limestone was recrystallized later to dolomite, which is calcium-magnesium carbonate. The sea level dropped at the end of Cambrian time, and all of Wisconsin was above sea level for several million years. The land was eroded by whipping wind and rain, creating an unconformity before the next episode of deposition. Some

Figure 7. Cambrian trilobite fragments in sandstone from northwestern Wisconsin. Largest fragments are less than 1 inch across. Dime shows scale. (Photo by Kent M. Syverson.)

of the western Green Bay Lobe, Chippewa Lobe, and Superior Lobe IAT segments cross Cambrian sandstone (fig. 4).

In the Ordovician Period, beginning about 490 million years ago, the sea again covered the land that had been eroding since the Cambrian sea retreated. Limestone was deposited on the unconformity that had developed. Prairie du Chien dolomite is the oldest such Ordovician layer, and the IAT crosses Prairie du Chien dolomite in part of Dane County.

Then, about 480 million years ago, still in the Ordovician, the sea level dropped again, exposing Wisconsin to intense stream erosion. Not only did the water erode the recently deposited limestone, but it also carved deep valleys into the Cambrian sandstone. The next rise in sea level produced the St. Peter sandstone, an almost pure quartz sand that was deposited by waves in the shallow sea and by wind on land. Look for St. Peter sandstone in several places along the IAT in southern Dane County. The sea level continued to rise and the supply of sand was reduced, resulting in deposition of limestone on top of the St. Peter sandstone. These Sinnipee Group limestones form some of the hilltops along the southern Green Bay Lobe IAT segments. In far southwestern Wisconsin these rocks are very fossiliferous (fig. 8), a colorful word meaning "fossil rich," but in south central Wisconsin you won't find many of these fossils, because recrystallization of the dolomite destroyed them. In some of these rocks, recrystallization left behind small openings that later partly filled with quartz crystals. Before the sea level fell near the end of the Ordovician, shale was deposited on top of the Sinnipee Group dolomite. You'll rarely see this Maquoketa Shale in outcroppings, because it is soft and easily eroded and therefore doesn't last. Glaciers flowing down the Green Bay–Lake Winnebago lowland presumably excavated this shale to produce features like Lake Winnebago and Horicon Marsh.

The youngest rocks along the IAT are of Silurian age, ranging from about 445 to 416 million years old (table 1; fig. 4). This dolomite is informally called the Niagara dolomite because it is the same rock unit over which Niagara Falls thunders. This relatively resistant rock forms the Niagara Escarpment (SB 19) in eastern Wisconsin. It forms the backbone of Door County and underlies nearly all of easternmost Wisconsin. Most IAT segments in the northeast, northern Kettle Moraine, middle Kettle Moraine, and southern Kettle Moraine Lobes are on Silurian dolomite (figs. 3, 4). As you hike, you're likely to see dolomite exposures

Figure 8. Ordovician-age fossiliferous dolomite from southwestern Wisconsin. Recognizable shells are brachiopods. Largest shell is about 1 inch across.

in a few places. You may also pass one of many large quarries in the Silurian dolomite. People extract the rock because it is a good building stone, an excellent aggregate for concrete and asphalt, and sometimes a decorative stone.

Corals were present during the Ordovician, but large coral reefs made their first appearance during the Silurian, and because the reef dolomite tends to be more resistant than the dolomite around it, many of the bedrock hills that still stand today in far eastern Wisconsin are fossil coral reefs. The relatively high topography of this dolomite Niagara Escarpment is responsible for splitting the flow of the glacier into the Green Bay and Lake Michigan lobes.

Sediments were undoubtedly deposited after the Silurian in Wisconsin, but there is little trace of them. A small area of Devonian rock occurs along the Lake Michigan shoreline north of Milwaukee and in Sheboygan (fig. 4). All of the younger rocks have been eroded away, resulting in an unconformity between all of the older rocks in Wisconsin and the much younger glacial deposits of the Quaternary Period (table 1).

## QUATERNARY WISCONSIN

During the Quaternary, Wisconsin experienced its great Ice Age. But first, some terminology! The Quaternary Period, the last period of geologic time—and the period we are now in—began about 2.6 million years ago (table 1). By this time ice sheets had already formed in the Antarctic, Greenland, and probably in the Canadian Arctic but to our knowledge had not grown large enough to cover any of Wisconsin. All but the last 11,700 years of the Quaternary are known as the Pleistocene, a name that is generally synonymous with the Ice Age. Based on the climate record preserved in ocean cores, there may have been as many as 15 glacial episodes during the Pleistocene when ice sheets grew large enough to enter what is now the central and eastern United States. Between those episodes, the large ice sheets disappeared from North America entirely. Ice sheets in Greenland and the Antarctic shrank in these warmer times but in all likelihood did not disappear. The early glacial record of Wisconsin is poorly known. Because each glacial advance erodes, much of what the early glaciers deposited disappears. With that, we lose more than 90 percent of the record of earlier glaciations. Figure 1 shows where you can see some of the remaining signs of these pre–late Wisconsin glacial deposits at the land surface today.

The most recent glacial episode was the Wisconsin Glaciation. This began about 100,000 years ago, and in its early stages, the glacier probably entered northern and central Wisconsin. We have no way to date these deposits, so we can only estimate their extent. The major glacial event, which created the landforms that we see in about two-thirds of Wisconsin, was the *late* Wisconsin Glaciation (fig. 1). Nearly all of the IAT traverses deposits of this glaciation, with the exception of a few trail segments in the Southern Green Bay Lobe (figs. 1, 3).

The time since the last glaciation is known as the Holocene, which we define as beginning 11,700 cal. years ago. However, the climate began warming about 18,000 years ago. By 14,000 cal. years ago, the climate had warmed, the glaciers had retreated far enough that humans hunted mammoths and other now extinct large mammals in southern Wisconsin, and spruce forests grew in northeastern Wisconsin (Two Creeks Buried Forest Unit of the Ice Age National Scientific Reserve; SB 18).

## The Driftless Area

The Driftless Area of southwestern Wisconsin (fig. 1) was not glaciated, but it nonetheless experienced bitterly cold climates at times when glaciers sat just a few miles away in the east and north. We have recognized for well over 100 years that the southwestern area was not covered by glaciers. There are a few ways to tell: this part of the state has deeper valleys, steeper hill slopes, and narrower ridge tops than the landscape in most of the glaciated area. Elsewhere glaciers have blunted hilltops and softened the steep, deep valleys with their deposits, reducing the overall relief in the landscape. If you can imagine it, much of southern and southeastern Wisconsin looked like southwestern Wisconsin before the last glaciation.

Another sign that the southwest remained glacier free: we don't find any erratics (SB 5), out-of-place rocks, on its uplands (except those that landscapers are bringing in!). There are erratics in some riverbeds here, which were carried by water draining from ice sheets to the north and east. Finally, the Driftless Area has many sandstone chimneys, or buttes, some of which are quite fragile. These formed because slightly harder layers of rock (mostly in sandstone) protected the softer sandstone beneath from erosion. The Mill Bluff Unit of the Ice Age National Scientific Reserve is one of these sandstone monuments. It seems likely that if glaciers had overridden them, most would have been pushed over and broken. The fact that they stand tall just adds to the evidence that glaciers left this part of the state untouched.

In the absence of glaciers, weathering and erosion have slowly shaped the Driftless Area landscape for tens of millions of years. Streams have alternately eroded their beds and deposited sediment as changes in climate and vegetation cover affected how much sediment they carried. As rainwater slowly washed away grains of sand from the hillslopes of sandstone, dolomite was dissolving. Most of the dissolved elements were washed away, but what didn't dissolve accumulated as reddish-brown clay. This clay is a common residual soil in places where dolomite bedrock is the surface rock. Dissolving dolomite beneath the land surface has even produced some caves! Another product of the weathering of dolomite is the accumulation of pieces of a rock called chert, a form of the mineral quartz ($SiO_2$) that has tiny crystals or no crystals at all. It is resistant to weathering and is a common component of gravel in Driftless Area streams.

During glacial times, the climate in the Driftless Area was subfreezing throughout the year. There were no towering trees or grasslands to be found; instead, tundra vegetation covered most of the landscape. The area had permafrost similar to that on the north slope of Alaska today. In these areas, the top few feet of soil can thaw and slip over the still-frozen deeper soil, resulting in mudslides and flows wherever slopes are steep. Much of the chert created during the dolomite weathering discussed above accumulated on stream bottoms, diverting the streams into rivulets and creating braided streams (SB 8) much like outwash streams in the glaciated area.

A widespread sediment in the Driftless Area, and in parts of the glaciated area, was caused by glaciation but was not deposited by glaciers. That sediment is called loess (SB 8) and was deposited by wind blowing silt from outwash plains and other vegetation-free surfaces. The loess layer is thick in the Driftless Area but generally thin to the east.

Only a few segments of the IAT touch the Driftless Area: part of the Devil's Lake Segment, and the entire Cross Plains and Table Bluff segments. The Mill Bluff Unit of the Ice Age National Scientific Reserve is entirely in the Driftless Area, as are the western parts of Devil's Lake State Park and the Cross Plains Ice Age complex (figs. 1, 2).

# Science Briefs

## Science Brief 1: Why Glaciers? Why Not?

Glaciers can exist only when and where more snow falls in the winter than melts in the summer. When snow that has fallen one winter is covered by new snow in the following winter, it compacts under the added weight and slowly changes to ice. As more weight from more seasons of snow is added, the snow particles compact and freeze together to form ice crystals. Then the ice crystals merge and eventually become solid ice with only a few air bubbles remaining. The process could take as few as 5 years under warm conditions with lots of snow or more than 100 years where it is very cold. The part of the glacier where more snow accumulates in the winter than melts in the summer is called the accumulation zone.

Figure 9 shows a longitudinal profile of a hypothetical glacier. Although the cross section used is that of a valley glacier, the same concepts of ice movement apply to a large continental ice sheet. The glacier is divided into two zones: the accumulation zone mentioned above, where more snow falls in the winter than can melt in the summer, and the ablation zone, where more snow melts in the summer than falls in the winter. In accumulation zones the glacier gains in mass, and in the ablation zone a glacier loses mass over the year. The boundary between these two zones is called the equilibrium line. Figure 9 shows that unless there were some way to transfer ice from the accumulation zone to the ablation zone, the glacier would thicken in its upper part and disappear in its lower part. In fact, a glacier transfers its mass in three ways: ice moves from the accumulation zone to the ablation zone by bed deformation, by sliding at the glacial bed, and, if the ice is greater than about 100 feet thick, by internal flow (SB 2). The ice always moves from the accumulation zone to the ablation zone; that is, it always moves down the ice-surface slope. The glacier loses ice (ablates) through gradual melting or sudden release of icebergs.

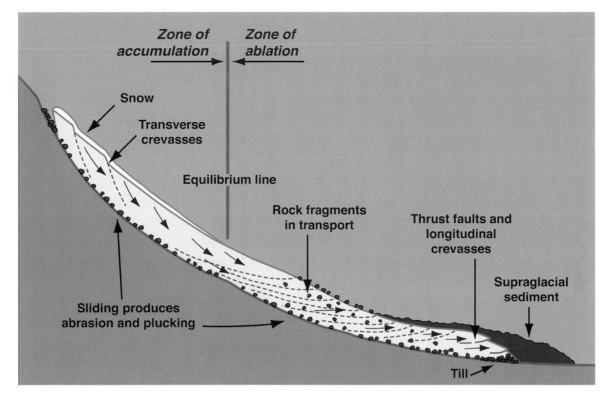

Figure 9. Important parts and processes of a glacier. Several terms are discussed in the science briefs that follow. (Drafted by Mary Diman.)

"Ice advance" and "ice retreat" refer to changes in the position of the ice edge, or to what is often called the ice margin. Glacier ice always flows down the ice-surface slope toward the edge of the glacier, and changes in the position of the ice margin are due to changes in the balance between the amount of ice coming in from the accumulation zone and the amount of ice being lost by ablation. The glacier is analogous to the balance in your bank account. If you take out more money for your mortgage and groceries than you deposit, your bank account will shrink. Similarly, if the volume of ice lost is more than the volume of ice gained, the glacier shrinks and the ice margin retreats. With a somewhat delayed response, a glacier retreats if snowfall decreases (less accumulation) or summer temperatures increase (more ablation). Likewise, a glacier advances if snowfall increases (more accumulation) or the summer temperatures decrease (less ablation).

## Science Brief 2: How Do Glaciers Move?

If glacier ice did not move, snow in the accumulation zone would pile thicker and thicker, stopping only when it touched the clouds that fed it with snow. As we know, this does not happen. Even though ice sheets can be over 2 miles thick in their central areas, there is a limit to how thick they can become. As more snow piles on top, gravity forces the ice beneath to move outward from the area of accumulation.

How does the ice move? We think of ice as a brittle solid. If we drop an ice cube on the floor it shatters. In fact, ice in the top 100 feet of a glacier *is* brittle. But ice under thicknesses of accumulated snow and ice greater than about 100 feet behaves differently. This ice deforms, much like the way cold molasses or *very* thick pancake batter flows slowly, but the processes involve different mechanisms. The ice crystals actually melt at their edges, where the pressure is high, and refreeze in areas where the pressure is lower. Slippage within individual crystals also contributes to the flow of the glacier as a whole. Most glacier ice moves at velocities of a few feet to a few hundred feet per year, although so-called surging glaciers move much faster. As ice deep within a glacier flows, the brittle ice at the top of the glacier cracks with the movement, forming deep surface fissures called crevasses (fig. 10).

Ice in a glacier can also be transported in two other ways. If the sediment below the glacier is soft and wet, the ice can ride "piggyback" on the deforming sediment beneath. More important, for most glaciers, is sliding. Whenever water is present at the bed of the glacier, there is a very weak bond between ice and the bed below, so glaciers can slide over rock or sediment. The latter two processes shaped the landscape and produced the landforms and sediments that we see today in Wisconsin.

Because ice flow is controlled by gravity, its path is partly determined by the shape of the bed beneath. As the ice sheet expanded into the Great Lakes area, ice funneled down broad lowlands such as the Great Lakes basins, creating glacier lobes that are now named for the basins they occupied (fig. 1). Exceptions in Wisconsin are the Chippewa, Wisconsin Valley, and Langlade lobes. That ice flowed toward the southwest out of the east-west-trending Lake Superior basin.

The Ice Age Trail follows the outer boundary of the Superior, Chippewa, Wisconsin Valley, Langlade, and Green Bay lobes, as well as the Kettle Moraine, which formed at the junction of the Lake Michigan and Green Bay lobes.

## Science Brief 3: Are All Glaciers Cold?

Relative to temperatures at which most humans are comfortable, yes, glaciers are cold! However, some glaciers are colder than others. In fact, whether or not the bed of the glacier is at the melting point (32°F, or 0°C) makes a huge difference in the behavior of the glacier. If the bed is at the melting point, water and ice can coexist for a long time, allowing the bulk of ice to "float" on the layer of water and slide across the landscape. When the temperature is below the melting point, the glacier freezes to the bed. In that case, there is no water present to lubricate the bed and allow sliding. These cold glaciers with a frozen base do little work on

Figure 10. Crevasses on the surface of Reid Glacier in southeastern Alaska. Ice flow is toward the water in the foreground.

the landscape because they are frozen solid to the ground. In parts of Scandinavia it is well documented that the landscape was glacier covered for thousands of years, yet the glacier left hardly any trace because its bed was below the melting point. There was little or no erosion from this cold glacier.

Under warmer conditions, like those of most Alaskan glaciers today, the base of the glacier is at 32°F (0°C) and both water and ice are present. Melting and refreezing often take place in spots close to each other. Even at the peak of the Wisconsin Glaciation, the ice sheet in the Great Lakes area had zones where the glacier bed was "warm." In much of what is now Wisconsin, these zones looked like those in figure 11. Along the southern edge of the ice sheet in Illinois, Indiana, and Ohio, there was little or no frozen zone near the ice edge because the climate was even warmer. Later, as the climate in Wisconsin warmed, the glacier bed also warmed and water was present everywhere.

A warm glacier leaves many traces. It can slide over its bed, scraping and gouging the earth beneath, forming, among other features, striations. Grains of rock and sediment frozen onto the base of the glacier are simply dragged along by the glacier's forward plowing.

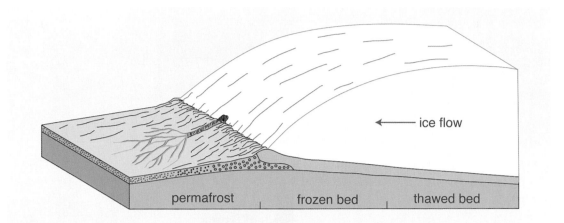

Figure 11. Cross section of the ice sheet at the glacial maximum in what is now Wisconsin. Note zones of permafrost, frozen, and unfrozen bed. (From Attig et al. 1989; drafted by Mary Diman.)

Water moving around and through the glacier creates other glacial remnants on the landscape. Water can flow at the base of a warm glacier to its edge, sometimes as a thin sheet and sometimes in ice tunnels. When the glacial bed is at the melting point, water-filled tunnels do not freeze shut and they can carry water for several years. In some tunnels, flowing water deposits sand and gravel, leaving long deposits on the land called eskers (SB 13). If the tunnel isn't filled with sand and gravel and the flow of water in the tunnel stops, air fills the tunnel (as in figure 12), and the tunnels melt rapidly.

Figure 12. Tunnel in the wet-bed Burroughs Glacier in Alaska. Tunnel is about 15 feet high and more than 300 feet long.

Even in areas where the glacier bed near the outer edge of the ice is below the melting point, trapped water behind that zone escapes under the frozen glacier margin (fig. 11). When it does, a tunnel channel forms (SB 17).

## Science Brief 4: Scratching the Surface

The most common evidence of glacier erosion on bedrock surfaces is striations, or striae. These scratches are often less than 0.1 inch wide and deep, are generally straight, and can stretch for tens of feet on smooth bedrock surfaces (fig. 13). They form parallel to ice flow, and so for more than 100 years, glacial geologists have used striations on bedrock to interpret the direction of ice movement. It is common to see more than one direction of striations represented on a freshly exposed bedrock surface. This can be evidence of more than one glacial advance or of changes in the direction of the ice flow during a single advance.

Figure 13.  Striations on limestone at Valders, Wisconsin. Ice moved from upper left to lower right. Compass shows direction of ice flow.

Because striations are usually symmetrical, they give a compass bearing but don't show the unique direction the glacier was moving along that line. For instance, with only north-south striations, one can not tell whether a glacier moved from north to south or the reverse. But sometimes asymmetrical shapes, such as crag-and-tail features, can be seen. These occur where the bedrock has spots on its surface that differ in their resistance to abrasion. In figure 14, the crag is a hard fossil and the tail is a ridge of soft limestone on the lee (downstream) side of the harder nodule. Shallow grooves on either side of the tail were produced by increased erosion as the ice went around the obstruction. Crag-and-tail features on bedrock are useful because they reveal the unique direction of the ice flow.

Striations also occur on rocks carried in the glacier itself (fig. 15). Look at almost any exposure of till (SB 5), and you will find striations on some of the rocks, particularly on limestone or dolomite, which are softer than other rocks.

Freshly exposed bedrock surfaces and rocks in till often have a shiny surface much like that of polished jewelry or the faces of smooth gravestones. Glacially transported rocks are polished the same way we smooth wood. When a glacier moves over the land, fine-grained

Figure 14. Crag-and-tail feature on dolomite in Middleton, Wisconsin. Ice moved from left to right, in direction pencil is pointing. Crag is resistant chert (a hard rock) and tail is softer dolomite. (Photo by Kent M. Syverson.)

Figure 15. Striations on limestone fragment from till (SB 5) in Iowa.

sediment frozen in the base of the ice slides across the bedrock surface, abrading and polishing the rock. Just as finer sandpaper is used to produce smoother finishes on wood, fine-grained sediments in the debris-rich layer at the bed of the ice produce scratches so fine the resulting rock surface appears polished, whereas coarser particles produce striations.

## Science Brief 5: Scraping Up the Rubble: The Making of Till

In a way, a glacier acts like a conveyor belt. Sediment is picked up at one location, carried, and deposited closer to the ice margin. If the glacier is advancing, the sediment can be eroded again and again as the glacier advances to its outermost advance position. It is a complex system, however, because what happens to the sediment during the transport process differs depending on where the sediment is carried within the glacier. When debris is carried in dirty ice near the glacier bed, the sediment particles interact with each other, abrading and breaking themselves. In other cases, rock particles are carried high in the ice for hundreds of miles without being modified at all. Sediment deposited directly by the glacier is called till. Figure 16 shows till at the edge of a retreating glacier in Alaska. Note the very fine to boulder-sized particles and the lack of the layering that is typical of water-deposited sediment.

As the glacier thins, debris at the base no longer moves with the glacier. If the debris is frozen, ice in the pore spaces melts, leaving this sediment behind. Figure 17 is from Langlade County, but it shows till similar to that throughout southern, east central, and much of northern Wisconsin. The till in a narrow belt south of Lake Superior, in the Lake Winnebago–Green Bay lowland, and along the shore of Lake Michigan contains much more clay than does till in the rest of the state. The clay was derived from lake sediment overridden by the glacier.

Figure 16. Till being deposited from the base of a glacier in southeastern Alaska (ice edge is in upper left). Shovel handle is about 20 inches long.

Figure 17. Till exposure in Langlade County. Exposed face is 15 feet high.

Because of the wide range of its grain size, till is not useful as an aggregate for concrete or asphalt. It is, however, commonly used as fill. Till often contains erratics—rocks that are out of place and different from the bedrock beneath (fig. 18). Erratics in eastern Wisconsin have been carried from areas east and north of the Lake Superior basin. In northern Wisconsin, erratics are from the Lake Superior basin or areas to the north. Most big boulders on the ground surface in the glaciated part of Wisconsin are erratics. Often farmers have to plow around or try to move these huge boulders; although till generally makes good farmland, a landscape studded with erratics is hard to farm.

## Science Brief 6: End Moraines

Moraines are ridges of glacial debris. They form wherever till (SB 5), and to a lesser extent sand and gravel, are piled up either on or adjacent to a glacier margin. End moraines are

Figure 18.  Erratic in the Kettle Moraine in Sheboygan County. Boulder is about 4 feet across.

ridges that form at the edge (end) of the glacier when debris is carried to the ice margin and released as the glacier ice melts (fig. 19). The Ice Age Trail follows morainal topography along most of its path across Wisconsin (fig. 3).

Debris that is picked up by the glacier is carried as if on a conveyor belt to the ice-marginal zone, where it melts out either at the base of the ice or on the ice surface. The conveyor belt is always operating, thus sediment is always carried toward the marginal zone. If the glacier margin remains in one place, debris is piled up into a moraine ridge just as it would if it fell off the end of a conveyor belt. If the marginal zone advances or melts back across the landscape at a more or less constant rate, then no moraine is formed and a more or less uniform thickness of till is deposited.

If the glacier is advancing, the sediment melting out of the ice is typically reincorporated and carried farther along the flow path as the glacier advances over it. Only at the position of maximum advance does the sediment reach its final resting place during the advance stage.

Figure 19.  Glacier in Greenland with debris-covered ice (about 500 feet wide) in marginal zone. An end moraine lies between the ice edge and the river on the left. (Photo by R. P. Goldthwait.)

This is the outermost end moraine, sometimes called a terminal moraine. When the glacier edge retreats, sediment deposited near the ice margin is left behind by the retreating glacier. If the glacier edge stays in place for some time during ice retreat, then the sediment accumulates as an end moraine. These moraines are often called recessional moraines.

The volume of sediment in a moraine depends on how much debris was being carried by the glacier and how long the glacier edge sat in one position. In some places, for instance in northern Dane County, no moraine formed at the position of the outermost late Wisconsin advance. This means that the ice was relatively clean, with very little debris being carried to the ice margin. In much of southern Wisconsin, moraines are only about 50 feet high and a quarter to a half mile wide. Figure 19 shows a glacier in Greenland with a moraine in front that probably looks much the same as the glacier did along the southern edge of the Green Bay Lobe at the time of the glacial maximum. In some places moraines are even narrower

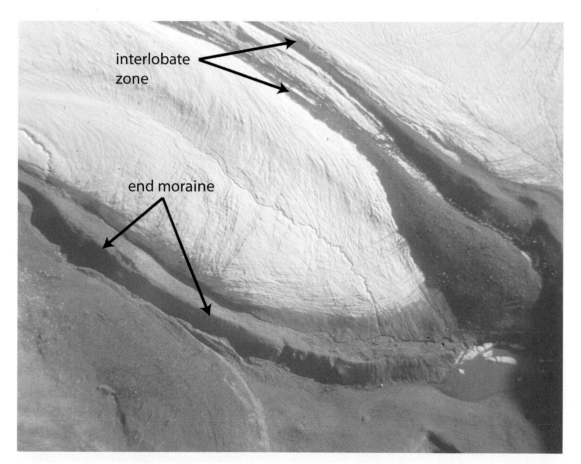

Figure 20. Glacier in Greenland with end moraine and interlobate zone indicated. View is several miles across. (Photo by R. P. Goldthwait.)

and consist of a single slender ridge. Figure 20 shows another example from Greenland where the moraine is a single-crested ridge. The moraine at the northern entrance to Devil's Lake State Park is a good example of a narrow, single-crested ridge.

Many moraines have a significant component of supraglacial sediment that melted out on the top of the ice. The ice beneath this sediment melts slowly and unevenly, producing hummocky topography (SB 11). Kettles (SB 9) produced by this buried ice are more common in moraines than in the till areas between moraines. In places where the supraglacial sediment was relatively thin, the kettles are shallow. Figure 21 shows the edge of the Johnstown Moraine in southern Dane County. Here kettles are only about 30 feet deep, implying that the sediment cover on the ice was about 30 feet thick.

In contrast, in much of northern Wisconsin a huge moraine formed. Figure 22 shows an aerial view of the moraine in northern Wisconsin. Note the high-relief hummocky topography (SB 11) in the moraine. The Chippewa Moraine Unit of the Ice Age National Scientific

Figure 21. View of Johnstown Moraine (bottom) and outwash surface (top) in southern Dane County.

Reserve and most of the Ice Age Trail in northern Wisconsin exhibit excellent high-relief hummocky topography in the moraine.

The name Kettle Moraine is actually a misnomer, but it has been used for more than 100 years and it is unlikely that this area will ever be known as anything else. Most moraines are made up of till, but the Kettle Moraine contains mainly sand and gravel deposited by streams flowing in the interlobate zone between the Green Bay Lobe and the Lake Michigan Lobe. Although on a smaller scale, figure 20 shows an interlobate zone with ridges of debris upon the ice surface. More details about the formation of moraines are given in the regional descriptions.

## Science Brief 7: Water, Water Everywhere

Think of all of the ice that lay over much of Wisconsin 25,000 years ago. It was thin at the edge but thickened rapidly toward the north and east. The glacier was probably about 800 feet thick over what is now the capitol building in Madison. It was probably more than half a mile thick over what are now the cities of Green Bay and Superior. Even in the coldest years, there would have been some melting in summer, with the flow of ice from the north

Figure 22. Kettle lakes in the end moraine of northern Wisconsin.

replacing that which melted. As the climate warmed about 18,000 cal. years ago, the ice melted rapidly, and torrential rivers flowed away from the ice sheet in many places, particularly down major river valleys such as the Rock, Wisconsin, Chippewa, St. Croix, and Mississippi. In addition to the water from all of that ice melting, the normal precipitation levels were similar to those we have today.

Some meltwater entered the groundwater system, but most ran into the Mississippi and, later, into the Great Lakes' basins. Because of shrinking ice sheets and smaller glaciers all over the world, the global sea level has risen about 360 feet since the last glacial maximum. The rise itself did not affect Wisconsin, but the huge ice sheet in Wisconsin and to the north contributed greatly to the change.

Other than the glacier itself, water has produced more of our landforms than any other agent. Significant amounts of sediment were carried by streams flowing beneath the glacier (fig. 23). Much of the subglacial sediment that remains is in the form of eskers (SB 13) and

Figure 23.  Water rushes from beneath a glacier in Iceland. Ice cliff is about 30 feet tall.

hummocks in tunnel channels (SB 17). In front of the ice margin, rivers deposited outwash gravel and boulders in fan-shaped deposits near the ice edge and finer sand farther from the ice. Streams also eroded the land. Because of topography, water flowed along the ice edge, leaving behind ice-marginal channels. Normally the flow of water beneath the ice was toward the ice margin, but locally some water flowed off or from under the ice, and then back under the ice edge. These streams formed submarginal chutes. A good example of one is at the Cross Plains Unit of the Ice Age National Scientific Reserve.

In other places where drainage was prevented from flowing away from the ice because of the slope of the land, lakes developed. These ice-dammed lakes were sites of accumulation of thick sand, silt, and clay. These lakes were especially common in the area in front of the retreating Green Bay Lobe because that land sloped toward the northeast, trapping water along the ice front. Figure 24 shows a small ice-marginal lake in Iceland.

Unlike glaciers, which collect different sizes of sediment into a complete mix of boulders, gravel, sand, silt, and clay and deposit till (SB 5), water tends to separate particles by size before and during deposition. Because the size of the particles that water can carry is a function of velocity and turbulence, coarse particles are deposited first as the velocity decreases.

Figure 24.  Ice-marginal lake (Jökulsárlón) in Iceland. The glacier itself is across the lake (rear). (Photo by John Chapman.)

Figure 25. Stratified sand and silt. Exposed part of knife blade is 1 inch long.

As a result, the deposit is layered, or stratified (fig. 25). The stratification is produced by differences in the grain size of the particles in various layers. Fast-moving streams deposit boulder- to gravel-sized rocks, and slower currents deposit sand. Even finer particles, such as silt and clay, are typically deposited in very-slow-moving streams or standing water.

## Science Brief 8: Outwash and Loess

Outwash is the term used to describe sediments deposited by water flowing away from a glacier. Like other water-deposited sediment (SB 7), outwash is sorted and stratified, meaning that particles are sorted by size before they are deposited (fig. 26). Outwash that accumulates very close to the ice edge is typically coarse and bouldery because water is moving off the steep ice slope. As the streambed becomes less steep, the velocity of the water slows, so outwash even a mile or so away from the ice edge is already mostly sand and gravel. The outwash continues to get finer in the downstream direction, and tens of miles downstream it typically consists of sand. Water flow will carry the fine-grain silt and clay even farther downstream.

Because of this sorting, sediment at any one site is of a relatively uniform size, so outwash sand and gravel deposits are a common source of aggregate for concrete and asphalt. The sorting of grains takes place in braided streams (fig. 27), which typically have large flows of water late in the day, after hours of melting, and much less water late at night or early in the morning. Deposition takes place each time the water velocity decreases, and erosion takes place each time the water velocity increases. This causes channels to shift position on a daily

Figure 26. Stratified sand and gravel in a Wisconsin gravel pit.

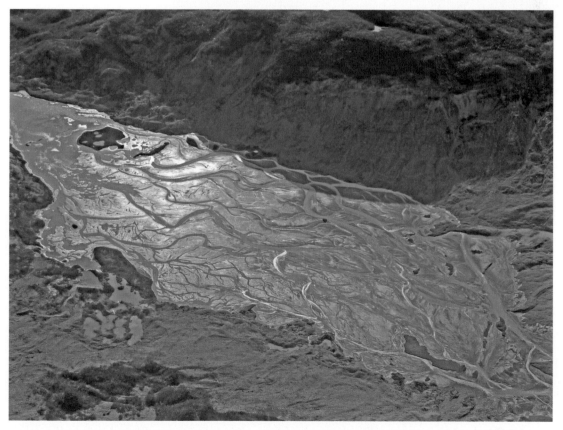

Figure 27. Oblique aerial view of braided outwash stream in front of the Burroughs Glacier in southeastern Alaska. Stream floodplain is about a quarter mile wide. Note kettle holes forming in lower part of photo. (Photo by Richard Becker.)

basis as sediment is deposited with slowing velocity, clogging the active channels. Although some of the sediment is picked up again as velocity increases, much of it remains in gravel bars and islands. Outwash streams fed by today's glaciers can deposit several feet of sand and gravel each year across the entire floodplain.

Outwash streams occupied many of the large river valleys of Wisconsin during the last glaciation. In some valleys the outwash sand and gravel is over 300 feet thick. This means that at some time, probably during or just before the glacial maximum, these valleys were 300 feet deeper than they are today. Some of the biggest outwash rivers were in the valleys of the Chippewa, Wisconsin, Rock, and, of course, the Mississippi rivers. Rivers also formed broad, gently sloping outwash plains in front of the ice margin, both when the ice was at its maximum extent and also during its retreat (fig. 28). Excellent examples of outwash plains occur in many places in front of moraines. Outwash fans occur along moraine fronts where tunnel channels or other outwash sources exited the glacier. Commonly, outwash is deposited on and against an ice margin. Where the ice has melted out, a collapsed, often kettled, ice-contact face is left behind, which is known as an outwash head (SB 10).

The grinding action at the base of glaciers produced a huge amount of silt-size grains (SB 21) that were washed down these braided rivers. On warm days much of the riverbed flooded, then at night, when melting slowed and the water level dropped, silty water coated the exposed gravel. Wind dried the silt grains, and they blew into the air to be deposited somewhere downwind as a sediment called loess. Eventually this loess blanketed much of Wisconsin. In some areas, much of it has been washed off since the glaciers departed, but in other areas, it remains thick: in places along the Mississippi River the loess layer is over 50 feet thick!

As the glaciers retreated, the rivers that deposited sand and gravel changed their behavior. They carried less overall sediment, so they deposited less sediment. They also had a more constant water flow because they were farther from the glacier's edge and the rush of afternoon meltwater. Both of these changes caused the rivers to be less braided. Many rivers completely changed from braided streams to ones with a single, deeper channel. As a result, rivers began to erode their beds, leaving the former braided streambed well above the modern floodplain. These abandoned streambeds are called terraces (fig. 29). Most Wisconsin outwash deposits have terraces that represent the streambed at a time when the river was flowing at a higher level.

The formation of terraces on some of Wisconsin's largest rivers, such as the Wisconsin and Mississippi, also occurred relatively abruptly during deglaciation in response to cataclysmic floods. These erosive floods resulted from rapid drainage of lakes that had formed between the former ice margins and the retreating ice front.

Other forms of outwash are described in SB 10.

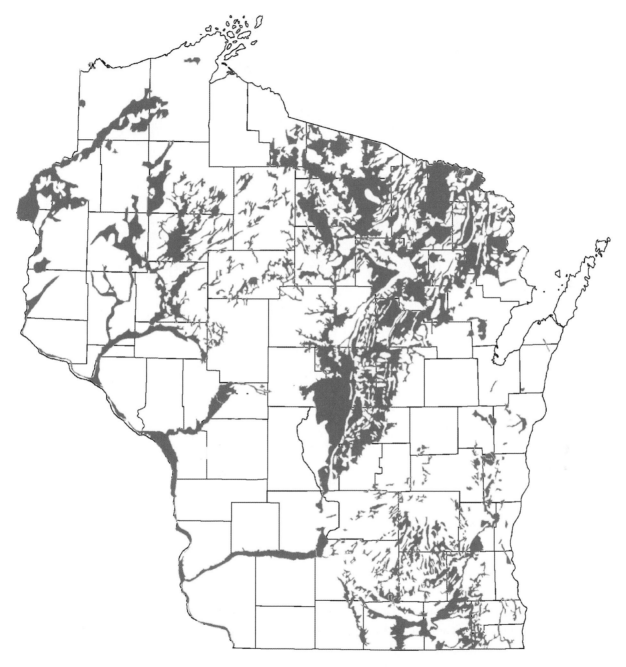

Figure 28. Distribution (shown in red) of outwash and other sand-and-gravel deposits in Wisconsin. (From Farrand et al. 1983, Goebel et al. 1984, Lineback et al. 1984, Hallberg et al. 1991.)

Figure 29.  Several terraces along an outwash stream in front of the Burroughs Glacier in Alaska. Stream in center of photo is about 25 feet wide. (Photo by Kent Syverson.)

## Science Brief 9: Kettles

A kettle, or kettle hole, is a depression formed when a mass of glacier ice that is buried in sand and gravel or till (SB 5) melts away (fig. 30). As the buried ice block slowly melts, the sediment above collapses into the void and produces a hole that may range from only a few tens of feet across to several miles across (fig. 31). If an ice block is simply dropped on a gravel surface, it will melt without leaving a trace. Formation of kettles, however, requires that the sediment be deposited on top of, or at least around, the block of ice, so they form where substantial thicknesses of sand, gravel, or till end up in a supraglacial position (on top of glacier ice). Kettle formation often takes place along glacier margins where the flow of the ice carries sediment upward from the base of the glacier to the ice surface, where it accumulates. This process typically produces hummocky topography (SB 11). Kettles also form where outwash streams (SB 8) flowing from the glacier bury blocks of ice, eventually resulting in

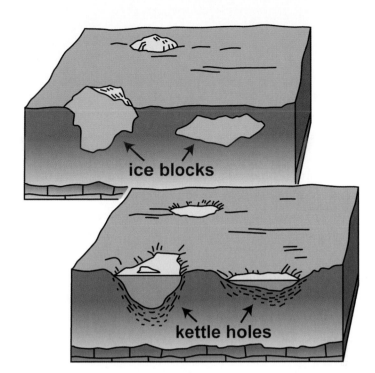

Figure 30. Kettle formation in outwash. (Modified from Mickelson 2007; drafted by Mary Diman and Susan Hunt.)

Figure 31. Ice block melting out of sand and gravel in Iceland. Several students shown for scale.

pitted outwash (SB 10). Typically moraines (SB 6) contain more and deeper kettles than the till surfaces between moraines. Thus, kettles are common throughout the glaciated area wherever circumstances allowed burial of glacier ice at the time of the glacier's retreat.

Kettles may be dry, seasonally wet, or nearly water filled on a permanent basis. Where the water table is high or where clay lines the bottom of the kettle, lakes commonly form. Nearly all of Wisconsin's lakes (over 10,000 of them) are kettle lakes formed in this way. It is interesting to note that there are no natural lakes in the Driftless Area of southwest Wisconsin because in this area there was no glaciation and therefore no kettle formation.

Although some water-filled kettles remain clear and deep lakes today, peat and other organic deposits have slowly accumulated in many of the shallow kettles. There are thousands of peat bogs throughout the glaciated part of Wisconsin (fig. 32), some with open water in the middle and some entirely filled with organic sediments. Some are even tree covered.

Figure 32.  Oblique aerial view of Spruce Lake Bog, a Wisconsin State Natural Area at the edge of the northern Kettle Moraine. Open water is surrounded by a sphagnum bog. Access to the lake is from Airport Road beyond the lake. Lake is 0.3 miles across.

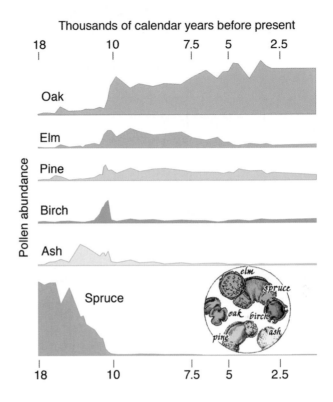

Figure 33. Abundance of pollen in Hook Lake in Dane County since deglaciation. (Modified from Mickelson 2007; drafted by Mary Diman and Susan Hunt.)

Kettles contain a record of the changes in vegetation that have taken place since the last glaciation. Pollen grains from plants in surrounding areas accumulate in lakes, with the oldest sediment at the deepest layers of the lakebed, and the youngest at the present lake bottom.

Pollen grains can be examined, counted, and used to reconstruct how the vegetation has changed. Figure 33 shows the change in the abundance of pollen with time since the disappearance of glacier ice from southern Dane County. (A more detailed pollen record is illustrated in the description of the Devil's Lake Segment.) All of these studies show that spruce was the dominant tree for several thousand years after deglaciation.

## Science Brief 10: Pitted Outwash

Often during ice retreat, sand and gravel outwash will cover glacier ice masses (SB 8). When the buried ice masses eventually melt out, they leave depressions in the earth's surface called kettles (SB 9). Figure 34 shows the edge of a retreating glacier in Alaska. Note the braided stream (SB 8) flowing away from the glacier's edge and the kettles developing as ice blocks beneath the outwash melt. Pitted outwash is a landscape that is on a continuum between completely unpitted (with no kettles) outwash and kettle-filled hummocky topography (SB 11). Pitted outwash preserves a good bit of the original streambed surface, but kettles

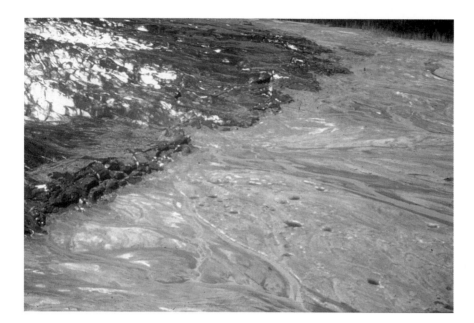

Figure 34. Retreating glacier margin of Scott Glacier in Alaska. Note the kettles beginning to form in the outwash surface. Assuming that these are isolated masses of buried ice, the surface will be pitted outwash after meltout. (Photo by Gail Ashley.)

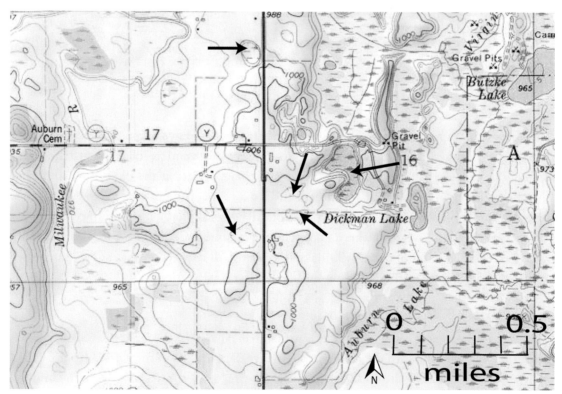

Figure 35. Topography of pitted outwash in an area just west of the Kettle Moraine near Campbellsport. Arrows indicate a few of the kettles. Also note the almost flat outwash surface surrounding the kettles. (Map is part of the USGS Campbellsport Quadrangle and was created with TOPO! © 2011 National Geographic Maps.)

interrupt it. Hummocky terrain is a surface that retains little or no original depositional surface. It has all collapsed. Figure 35 provides an example of pitted outwash. Note that the gently sloping outwash surface is interrupted by kettles. Examples of hummocky terrain are shown in SB 11.

Another outwash landform is called an outwash head (fig. 36). These features are common along the west side of the Green Bay Lobe where the glacier flowed up the regional slope of the land. They formed as the ice margin retreated from a stable position. Water could not flow directly away from the ice because of the land slope, so streams carried outwash along the ice margin, sometimes burying the ice near the edge of the glacier. As the glacier continued to melt, it left behind outwash with a steep ice-contact slope facing the glacier. Typically, the surface of a steep outwash head is pitted.

## Science Brief 11: Hummocky Terrain

Sediment of all sizes, from very small particles of clay to very large boulders, may accumulate in varying thicknesses on the top surfaces of glacier ice. These are called supraglacial deposits. When the underlying ice melts out, this sediment collapses into hills and kettles.

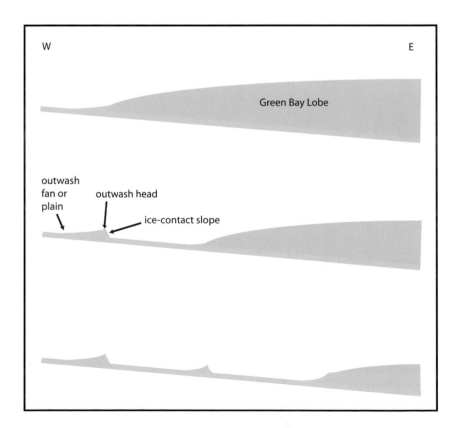

Figure 36. Development of outwash features along the west side of the retreating Green Bay Lobe. After ice retreated from the outwash head, water carried sand and gravel parallel to the ice margin because it could not flow uphill away from the ice margin. Not to scale.

Sometimes these ice-disintegration ridges can be confused with eskers or other landforms. This accumulation is a common process particularly when the glacier moves up a steep slope before the ice gets to its outer margin. It also occurs if the outer zone of the glacier is frozen to its bed (SB 3). In both cases, more rapidly moving, debris-rich ice from the base of the glacier rides up over the slower-moving ice near the margin. As the ice melts in the marginal zone, the debris is released and accumulates on the surface of the glacier. In other cases, streams redistribute and deposit this sediment as sand or a sand-gravel mix on glacier surfaces. Thus, sediment on the glacier's surface can be a sorted stream sediment or a very poorly sorted, boulder-rich mix. Figure 37 shows the edge of an Alaskan glacier with a thick cover of debris. As the glacier ice beneath the debris slowly melts, a landscape of closed depressions (kettles) and intervening irregular hills (hummocks or ice-disintegration ridges)

Figure 37. Debris-covered Kennicott Glacier in Alaska. Upper ice-walled lake in right of photo is about 500 feet across. The entire area is underlain by glacier ice. (Photo by Richard Becker.)

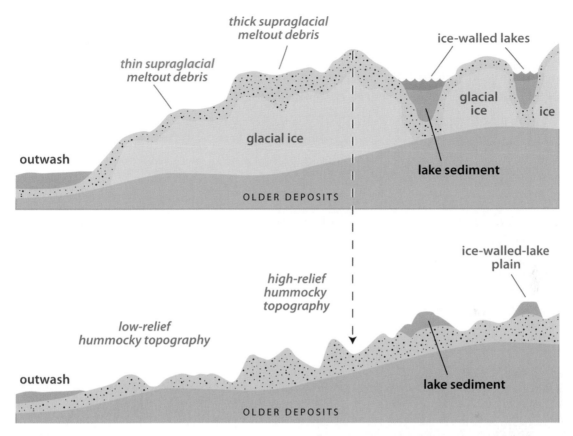

Figure 38. Accumulated supraglacial debris on the glacier surface and the resulting hummocky topography after the glacial ice melts out. Note that thickness of supraglacial sediment determines the relief in the present landscape. Arrow indicates reversal of topography. High points on the debris-covered ice often become low points in the landscape. Two ice-walled-lake plains are shown forming on the right-hand side of the figure. Note also that they are low spots on the buried ice, but high spots on the landscape. (Modified from Clayton et al. 2001; drafted by Mary Diman.)

is the result (fig. 38). Often what is a hill on the buried ice ends up being a kettle after all of the underlying ice has melted out, a process called topographic reversal. As figure 38 illustrates, thinner supraglacial sediment produces low-relief hummocky topography, and thick supraglacial sediment results in high-relief hummocky topography.

Such hummocky terrain is very strikingly apparent along much of the Ice Age Trail in northern Wisconsin. The high-relief hummocky topography is probably there because ice had to flow up and out of the Lake Superior basin and also because there was a fairly wide zone near the ice margin with a frozen bed (SB 3). There was apparently less debris on the glacier's surface along the west edge of the Green Bay Lobe in central and southern Wisconsin because the hummocks are not nearly as well developed as they are in the north. We do

Figure 39. Hummocky topography along the Kettle Moraine in eastern Wisconsin. (Photo by Donna Harris.)

see hummocky terrain in eastern Wisconsin: the area of the Kettle Moraine has high-relief hummocky topography (fig. 39) because rivers on top of the ice deposited sand and gravel that collapsed as buried ice slowly melted. A reader might wonder what distinguishes pitted outwash (SB 10) from hummocky topography: in hummocky topography, very little or no original outwash surface is preserved.

A landform often associated with hummocky terrain is the ice-walled-lake plain (SB 15). Meltwater and mud flow or slide into the lake from the ice; coarse sediment falls near the lake's edge, and finer sediment travels farther to settle in deeper water. After the surrounding ice melts out, the fine-grained lake sediment is often preserved on the tops of hummocks, and the coarser bouldery sediment accumulates in the depressions between the hummocks (fig. 38).

## Science Brief 12: Moulin Kames

The word "kame" has been used for well over a hundred years to describe hills, usually those composed of gravel of various shapes and sizes. In fact, it has been used in a general sense for hills with so many different origins that the word has little specific meaning in itself. Associating it with the word "moulin," however, narrows its meaning. Moulins are vertical shafts through the glacier that often extend all the way to the bed. They only occur where the ice is relatively thin and where the glacier is at the melting point (SB 3), so water can coexist with ice. Figure 40 shows the top of a moulin in Iceland. This glacier is relatively clean at the surface, but where debris is available, it spills with meltwater down the moulin to the bottom. Figure 41 shows a small kame at the bottom of a moulin in southeast Alaska. Note the water and sediment falling onto the top of the kame and then sliding and flowing down its sides.

Very large moulin kames are common in the Kettle Moraine of eastern Wisconsin (fig. 42). They are exceptionally large and almost unique in their size and number. Garriety Hill is one of these large kames in which deposits have been exposed (fig. 43). Much of the hill is

Figure 40.  Surface of a glacier (Gigajökul) in Iceland showing the top of a moulin in center foreground. Note thin layer of debris on the ice surface and people for scale. (Photo by John Chapman.)

made up of poorly sorted debris-flow sediment with boulders, suggesting that the sediment was dumped from the ice surface above without much sorting (SB 8) by water. Thin pond sediment at the very top suggests that the moulin filled with sediment to the present height of the kame while the kame was still surrounded by ice. There is no exposure of the deposits contained in most of Wisconsin's very large moulin kames, so some may have formed in somewhat different ways—and certainly at a different scale—than the one shown in figure 41.

Moulin kames are present here and there throughout much of the part of Wisconsin covered by the last glaciation. For the most part they are not as large and not as numerous as in the Kettle Moraine. Note that most hills in hummocky terrain (SB 11) are not moulin kames. Typically moulin kames are more isolated and are commonly surrounded by flat or rolling topography. Some of the large kames have small eskers (SB 13) connected to them.

Figure 41. Moulin kame forming at the base of a moulin in southeast Alaska. Note water and sediment falling onto top of kame. Kame is about 10 feet high.

Figure 42. Oblique aerial view of Garriety Hill, a partly excavated moulin kame in the northern Kettle Moraine. See figure 43 for scale and location.

## Science Brief 13: Eskers and Crevasse Fillings

Eskers are long, winding ridges of stream sediment deposited in tunnels (fig. 12) beneath the ice. They form in warm glaciers (SB 3) where flowing water can melt upward into the ice (fig. 44) instead of downward into the bed as tunnel-channel rivers do (SB 17). Many of the tunnels in a modern Alaskan glacier leave no trace because they disappear when the ice melts away.

Under some conditions, however, the stream flowing through the tunnel fills the tunnel with sand and gravel (fig. 45). When the surrounding ice melts away, it leaves a long, sinuous ridge called an esker (fig. 46). Eskers range from a few to over 150 feet high, from hundreds of feet to many miles long, and up to as much as a half mile wide. They have steep sides

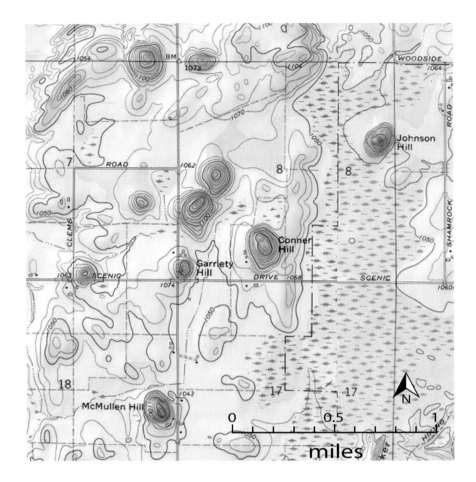

Figure 43. Topography of very large moulin kames in the northern Kettle Moraine. Garriety Hill (fig. 42) is close to the center of the map. (Map is part of the USGS Dundee Quadrangle and was created with TOPO! © 2011 National Geographic Maps.)

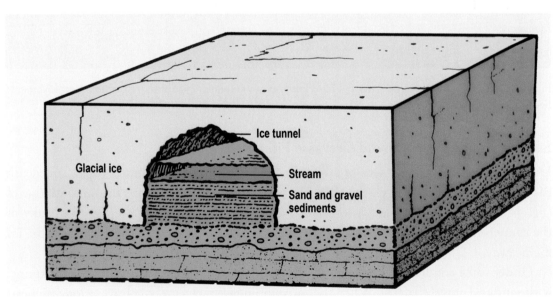

Figure 44. Subglacial tunnel filling with sand and gravel. (Modified from National Park Service; drafted by Mary Diman.)

Figure 45.  Gravel-filled tunnel in warm-bed Burroughs Glacier in southeastern Alaska. Tunnel is about 10 feet high.

(a little over 30 degrees) caused by collapse. Eskers are distributed throughout much of the glaciated area but seem to be most abundant in areas where the glacier was slow moving or stagnant and where there was a supply of debris at or near the base of the glacier, where debris for stream transport is abundant.

Pressure from upstream water in the tunnels pushes water through the lower sections, so water flows from areas of thick ice to areas of thinner ice. The mechanism is much like water in a garden hose with one end held higher than the other. Because of this pressurized system, the water in tunnels can flow uphill in the marginal areas of the glacier. For example, streams that formed eskers along the west side of the Green Bay Lobe flowed toward the west, up the land-surface slope.

Eskers are generally composed of fairly coarse sand and gravel, although in extreme cases they are made up almost entirely of boulders and in other cases of fine-grained clay, silt, and sand. Often the gravel contains far-traveled erratics. People mine eskers for sand and gravel,

Figure 46. Oblique aerial view of meandering Parnell Esker in the northern Kettle Moraine.

and in southern Wisconsin many have been mined away entirely. Eskers are fairly common in northern Wisconsin and in the Kettle Moraine.

Another closely related sign of glaciers on the landscape is the crevasse filling (SB 2). These short, ridge-shaped deposits form in open-to-the-sky cracks in the ice, often along the ice margin. Generally there are several parallel ridges that are closely spaced. Often it is impossible to say whether a single short ridge of sand and gravel is an esker or a crevasse filling.

## Science Brief 14: Drumlins

A drumlin is an elongate hill that forms at the bed of a glacier and parallels the direction of the ice flow. Normally described as a teardrop- or inverted-spoon-shaped hill, the classic drumlin is steeper on the up-ice side, where the glacier ice was forced against the hill, and gentler on the down-ice side, where the ice was less erosive (fig. 47). Drumlins do not occur as individual landforms but rather are present in groups called drumlin fields, evidently

Figure 47. Inverted-spoon-shaped drumlin in Jefferson County. Ice flowed from upper right to lower left. (Photo by Donna Harris.)

because their formation requires numerous hard and soft spots on the glacier bed. The Green Bay Lobe drumlin field, which includes the Campbellsport Drumlins Unit of the Ice Age National Scientific Reserve, contains more than 5,000 hills. A small drumlin field formed below the Lake Michigan Lobe in eastern Wisconsin, and drumlins also formed beneath the ice of the Langlade, Chippewa, and Superior lobes in northern Wisconsin.

Although drumlins in a single area are relatively the same shape, drumlins from different areas of the same drumlin field vary in their length-to-width ratios. Drumlins deposited near the center line of a moving glacier, under the thickest ice, are long and narrow. Toward the glacier's former margins, where the ice was thinner, drumlins have a rounder outline in map view.

You can find examples of different drumlin shapes throughout glaciated Wisconsin. Near Beaver Dam and Juneau, areas once under thick glacier ice, drumlins are narrow ridges across the landscape (fig. 48), whereas drumlins near Wautoma, Madison, Fort Atkinson, and Sullivan, areas at the glacier's edges, are almost equidimensional. They disappear entirely at the glacier's outermost edges, within 5 to 10 miles from the outermost extent of an advance. A number of IAT segments pass over or between drumlins. The Campbellsport Drumlins Unit of the Ice Age National Scientific Reserve also preserves excellent examples of very high, nearly equidimensional drumlins.

Although tens of thousands of drumlins formed during the last glaciation, we find drumlin fields only in certain areas. In the United States and southern Canada, drumlins are present in New England, southern Ontario into New York, Michigan, Wisconsin, Minnesota,

Figure 48.  Elongate drumlin in Dodge County. Ice flowed from lower left to upper right. (Photo by Donna Harris.)

and areas to the west. Yet they are absent from Ohio, Illinois, and Indiana, the southernmost reaches of the glaciers, where flat till plain covers the land instead. Geologists have written much about drumlin formation, but because they cannot see these features form beneath thick ice, they still have more to learn. Drumlins are composed of nearly all bedrock in places, nearly all sand or sand and gravel in other places, and all till in others. It seems most likely that drumlins formed when the bed of the glacier was partly frozen and partly thawed (SB 3). Frozen spots resisted erosion and remained behind as high spots, but areas where the bed was thawed slid, and the sediment beneath deformed, carrying sediment away toward the ice margin and streamlining the more resistant hills.

## Science Brief 15: Ice-Walled-Lake Plains

Ice-walled lakes form when water is dammed in a complex of debris-covered ice. If the lake is on ice, then later melting causes the collapse of the lake sediment, forming hummocky topography. But in cases where a lake accumulates sediment that is on solid ground below, then the flat lake sediment surface is preserved as an ice-walled-lake plain (figs. 38, 49). Glacier ice acts as the container that holds the water. Fine-grained sediment is carried to the more central, deeper parts of the lake. Coarse sediment—typically sand, gravel, and debris-flow sediment—accumulates around the edge of the lake as sediment flows and slides off the surrounding ice.

When the ice walls containing the lake have completely melted away, the lake sediment stands as a relatively flat-topped hill that is higher than the surrounding area. Thus, in this environment it is not unusual to find fine-grained lake sediment on hilltops and bouldery

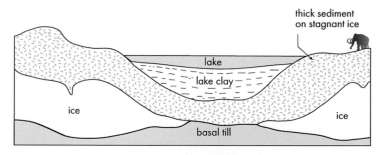

Figure 49. Formation of an ice-walled-lake plain. Note the disappearance of the ice and the collapse of overlying sediment. (Modified from Syverson 2007.)

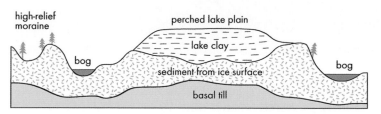

supraglacial sediment in low parts of the landscape. Figure 50 shows a lake on the surface of a modern glacier in Alaska. This is similar to what existed during deglaciation in northern Wisconsin, where ice-walled-lake plains are relatively common in the high-relief hummocky topography (SB 11) of the outermost moraines of the last glacial advance. Several northern Wisconsin IAT segments and the Chippewa Moraine Unit of the Ice Age National Scientific Reserve have excellent examples of ice-walled-lake plains and hummocky topography.

## Science Brief 16: Glacial Lakes

Lakes were much more extensive on the Wisconsin landscape during deglaciation than they are today. Certainly kettle lakes (SB 9) formed as the glacier retreated, and many of these remain as lakes today. Many other lakes formed along the ice margin and either shrunk

Figure 50. Ice-walled lake in the Malaspina Glacier in Alaska. Ice underlies the entire area surrounding the lake. If there is no glacier ice under the lake, the lake sediment being deposited will become an ice-walled-lake plain. (Photo by Thomas Gustavson.)

or completely disappeared after the retreat of the glacier. Figure 51 shows an ice-dammed lake in Greenland. In this case, the lake drains along the glacier edge to a lower lake on the right side of the photo. If the glacier were to advance, this outlet would be blocked, and the lake level would rise to the next lowest place for water to escape. Although we have no idea what topography beneath the glacier in figure 51 is like, if the ice were to retreat and expose lower ground, then the outlet along the present ice margin would be abandoned, and the lake level would drop. Wherever the land near the ice margin sloped toward the glacier, water was dammed by the glacier. In this way, ice advance usually raised lake levels by blocking the lake outlets. Likewise, ice retreat lowered lake levels by opening new lower water outlets.

The largest ice-marginal lakes in Wisconsin were what are now Lakes Michigan and Superior (SB 20). Another large lake was glacial Lake Wisconsin (fig. 52), which formed

Figure 51. Ice-dammed lake at the edge of the Greenland Ice Sheet. Note the lake outlet is against the ice margin at the right side of the lake. (Photo by R. P. Goldthwait.)

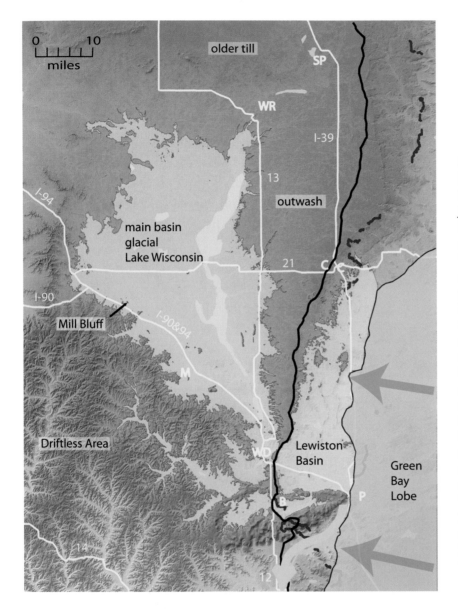

Figure 52. Shaded relief of glacial Lake Wisconsin after the Green Bay Lobe withdrew from the Johnstown Moraine (heavy black line). Blue arrows show ice-flow direction. Basins of glacial Lake Wisconsin and associated lakes are shown in gray. Light blue is present-day water. Red lines are IAT segments labeled in figures 137 and 183. Green arrow in southern part of the map shows the position of the outlet of glacial Lake Wisconsin when the lake rapidly drained around the east end of the Baraboo Hills. Yellow lines and numbers indicate highways. Cities shown: (B) Baraboo, (C) Coloma, (M) Mauston, (P) Portage, (SP) Stevens Point, (WD) Wisconsin Dells, (WR) Wisconsin Rapids. (Base map constructed from USGS National Elevation Dataset and modified by WGNHS.)

when drainage from the north (pretty much what is the Wisconsin River today) was blocked by the Green Bay Lobe near the city of Portage. Water level in the lake rose until it could flow westward into the Black River and then the Mississippi. The history of this lake is described further in the introduction to the western Green Bay Lobe and the description of the Mill Bluff Unit of the Ice Age National Scientific Reserve.

Another large ice-dammed lake was glacial Lake Scuppernong (fig. 53), which formed in front of the retreating Green Bay Lobe in southern Wisconsin. At that time land to the north

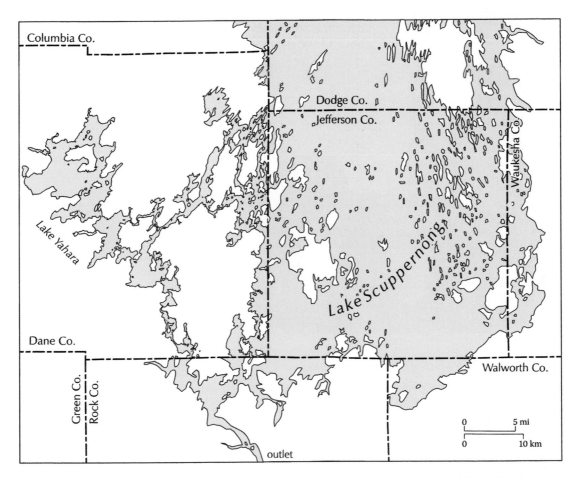

Figure 53. Approximate boundaries of glacial Lake Scuppernong and glacial Lake Yahara. These lakes began to form as the Green Bay Lobe retreated from the Milton Moraine, and they grew larger as ice continued to retreat almost as far north as Fond du Lac (not on map) before the lakes drained. (Modified from Clayton 2001.)

was isostatically depressed (SB 20) because of the weight of the overlying retreating ice. Thus, much of the land sloped toward the glacier, and this large, and probably shallow, lake grew in size as the glacier retreated then shrank over several thousand years as rebound took place. Much of this lake bottom was poorly drained peat before European settlers drained the land for agriculture. It appears that glacial Lake Scuppernong was connected for at least a short time with glacial Lake Yahara, which formed because drainage in the present Yahara River valley was dammed south of Stoughton by debris-covered ice in the Milton Moraine.

Another large lake formed in the Lake Winnebago basin as the glacier continued to retreat. This lake drained and redeveloped several times during successive late glacial ice retreats and advances (SB 18). Figure 54 shows glacial Lake Oshkosh at one stage when it

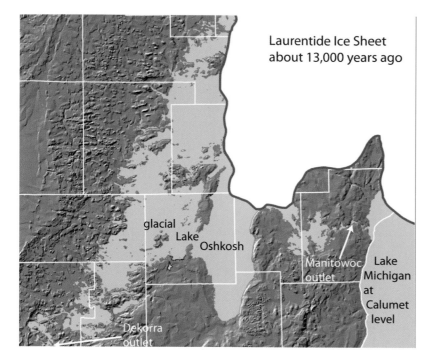

Laurentide Ice Sheet
about 13,000 years ago

glacial Lake Oshkosh

Manitowoc outlet

Lake Michigan at Calumet level

Dekorra outlet

Figure 54. Glacial Lake Oshkosh (blue) in eastern Wisconsin during retreat of the Green Bay Lobe (white) about 13,000 cal. years before present (BP). The Dekorra outlet near Portage is indicated in the lower left corner of the map, and the Manitowoc outlet is just south of the ice sheet in Calumet and Manitowoc Counties. A high stage of Lake Michigan (blue) is shown along the eastern edge of the map. (Modified from Hooyer 2007.)

drained into Lake Michigan through the valley of the present-day Manitowoc River. When the lake was at a higher level, the outlet was at the south end of the lake near Portage (see the description of the Marquette Segment). When the lake was lower, the outlets of glacial Lake Oshkosh included the present-day Kewaunee and Ahnapee river valleys (see the description of the Forestville Segment). In northern Wisconsin the land generally sloped away from the ice, so that except for Lake Superior, huge lakes were not present. Along the IAT there is evidence of many small, unnamed ice-marginal lakes.

## Science Brief 17: Tunnel Channels

We see the work of water everywhere: rivers usually slowly cut through rock over millennia, but sometimes they can rapidly erode channels into the land during a single flood. Deep under glacial ice, rivers also shaped the landscape. Subglacial rivers cut into the underlying glacial bed, forming tunnel channels that remain today. In Wisconsin, such channels occur only along the outermost edge of the late Wisconsin advance, where they cut through the moraines and end in large alluvial fans (fig. 55). The channel bottoms rise and fall in elevation along their length, indicating that the water carrying the sediment through the tunnels was under considerable pressure. In many areas the channels are partly filled with hummocky sand and gravel (SB 11), and in some places eskers (SB 13) are present along their floors. These hummocky deposits and eskers formed after the actual channel, when the ice warmed and the sediment collapsed into the tunnel from above as the overlying debris-rich ice melted.

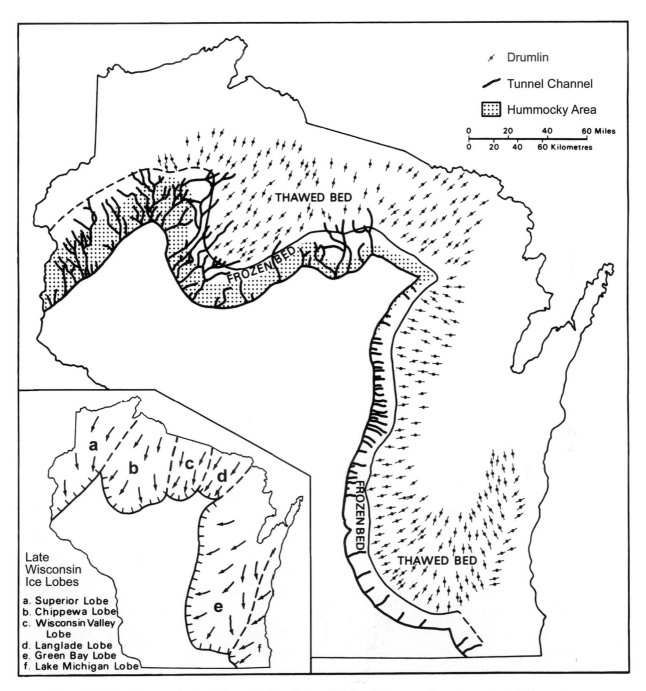

Figure 55. Distribution of tunnel channels, drumlins, and high-relief hummocky topography in Wisconsin. (Modified from Clayton et al. 1999.)

It appears that tunnel channels are restricted to areas where there was a cold glacier (SB 3) with a frozen bed near the ice margin and a large supply of water under thicker ice behind the margin (fig. 11). Because they form above a frozen glacial bed, tunnel channels, like drumlins, did not form in Illinois, Indiana, and Ohio, areas of the southernmost extent of the ice sheet (SB 14). The flows of water that cut the tunnel channels in Wisconsin were apparently sudden, short lived, and huge. Cutler et al. (2002) estimated that flows of thousands of cubic feet per second carried boulders up to 6 feet in diameter to the outwash fans!

Subglacial floods occur on many modern glaciers. They are called jökulhlaups, a term from Iceland, where they occur commonly. They produce enormous amounts of outwash, much of it bouldery, as water pours from the tunnel mouth for only a few days before slowing and finally ceasing (fig. 56). The tunnel then closes, partly filling with a mix of ice, boulders, and gravel that later collapses to produce hummocky topography.

Figure 56.  Oblique aerial view of flood from subglacial tunnel in Iceland. This 1996 photo shows the River Skeidara while its discharge was still rising in the huge, fast-flowing jökulhlaup that followed the Gjalp Eruption. The bridge is about 3,000 feet long. (Photo by Magnus T. Gudmundsson, University of Iceland.)

## Science Brief 18: When Was the Ice Here? How Do We Know?

We know that the last glaciation in the Great Lakes area, the late Wisconsin Glaciation, lasted from approximately 30,000 cal. years ago to 10,000 cal. years ago. Knowing the details of when the glacier advanced and retreated in Wisconsin requires being able to date some type of geologic material. Before the late 1940s there was no accurate way to establish when glaciers were here, and estimates ranged to as long ago as 35,000 years for the disappearance of the last glacier. About 1950, radiocarbon ($^{14}$C) dating became possible, allowing much more accurate estimates of age (Libby 1952). This method requires finding organic material—plant or animal remains, such as wood, bone, or peat—that was buried at the time of the event being dated and that has not been exposed to the atmosphere since.

Many thousands of radiocarbon analyses have been done in the last 50 years, and a chronology of events has been developed in many parts of the world. However, in Wisconsin it appears that the climate was cold enough that trees were not here during the time of ice advance. Although there are numerous radiocarbon dates for the advance of ice into central Illinois, there is not a single one for the advance of the Green Bay Lobe in Wisconsin. A radiocarbon date near Sheboygan suggests that the Lake Michigan Lobe advanced down that basin about 30,000 cal. years ago. In Wisconsin there are numerous radiocarbon dates on wood younger than about 14,000 cal. years because the climate had warmed and forests were widespread.

Scientists have long recognized that "years" in the radiocarbon chronology are not the same as calendar years. The discrepancy results from the cosmic ray flux (the abundance of energy-charged subatomic particles from space), which varies in time and causes the concentration of $^{14}$C in the atmosphere (and in living things) to vary as well. Using independent dating of objects that are also radiocarbon dated, it has been possible to develop a way to convert radiocarbon years to calendar years. Whenever events are based on the radiocarbon chronology, the dates should always be expressed as either radiocarbon, or $^{14}$C, years BP or calendar years (cal. years BP). In this book almost all radiocarbon dates are calibrated to calendar years. Another thing that can be confusing is that in the geologic community, radiocarbon dates are given in years before present (BP), with "present" defined as 1950. Archaeologists commonly refer to events before the year 0 in the Christian calendar as years BC or after as years AD.

In the last decade scientists have developed other approaches to dating glacial events. Cosmogenic dating, which uses isotopes of $^{10}$Be (beryllium) or $^{27}$Al (aluminum), is one such approach. In this case scientists do not need organic materials. Boulders deposited on the ground surface during the last Ice Age can be dated using cosmogenic dating if they have been continuously exposed to atmospheric radiation. Another new dating approach is called optically stimulated luminescence, or OSL. This method works on fine-grained sediment that was exposed to light at some time in the past and then buried. When sediment is buried, radioactive minerals in the sediment cause damage to the structure of some minerals. Scientists can sample sediment in the field, expose it to light in the laboratory, and measure the amount of luminescence of the very common mineral quartz grains to estimate the age of the sample. Neither of these methods has the accuracy of radiocarbon for dating the last glaciation, but they are still being developed and are crucial to aid scientists in dating glacial events in areas lacking organic materials for radiocarbon dating. Unlike the $^{14}$C method, cosmogenic and OSL dating produce calendar-year dates.

So, how do the timescales of these three dating methods compare? There are two calibration programs available on the web that translate dates from radiocarbon years BP to calendar years and vice versa. In table 2 we list some examples of the relationship between the two timescales. Note that the relationship is not linear; for an accurate conversion you really need to visit one of the websites listed below and download or use the online calibration program. What is important to remember is that the radiocarbon time scale is an "elastic" timescale. Be careful when you use these dates! For further information visit these websites: http://www.calpal-online.de/ or http://calib.qub.ac.uk/calib/.

Table 2. Examples of the difference between $^{14}$C dates and calibrated calendar-year dates using the CalPal program.

| $^{14}$C age (BP) | Calibrated age (cal. year BP) |
|---|---|
| 10,000 | 11,568 |
| 11,000 | 12,927 |
| 12,000 | 14,016 |
| 13,000 | 15,967 |
| 14,000 | 17,390 |
| 15,000 | 18,270 |
| 16,000 | 19,220 |
| 17,000 | 20,291 |
| 18,000 | 21,520 |
| 19,000 | 22,876 |
| 20,000 | 23,908 |
| 21,000 | 25,131 |
| 22,000 | 26,589 |
| 23,000 | 27,819 |
| 24,000 | 28,922 |
| 25,000 | 29,947 |

In summary, glaciers advanced into Wisconsin about 25,000 [14]C years BP (30,000 cal. years ago). The glacier edge probably remained close to its outermost advance position for more than 6,000 to 8,000 years before it began a slow retreat interrupted by many minor readvances. There are very few radiocarbon dates of this retreat in north central and northwestern Wisconsin, but it was probably later than the retreat recorded in eastern Wisconsin, which scientists think is more accurately dated. There the glacier had retreated out of the state by just after 14,500 [14]C years BP (17,300 cal. years ago). Ice readvanced into northeastern and east central Wisconsin by 13,700 [14]C years BP (16,300 cal. years ago), extending to the south end of Lake Winnebago and to Milwaukee, and again at about 12,900 [14]C years BP (15,300 cal. years ago). The Two Creeks Forest grew between 12,100 [14]C years BP (13,900 cal. years ago) and about 11,600 [14]C years BP (13,400 cal. years ago); it was then buried by the Two Rivers ice advance, which extended to the north end of Lake Winnebago in the Green Bay lowland and to Two Rivers in the Lake Michigan basin. Ice had disappeared from eastern Wisconsin by about 11,000 [14]C years BP (12,900 cal. years ago), but a late, short-lived advance out of the Lake Superior basin took place about 9,900 [14]C years BP (11,400 cal. years ago).

## Science Brief 19: The Niagara Escarpment

A prominent landscape feature in eastern Wisconsin is the Niagara Escarpment. The escarpment is the steep face of a cuesta, or asymmetrical ridge with one steep side and one gently sloping side, that extends from east of Niagara Falls, New York, forms the backbone of Door County, and continues southward through Walworth County (fig. 57). It extends southward into Illinois and Iowa as a less pronounced ridge. Cuestas form where bedrock dips at a low angle in one direction, exposing rocks of differing hardness to erosion on the land surface. Relatively hard Silurian dolomite caps the Niagara Escarpment and is more resistant to erosion than the softer Maquoketa shale and dolomite below (fig. 58). The top of the escarpment was probably higher than the land to the west or east for millions of years before glaciation, but the erosive power of the Green Bay Lobe enhanced the escarpment's height relative to the floor of the Green Bay–Winnebago basin.

## Science Brief 20: The Ups and Downs of the Great Lakes

We have known since the late 1800s that water in the Great Lakes has been much higher and much lower during different periods in the past than it is now. Beaches from the high water levels—sometimes called strandlines—are preserved in places around both Lake Superior and Lake Michigan. We concentrate on Lake Michigan history here because the east end of the Ice Age Trail climbs across some of these old beaches and wave-cut cliffs. For a detailed discussion of the history of Lake Superior and its effect on the St. Croix River valley, see the

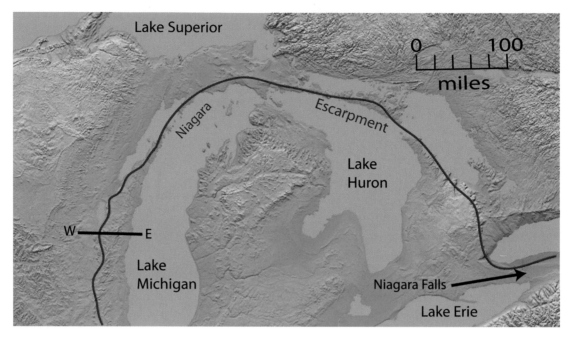

Figure 57. Approximate location of the crest (red line) of the Niagara Escarpment between Wisconsin and Niagara Falls, New York. W–E black line shows approximate location of cross section in figure 58. (Digital Elevation Model from NASA/JPL website: http://photojournal.jpl.nasa.gov/catalog/pia03377.)

last section of this book—Interstate State Park. Past changes in the lake level—which we can see evidenced by beaches above the present lake level and tree stumps below it—happened for several reasons: isostatic depression and subsequent rebound, changes in the amount of water flowing into the lake from precipitation and glacial meltwater, changes in evaporation, ice advance covering lower outlets and causing the lake level to rise to a higher outlet, and ice retreat opening lower outlets and allowing water level to drop. Let's look at them one at a time.

### ISOSTATIC DEPRESSION AND REBOUND

Isostatic depression of the earth's crust takes place because of the load of the ice sheet. Given sufficient time, the massive ice sheet can depress the land surface a distance equal to about 30 percent of the ice thickness. During and shortly after the last glaciation, when ice sheets well over 1,000 feet thick pressed on the earth's surface, the elevation in what is now northern and eastern Wisconsin was substantially lower (more than 300 feet) than it is today. Isostatic rebound takes place as the ice thins and retreats, and continues at an ever slower rate for thousands of years after the ice is gone, until the earth's surface returns to about the elevation it had before the glaciation. Rebound explains why beaches that formed when the

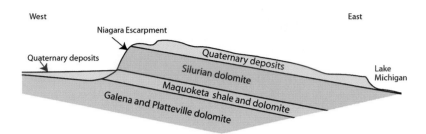

Figure 58. Diagrammatic "slice" through the Niagara Escarpment showing the distribution of deposits. This is generalized and could be anywhere from Door County to Walworth County along a line similar to the W–E line in figure 57. The dip of the rock is greatly exaggerated. The actual angle of dip is only a few degrees. Unit thicknesses vary from place to place. Not to scale. (Drafted by Mary Diman.)

ice sheet was just retreating have been uplifted and tilted. This is especially noticeable in Door County, where shorelines have been uplifted as much as 75 feet relative to beaches of the same age in southern Manitowoc County. This tilting of beaches across the landscape occurs because there was more rebound in Door County than there was farther south after the beaches had formed. This is partly due to the thicker ice over Door County and also the fact that it was deglaciated later and therefore rebounded later than land to the south.

CHANGES IN PRECIPITATION AND EVAPORATION

In the last 150 years, the water level in Lakes Michigan and Superior has fluctuated by about 3 feet in either direction because of changes in precipitation and evaporation. Over the last 4,000 years, the lakes rose and fell 6 to 8 feet. Most of these fluctuations were the result of changes in the balance between precipitation and evaporation. Water level changes due to these effects were small compared to those caused by rebound and changing lake outlets.

CHANGES IN LAKE OUTLETS

Today water in Lake Michigan flows into Lake Huron, which is at the same level. The outlet at Port Huron, Michigan, is the outlet of Lakes Michigan and Huron today, and it has been for about the last 5,000 years. However, Lake Michigan has had several other outlets at times in the past. Whenever the northern end of Lake Michigan was covered by glacier ice, the present-day path into Lake Huron was blocked, and the lake rose to drain down the Illinois River at Chicago (fig. 59). It was a typical ice-dammed lake (SB 16). Scientists recognize three distinctly higher levels of Lake Michigan (fig. 60): Glenwood (present elevation about 640 feet), Calumet (present elevation about 620 feet), and Toleston (present elevation

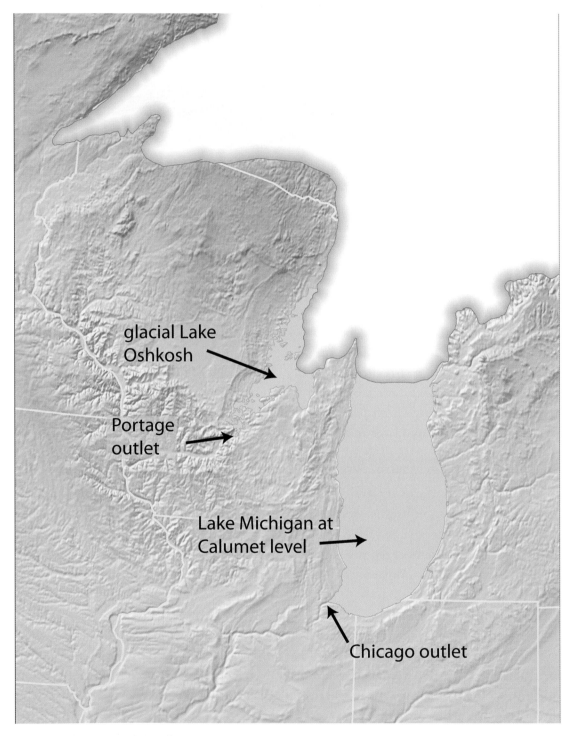

Figure 59. Glacier ice and ice-marginal Lakes Oshkosh and Michigan about 13,000 cal. years BP, when ice covered the northern end of the Lake Michigan basin. (Map by Michael Bricknell, Wisconsin State Cartographer's Office.)

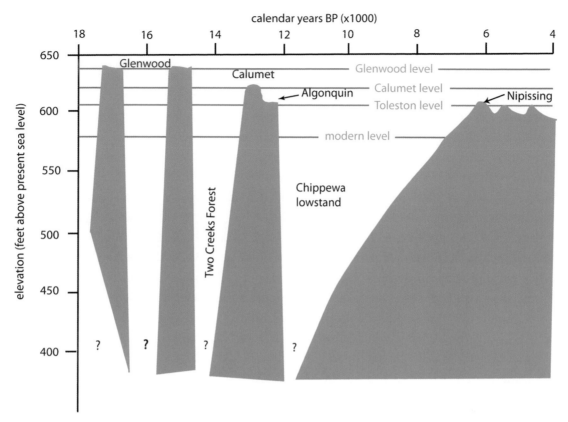

Figure 60. Fluctuations in the level of Lake Michigan (top of solid blue areas) between 18,000 and 4,000 cal. years BP. Lake levels are labeled in blue, lake stages in black. Elevation of low water stages is unknown but probably more than 300 feet below the modern level. (Modified from Hansel and Mickelson 1988.)

about 605 feet). Now Lake Michigan sits at a level of about 580 feet. These elevations of former levels are at the Chicago outlet, and minor differential rebound causes the Glenwood shorelines in Manitowoc County to be about 10 feet higher. The same Chicago outlet was used when the lake was at the successively lower Calumet and Toleston levels. Note that these names refer to levels and that the Glenwood and Toleston levels were attained more than once.

Whenever glacier ice retreated north of the northern end of Lake Michigan, the water level in the lake dropped as outlets lower in elevation than Chicago were uncovered. The land at the north end was much lower than it is today because it had not yet had time to rebound from the thick ice cover. Low northern outlets were progressively opened as the glacier retreated. They carried water into the eastern Great Lakes or across southern Ontario to the St. Lawrence River. Lake Superior now drains into Lake Huron at Sault St.

Marie. Whenever ice occupied the eastern part of the basin, the water level was much higher than it is today, and the lake drained to the southwest through an outlet in Minnesota and then through the St. Croix River valley in Wisconsin.

For a period of time after the ice retreat from Wisconsin, what are now Lakes Michigan, Huron, and Superior were at the same level; this large lake was called Lake Algonquin, and it formed a shoreline that is now at about 605 feet (Toleston level) in Manitowoc County but substantially higher in Door County. Subsequent glacier retreat in Ontario opened very low outlets, and Lake Algonquin shrank to three individual lakes. The level of Lake Michigan fell to as much as 300 feet below its present level. Then, as the northern outlets rose because of isostatic rebound, the level of Lake Michigan rose as well. By about 5,000 years ago the lake had again risen to the Toleston level at 605 feet, and water again drained through the Chicago outlet. This highstand of the lake is called the Nipissing phase of Lake Michigan, and features from this highstand are crossed by the IAT in several places (see especially the Point Beach Segment description). Because the land rebound in Door County was significant between Lake Algonquin and the later Nipissing phase, the Algonquin shoreline is higher than the Nipissing shoreline. The Sturgeon Bay Segment of the Ice Age Trail crosses both of these shorelines in Door County. Farther south in Manitowoc County, little or no differential rebound took place between the formations of the two shorelines, and therefore the beaches are indistinguishable.

The history of ice retreat from the Lake Superior Basin and the changes in the level of Lake Superior are detailed in the discussion of the Interstate State Park Unit of the Ice Age National Scientific Reserve.

## Science Brief 21: Rock Recognition 101

Rocks are generally divided into three categories: igneous, which are formed when a silica-rich magma cools and hardens; sedimentary, which are aggregations of weathered rock and mineral particles that have been deposited from water, wind, or ice; and metamorphic, which are formed from preexisting rocks that have been recrystallized upon being subjected to high temperature and pressure (see "An Overview of Wisconsin Geology" in the introduction). Some common rocks found along the IAT are defined below, and their distribution is shown in figure 4.

### IGNEOUS ROCKS

We classify igneous rocks on the basis of their mineral content, which we can often tell by their texture and color, and the size of their mineral grains. A rock with relatively coarse interlocking mineral grains that are predominantly orthoclase feldspar (pink to orange) and quartz (gray, sometimes clear), with lesser amounts of plagioclase feldspar (commonly white) and iron-magnesium silicates (black), is called granite. Mica is a common mineral in granite as well. The fine-grained (too fine to see individual grains) equivalent is rhyolite (often orange to purplish gray). A rock with minor amounts of orthoclase feldspar and quartz but with major amounts of plagioclase feldspar is called diorite. These rocks are somewhat darker than granite. Even darker in color is gabbro, which is composed mainly of plagioclase feldspar and dark silicate minerals; its fine-grained equivalent is called basalt.

### SEDIMENTARY ROCKS

Sediments and sedimentary rocks make up only about 5 percent of the volume of the outer 10 miles of the earth; however, they cover about 75 percent of the land surface. They might be considered "nature's dump," because they represent the debris left from the decay of preexisting rocks. Because these deposits are made of material from many different sources, the classification of sediments is perhaps less clear-cut than for igneous rock. As a first approximation, sedimentary rocks are often separated into three categories: clastic, chemical, and organic.

Clastic means "broken fragments." These broken pieces can be described by their shapes (angular to rounded) and degree of sorting (well sorted to poorly sorted). The size of the particles is used to name clastic sediments. For example, sandstone differs in the size of its particles from claystone (table 3). The sedimentary particles are often cemented together with mineral material, the most common of which are silica, iron, and lime (calcium carbonate).

Chemical sedimentary rocks include rocks that have formed by chemical precipitation. Certain types of limestone ($CaCO_3$) and dolomite ($CaMg(CO_3)_2$) are examples, as are rock salt, rock gypsum, and chert ($SiO_2$).

Table 3. Simplified particle size scale.

| Diameter of particles (mm) | Names of individual pieces | Names applied to aggregates | Name of rock |
|---|---|---|---|
| >4 | Rounded: pebbles, boulders | Gravel | Conglomerate |
| | Angular: blocks | Gravel | Breccia |
| 2–4 | Granules | Grit | Conglomerate |
| 1/16–2 | Sand grains | Sand | Sandstone |
| 1/256–1/16 | Silt particles | Silt | Siltstone |
| <1/256 | Clay particles | Clay | Claystone or mudstone |

Sedimentary rocks that owe their existence to organisms are called organic. Such rocks include limestone that is composed in large part of shells or coral debris. Coal, formed from compressed, ancient plants, is another example of organic sedimentary rock.

Some rocks are very difficult to put into this simple three-part classification scheme. A limestone made up of broken shell fragments can be said to be clastic, organic, and also perhaps of chemical origin.

Common sedimentary rocks along the Ice Age trail are sandstone and dolomite. Nearly all of the dolomite was probably deposited as limestone ($CaCO_3$) and later converted to dolomite ($CaMg(CO_3)_2$) as groundwater passed through the rock.

The most characteristic aspect of sedimentary rocks is their bedded, or layered, nature. Close study of the bedding and the features the beds contain can often provide useful information about the rocks. Some common features are cross-bedding (fig. 61), ripple marks (fig. 62), and mud cracks. Because sedimentary rocks are deposited layer upon layer, the

Figure 61.  Cross-bedding in sandstone in western Wisconsin. Rock hammer is about 1 foot long. Current was from left to right.

Figure 62. Ripple marks on Baraboo quartzite (top). Modern ripples on sand (bottom). Ripple spacing is about 2 inches, and both are at Devil's Lake State Park. (Photos by Diann Kiesel.)

rocks above are almost always younger than the rocks below. In sedimentary rocks that have been tilted or overturned by earth movements, geologists use a number of techniques for determining which direction was originally up, including the features mentioned above. There are many other methods; looking at fossils of rooted tree stumps is but one.

METAMORPHIC ROCKS

The term "metamorphism" means "change in form" and usually includes those processes that change rocks at higher temperature and pressure than we experience at the earth's surface. Minerals in equilibrium at the surface of the earth may no longer be stable when they are deeply buried and subjected to heat. New minerals usually form as existing ones disappear. Quartzite is metamorphosed sandstone, marble is metamorphosed limestone or dolomite, and schist and gneiss can be derived from the metamorphism of several sedimentary and igneous rocks.

In general, the older the rock, the more likely it is to be metamorphosed. Rocks that were deeply buried and subjected to heat and pressure over long periods of time tend to be metamorphic. A vast area of Precambrian metamorphic rock produced by this regional metamorphism is present in northern Wisconsin, Minnesota, the Upper Peninsula of Michigan, and Canada. Rocks can also be metamorphosed close to the earth's surface when they are intruded and baked by igneous magma (contact metamorphism).

In summary, the IAT crosses mostly Silurian-age dolomite in eastern Wisconsin and a mix of Ordovician sandstone and dolomite in southern Wisconsin (fig. 4). The trail goes over Cambrian sandstone in central Wisconsin and north of Eau Claire in northwestern Wisconsin. The remainder of the trail crosses metamorphic and igneous rocks that are described in more detail in the appropriate segment descriptions.

## Science Brief 22: Understanding Topographic Maps

All of the trail segments in this book include part of a topographic (or topo) map. Topo maps show you the lay of the land, including how much elevation you'll climb along a route, where and in what direction water runs, and the trail or road intersections you will come across. Such maps are very useful tools once you know how to read them, and they are fundamental to understanding the landscape features described in this book. The maps we show are limited by page size and readability. We recommend purchasing the full map or getting topographic map software that will allow you to see the landscape features over a larger area than we can show.

Here are a few step-by-step instructions to orient you to topographic maps and help you interpret contours to see the 3-D world around you in a whole new way.

SCALE

A scale tells you how much horizontal distance on the map equals a mile in the real world. A map that shows a small area close-up is called a large-scale map. On a map that is zoomed in to show only a small area, like that in figure 63, the scale bar may show that 1 inch equals 1 mile or even less. Conversely, a map that shows a large area is a small-scale map, and 1 inch may equal 10 or more miles. Most of our small-scale maps are shaded-relief digital elevation models. These are constructed from elevation data just as contour maps are, but shading is used to show changes in elevation.

CONTOUR LINES

The topo map looks etched with parallel lines. Each of these contour lines follows a constant elevation across the landscape, and the interval between the lines is a constant number of vertical feet. The contour interval on nearly all of the maps we use is 10 feet. If your map doesn't show you this in the legend, you can tell the contour interval by counting the number of contour lines between printed elevation numbers and dividing the change in elevation by the number of lines. In figure 63, we can see the 900-foot label (summit) just above the red line, and the 950-foot label is on the fifth contour line away, so we know each interval is 10 feet. To know how the land just around the next trail bend is going to change, you can look at how widely the lines are spaced. Look at the profile in figure 63 and compare the slopes with the appearance of the contours on the map. Widely spaced lines tell you the land is gently sloping. Closely spaced lines tell you elevation changes a lot over a short distance, so you should be ready for a steep climb or even a drop off ahead! The left end of the red path starts in a low spot at river level, crosses closely spaced contour lines next to the river (definitely steep!), then flattens out between 0.1 to 0.2 miles, where the contour lines are farther apart. At 0.3 miles, it climbs again through tightly spaced contour lines, showing that it covers a lot of vertical distance in a very short amount of horizontal space. Most trail builders have been well trained to construct trails with no more than an 8 percent sustained grade, with slightly steeper sections only for very short distances (Birkby 2005). Let your eye follow the rest of the profile line on the map and think about what the actual trail slope would look like.

LANDSCAPE FEATURES

You can also find landmarks waiting for you on your route using a topo map: look for blue lakes, rivers, creeks, and wetlands (see the symbol on fig. 63). A thin dashed-and-dotted blue line means the flow of a creek is intermittent. You can even spot valleys with no water by looking for a series of V shapes in the contours, with the V opening downhill (fig. 63).

Contours can tell you when you've finally reached the top—or the bottom—of a hill! Summits have a closed contour, meaning they are a complete oval or circle shape. In figure 63 the summit of the hill just above the profile line is higher than 900 feet but below 910 feet. Closed depressions like kettles often show up as closed contours with hachures as shown in figure 63. In that case, the bottom of the depression is below 900 feet but above 890 feet.

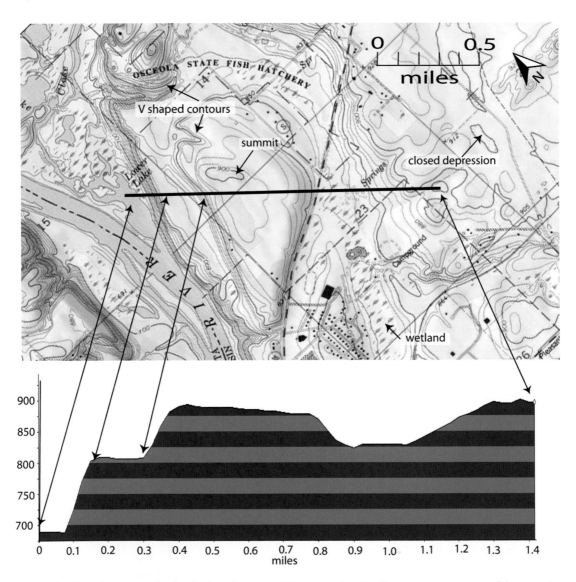

Figure 63. Sample topography (top) of northwestern Wisconsin showing features seen on many of the maps in this book. Compare the profile (bottom) with the pattern of contours on the map. Also note other features, such as scale, north arrow, water bodies, etc., described in text. (Map was created with TOPO! © 2011 National Geographic Maps.)

MAGNETIC DECLINATION

Most maps have a north arrow that points to true north. Across the state, the declination ranges from -4 degrees in northwestern Wisconsin to +1.5 degrees in Door County (http://www.sco.wisc.edu/maps/magneticdeclination.php), so your compass will be pointing very close to true north, and you don't have to worry about adjusting for it. Figuring out the declination isn't an issue for people who use a Global Positioning System (GPS) unit, either, because these automatically realign their reading so they point to true north. Note that although north is commonly toward the top of our maps, that is not always the case, as figure 63 demonstrates!

Whether you're using a compass or GPS with this book, take a few minutes to read the lay of the land before you stride onto the trail. And as you're hiking, check the book's maps and reorient yourself occasionally: the maps will tell you about the features just off the trail through the trees and help you see the land's shape under the vegetation that can cover it like a bulky coat. You'll learn to recognize the long, irregular furrows of tunnel channels and teardrop-shaped drumlins and to find a bluff-top lunch spot with a view of all you've hiked so far.

If you have special interest in finding routes, you may also want to take part in Earth-Caching, a GPS or compass activity to navigate to landmarks in many of the nation's protected lands. Participants are given the coordinates of a place, object, or living thing (such as a blooming tree), and they must plan their route—staying on trail—to find the "prize" at the end. There they take a picture that they submit to the organizers to show they found it. On the Ice Age Trail, hikers can try the glacial version of EarthCaching: ColdCaching! Neither one requires participants to leave any objects on the land (as Geocaching, a similar activity outside protected areas, often does). ColdCaching is a fun way to explore the landscape, and it complies with rules most parks and forests have about Leave No Trace land use. Contact the Ice Age Trail Alliance or visit www.iceagetrail.org for more information about ColdCaching on the Ice Age Trail. Have fun navigating!

# Northeast Ice Age
# Trail Segments

THE NORTHEASTERN SEGMENTS OF THE IAT are located between Potawatomi State Park, near Sturgeon Bay at the eastern end of the trail, and the city of Manitowoc (fig. 64). The landscape of this part of Wisconsin is dominated by the Niagara Escarpment (SB 19), the west face of an asymmetrical ridge, or cuesta, that forms the backbone of Door County. This escarpment affects much of the landscape of the Great Lakes area; it extends all the way to Illinois and Iowa to the south and to Niagara Falls to the east (SB 19). The Silurian dolomite (SB 21) that makes up the escarpment is more resistant to erosion than the softer rocks underlying Lake Michigan and Green Bay. Erosion before and during the Ice Age has increased the relief along this escarpment. Excellent exposures of the dolomite are present along other parts of the trail south to Waukesha County but are especially impressive along the Sturgeon Bay Segment in Potawatomi State Park.

Glaciers modified the underlying bedrock by eroding and shaping the rock surface. Much of the Door County landscape has a very thin cover of till (SB 5) or sand and gravel (SB 7) over the rock. When the edges of the Green Bay and Lake Michigan lobes were much farther south, the bed of the glacier was probably sliding right on the bedrock surface throughout much of this area. As it first retreated from here just over 18,000 cal. years ago, it deposited the light-brown sandy till of the Holy Hill Formation, similar to the material that the Green Bay Lobe deposited at its outermost margin. Ice retreat exposed the eastern Wisconsin landscape for almost 1,000 years before the glacier readvanced into this part of the state. As it advanced, it flowed over lake sediment that was reddish brown in color and quite clayey. A series of minor readvances then deposited this clayey, reddish-brown sediment as till as far south as Fond du Lac in the Green Bay basin and Milwaukee in the Lake Michigan basin. These reddish-brown till layers make up the Kewaunee Formation. The deposits of

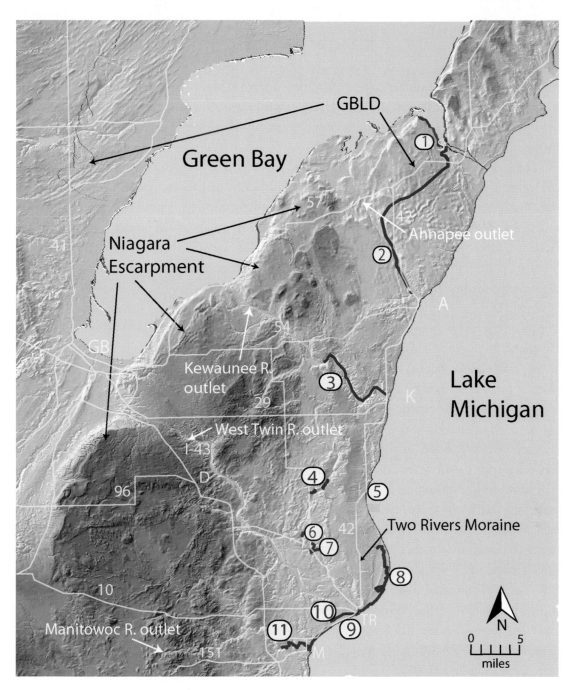

Figure 64. Digital elevation model showing the northeast IAT segments (red): (1) Sturgeon Bay, (2) Forestville, (3) Kewaunee River, (4) Tisch Mills, (5) Two Creeks Buried Forest Unit of the Ice Age National Scientific Reserve, (6) East Twin River, (7) Mishicot, (8) Point Beach, (9) City of Two Rivers, (10) Dunes, (11) City of Manitowoc. Yellow lines and numbers indicate major highways. Cities shown (yellow): (A) Algoma, (D) Denmark, (GB) Green Bay, (K) Kewaunee, (S) Sturgeon Bay, (TR) Two Rivers. GBLD: Green Bay Lobe drumlins. Outlets of glacial Lake Oshkosh are shown. Sturgeon Bay, at the City of Sturgeon Bay (S), was another outlet. Blue arrows show ice-flow direction during latest ice advance. (Base map constructed from USGS National Elevation Dataset and modified by WGNHS.)

the Kewaunee Formation are typically only a few feet to 20 feet thick and in places are missing entirely. Most of the glacial landforms that you can see at the scale of a topographic map are made up of Holy Hill Formation deposits and have only a thin cap of clayey, reddish-brown till on them.

The glacial advances that took place about 15,400 cal. years ago and again about 14,500 cal. years ago advanced across a cold tundra landscape. Evidence of fossil pollen suggests that only plants such as those that thrive on the north slope of Alaska today could survive in this environment. A dramatic warming, recorded in much of the world, including cores from the Greenland Ice Sheet, then caused rapid ice retreat about 14,000 cal. years ago. Very soon after this warming period, the glacier advanced again into northeast Wisconsin, depositing another layer of reddish-brown Kewaunee Formation till. This time, the glacier advanced into spruce forests. We know this because we find buried spruce wood beneath the youngest till in areas extending south to Appleton, in the Green Bay basin, and to Two Rivers, in the Lake Michigan basin. You can visit a good example of this buried land surface at the Two Creeks Buried Forest Unit of the Ice Age National Scientific Reserve, but unless there is further erosion on the bluff, it may be difficult to see.

The cooling that caused the resurgence of the glacial advance is known as the Younger Dryas cooling. It was short lived. By shortly after 13,000 cal. years ago, glacier ice had retreated from Wisconsin and, except for an advance into northwest Wisconsin from the Lake Superior basin, never returned to the state.

However, the presence of the glacier to the north continued to affect our landscape, especially that along the present Great Lakes shorelines. To understand the development of this landscape we must investigate the history of the Lake Michigan basin since the glacier began to retreat into it about 18,000 cal. years ago. As pointed out in SB 16, lakes were common around the edge of the retreating ice sheet. The Lake Michigan basin and the Green Bay–Lake Winnebago basins were no exceptions. The large lake that formed several times in the Green Bay–Lake Winnebago basins is called glacial Lake Oshkosh (fig. 59). The lake that formed in the Lake Michigan basin has been called glacial Lake Chicago, but in this book we prefer instead to consider the various levels of this lake as different past levels of Lake Michigan (SB 20).

## 1. Sturgeon Bay Segment
*Potawatomi State Park to CTH HH in Maplewood (13 miles)*

The eastern end of the Sturgeon Bay Segment is also the eastern terminus of the IAT (figs. 64, 65, 66). The trail begins at the base of an observation tower that sits high on the Niagara Escarpment (SB 19; fig. 67). The rock beneath is glacier-scoured dolomite with only a thin, discontinuous cover of till. Sturgeon Bay occupies a valley that cuts across the dolomite (SB

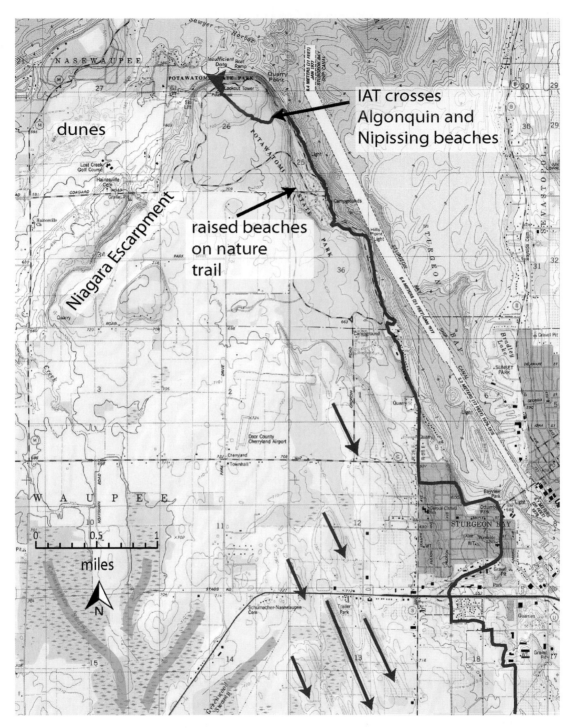

**dunes**

**Niagara Escarpment**

IAT crosses
Algonquin and
Nipissing beaches

raised beaches
on nature
trail

miles

N

Figure 65. Topography of the eastern part of the Sturgeon Bay Segment. Sturgeon Bay is to the east and Green Bay is to the west. The steep slope on the west is the Niagara Escarpment. Blue arrows are on drumlins and show ice-flow direction. Green arrows show path of the meltwater flow. (Map is part of the Idlewild and Sturgeon Bay West USGS Quadrangles and was created with TOPO! © 2011 National Geographic Maps.)

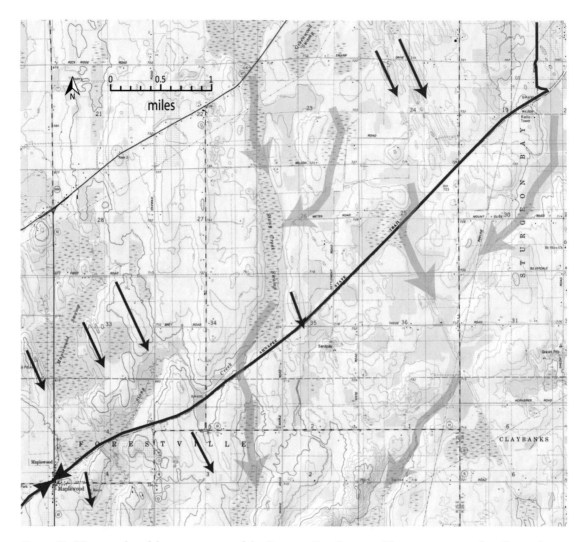

Figure 66. Topography of the western part of the Sturgeon Bay Segment. Blue arrows are on drumlins and show ice-flow direction. Green arrows show path of meltwater flow away from the retreating ice margin. (Map is part of the Sturgeon Bay West USGS Quadrangle and was created with TOPO! © 2011 National Geographic Maps.)

21) of the Door Peninsula. This may have been a preglacial valley, but it seems likely that a subglacial river further deepened it. Several of these valleys cut across the peninsula. As the glacier retreated, this valley acted as an outlet for glacial Lake Oshkosh (SB 16), which occupied the Green Bay–Lake Winnebago basins, into Lake Michigan. From the top of the tower you will have a great view to the north, including Sherwood Point, a sand spit that wave action built into the lake during the high-lake-level Nipissing stage (SB 20). A smaller

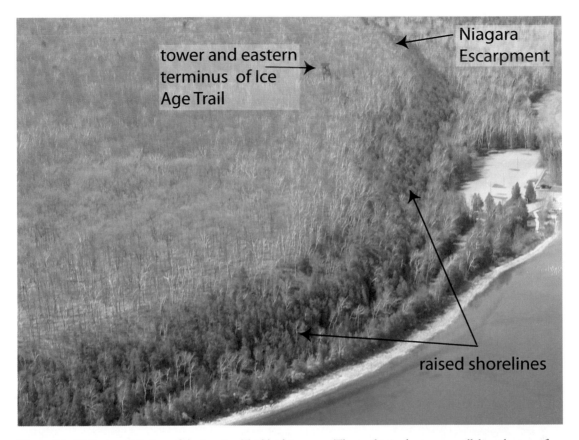

Figure 67. Oblique aerial view of the Ice Age Trail looking west. The trail is in the trees paralleling the top of the cliff. The base of the lookout tower is the eastern terminus of the IAT.

spit, Cabot Point, is a post-Nipissing spit. From the tower, look northeastward along the shore of the Door Peninsula. Directly across Sturgeon Bay there is a large dolomite quarry. Silurian dolomite is widely used for shore protection, as aggregate for concrete and asphalt, and as decorative stone.

The cliffs that tower above present lake level are not only the Niagara Escarpment but also wave-cut cliffs produced during higher levels of the lake (fig. 67). At the base, these cliffs have a gently sloping wave-cut platform that is analogous to the area just off shore of the present lake edge. As these cliffs were formed, breaking waves eroded the platform back to the base of the cliff. The highest platform is the Algonquin shoreline, which formed about 12,500 cal. years BP, as ice withdrew from the Lake Michigan basin for the last time. Rocks falling from cliffs above it have covered parts of it over the years, making it hard to see in places.

Notice as you walk down the escarpment on the IAT that there are two breaks in slope. The highest is the Algonquin and the lower one the Nipissing shoreline (SB 16; fig. 68).

You will find similar cliffs along much of the west side of the Door Peninsula. The elevations of the wave-cut platforms rise to the northeast, and the difference in elevation between them also increases. Both of these shorelines were horizontal when they formed. Because the Algonquin shoreline was formed first, it has been tilted more and raised higher than the younger Nipissing shoreline as the once-compressed earth rebounds (SB 20). The trail continues on the Nipissing wave-cut platform into the city of Sturgeon Bay.

About one and a half miles south of Sturgeon Bay, the trail begins to follow an old railroad grade, which is now the Ahnapee State Trail (fig. 66). This part of the trail crosses low drumlins (SB 14) that are oriented north-northwest to south-southeast and separated by marshes. Red, clayey till (SB 5) caps the drumlins. Dolomite bedrock is for the most part close to the surface.

Figure 68. Staircase down the 6,000-year-old lake shore bluff formed during the Nipissing high-stand of Lake Michigan. Actual beach was at base of staircase.

## 2. Forestville Segment

*CTH H at Maplewood to CTH M near Algoma (10 miles)*

From the east end of the segment at Maplewood, the IAT follows the Ahnapee State Trail across uplands and into the Ahnapee River valley just north of Forestville (fig. 69). Halfway between the two towns is a quarry in Silurian dolomite that is about a half mile east of the trail. The reddish-brown till of the Kewaunee Formation is only a few feet thick across much of this upland, with either dolomite or till of the Holy Hill Formation beneath. All of this area was last covered by the Green Bay Lobe. From Forestville to its western trailhead in this segment, the IAT remains in the Ahnapee River valley, through which the river meanders (fig. 70). Note how wide this valley is relative to the present size of the stream flowing through it (fig. 71). This is what is called an underfit valley; you will see many of these associated with glaciation along the IAT. One can surmise that a river much larger than the one flowing through it now produced the valley. At that time, water would have filled the valley from side to side. We know this happened when the valley was an outlet of glacial Lake Oshkosh (SB 16) during several late retreat phases of the glacier. Water flowed from the Green Bay basin into Lake Michigan, which was at a higher level than it is today. You can see evidence for this if you inspect the dissected delta—or wave-cut surface—through which the modern river valley passes (fig. 71). A delta top is generally just a little higher than the level of the water it forms in. The top of this delta is at about 620 feet above sea level, or at the Calumet level of ice-dammed Lake Michigan (SB 20). Most of the surrounding upland landscape was shaped by ice flow from the northwest (Green Bay Lobe), as indicated by the orientation of low drumlins (SB 14; fig. 71).

## 3. Kewaunee River Segment

*Hathaway Dr. access road to CTH A (11.2 miles)*

From the eastern trailhead the IAT follows an abandoned railroad grade across a broad wetland and the channel of the Kewaunee River (fig. 72). This valley was an outlet for glacial Lake Oshkosh (SB 16) whenever the retreating Green Bay Lobe allowed drainage into the valley through channels north of Luxembourg. Presumably water also flowed through this outlet during ice advance, but evidence of that was destroyed when the glacier advanced across the valley. Rivers cut this valley quite a bit deeper than it is today at times of ice retreat and low lake level (SB 20). At times of higher lake level, water flooded up the valley. Figure 72 shows the upstream extent of flooding during the Nipissing high stand. The trail continues on broad floodplain bounded by till-covered upland. It climbs onto a low terrace near the inland extent of the Nipissing high stand. This terrace is at the Calumet level (SB20). The terrace can be traced upstream (fig. 73), and the trail continues westward on and off this terrace to its western trailhead.

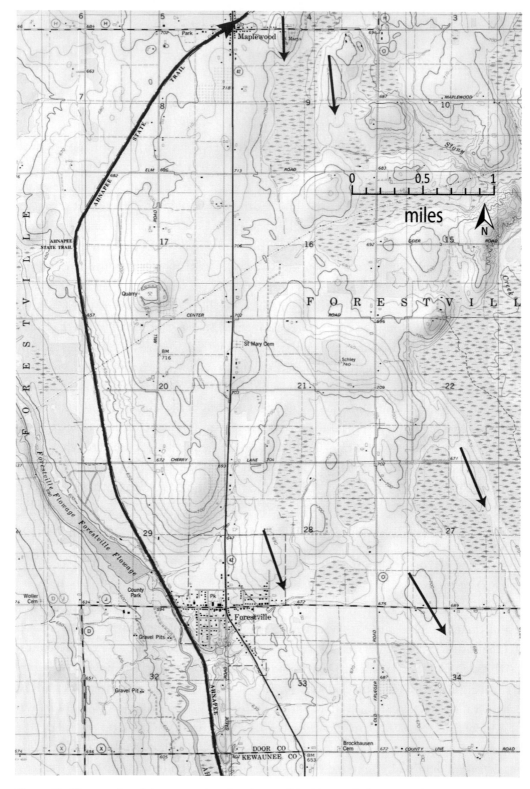

Figure 69. Topography of the eastern (northern) part of the Forestville Segment. Blue arrows show orientation of drumlins. Many hills are dolomite cored. (Map is part of the Forestville USGS Quadrangle and was created with TOPO! © 2011 National Geographic Maps.)

Figure 70. Oblique aerial view of the Ahnapee River looking toward the northeast. The intersection in the middle right is Hwy. 42 and Washington Rd., shown on figure 71. The Ice Age Trail is in the foreground.

## 4. Tisch Mills Segment

*Nuclear Rd. to CTH B (2.8 miles)*

When the glacier readvanced into Wisconsin about 13,400 cal. years BP, the climate had warmed and a spruce forest grew in this part of Wisconsin. Rising lake levels (SB 20) and the glacier itself killed the forest, which is now preserved as the Two Creeks Forest Bed (see the description of the Two Creeks Buried Forest Unit of the Ice Age National Scientific Reserve). The Two Rivers advance built the Two Rivers Moraine. The Tisch Mills Segment follows the East Twin River along the moraine front (figs. 64, 74, 75) to the village of Tisch Mills. West of the village the IAT is west of the river, but the moraine is clearly visible. When the glacier sat at the Two Rivers Moraine, water flowed toward the southwest in this valley. A braided stream (SB 8) choked with sand and gravel flowed into Lake Michigan, which was at the Calumet level at this time. Sand and gravel being deposited in the East

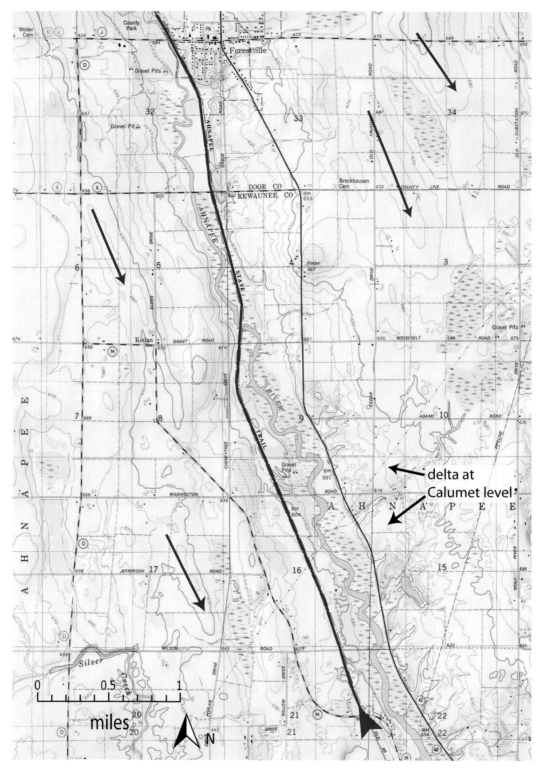

Figure 71. Topography of the western (southern) part of the Forestville Segment. The Ahnapee River valley passes through the central part of the map. Intersection that is shown in figure 70 is southwest of the top arrow that points to the Calumet level surface. Blue arrows show orientation of drumlins. (Map is part of the Forestville USGS Quadrangle and was created with TOPO! © 2011 National Geographic Maps.)

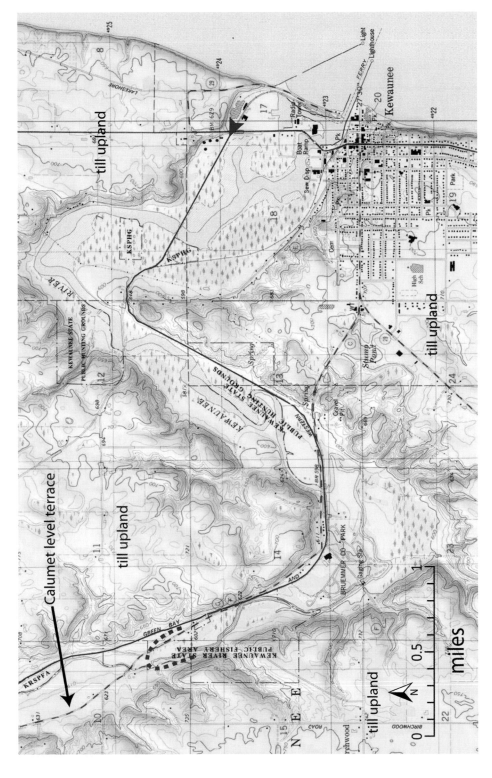

Figure 72. Topography of the eastern part of the Kewaunee River Segment. Dashed blue line shows the inland extent of Lake Michigan when it was at the Nipissing level. (Map is part of the Kewaunee USGS Quadrangle and was created with TOPO! © 2011 National Geographic Maps.)

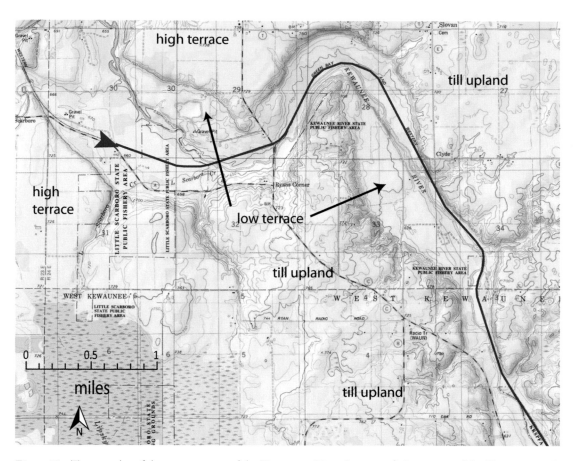

Figure 73. Topography of the western part of the Kewaunee River Segment. (Map is part of the Kewaunee and Casco USGS Quadrangles and was created with TOPO! © 2011 National Geographic Maps.)

Twin River valley by the outwash stream (SB 8) probably dammed Tisch Mills Creek, which the trail crosses near its western trailhead. Spruce forest would have covered much of the landscape to the west of this location.

## 5. Two Creeks Buried Forest Unit of the Ice Age National Scientific Reserve

The Two Creeks Forest Bed, a thin layer of spruce wood and needles, is among the most famous and studied geological sites in the Great Lakes area. In 1905, J.W. Goldthwait first described it as being about 2 miles south of the Village of Two Creeks. As far as we know, the forest bed has not been exposed there in the last 30 years, but it has historically been well exposed along the lake shore bluff just south of the Kewaunee County–Manitowoc County line (intersection of CTH BB with STH 42, between STH 42 and Lake Michigan). This site,

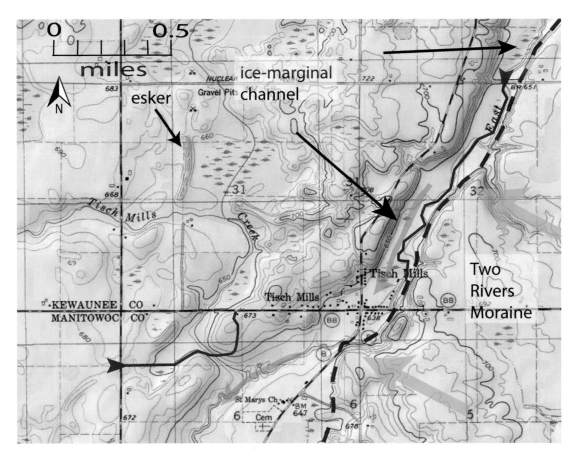

Figure 74. Topography of the Tisch Mills Segment. The East Twin River flows to the south along the front of the Two Rivers Moraine. When the glacier was at the Two Rivers Moraine, this valley was an ice-marginal channel, filled from side to side by meltwater carrying sand and gravel. Dashed blue line shows edge of the Two Rivers Moraine. Blue arrows show ice-flow direction when the moraine formed. (Map is part of the Two Creeks and Larrabee USGS Quadrangles and was created with TOPO! © 2011 National Geographic Maps.)

which is now part of the Ice Age National Scientific Reserve and is a State Natural Area, has Two Creeks wood, which is also present in several other places, for instance near the city of Denmark, Wisconsin, and in the Fox River valley between Green Bay and Appleton.

To reach the bluff containing the forest bed, drive east on CTH BB from the IAT at Tisch Mills to the Lake Michigan shore and park in the small lot there (fig. 76). Walk to the bluff edge. Public land extends from the obvious north property line southward for several hundred yards. To see the soil and in-place material, you have to crawl over the edge of the bluff (carefully—it is 30 feet high) and use a shovel to dig into the bluff's side. You can also

Figure 75. Low oblique aerial view of Two Rivers Moraine at Tisch Mills showing moraine front, East Twin River, the village of Tisch Mills, and the location of the Tisch Mills Segment. View is to the north. Ice flowed from east to west when depositing the moraine.

usually see pieces of logs from the forest bed scattered on the beach, but collecting the wood is *not* allowed!

From the beach you can see the layers of earth that underlie and cover the forest bed. (Be careful going down the bluff—it is very slippery!) In the base of the bluff, and generally covered by soil that slid from above, is till deposited by the Lake Michigan Lobe glacier when it advanced at least as far south as Milwaukee (fig. 77, Till A). Clayey silt and fine-sand lake sediment overlay the till, and above that is a layer of coarser sand and gravel that we interpret to be shallow-water and beach sediment. The forest bed lies on this sand and gravel and consists of a few centimeters of organic material that is mostly black spruce (*Picea mariana*) needles, twigs, and branches. We also find very limited amounts of white spruce, balsam fir, tamarack,

Figure 76. Oblique aerial view of the bluff at the Two Creeks Buried Forest State Natural Area. Lake Michigan is in the foreground. View is to the west.

and aspen. In addition to wood and needles, the forest bed contains shells of land snails that lived in the soil thousands of years ago, as well as some that lived in shallow water.

Note that the wood you see in this ancient forest is not petrified or decayed! At times, we even find tree stumps left upright, in the same positions they grew, at this level (fig. 78). Silty lake sediment, which is about 10 feet (3 m) thick, covers the forest bed, and atop it all we find Two Rivers till (fig. 77, Till B). This till contains numerous pieces of wood picked up as the glacier moved across the dead forest about 12,000 cal. years ago.

When ice that deposited the lower till retreated from this location, Lake Michigan was at the Glenwood level, about 60 feet above its present level (SB 20). Silt and clay particles settled through the deep water onto this spot, which was lake bottom, until the glacier retreated

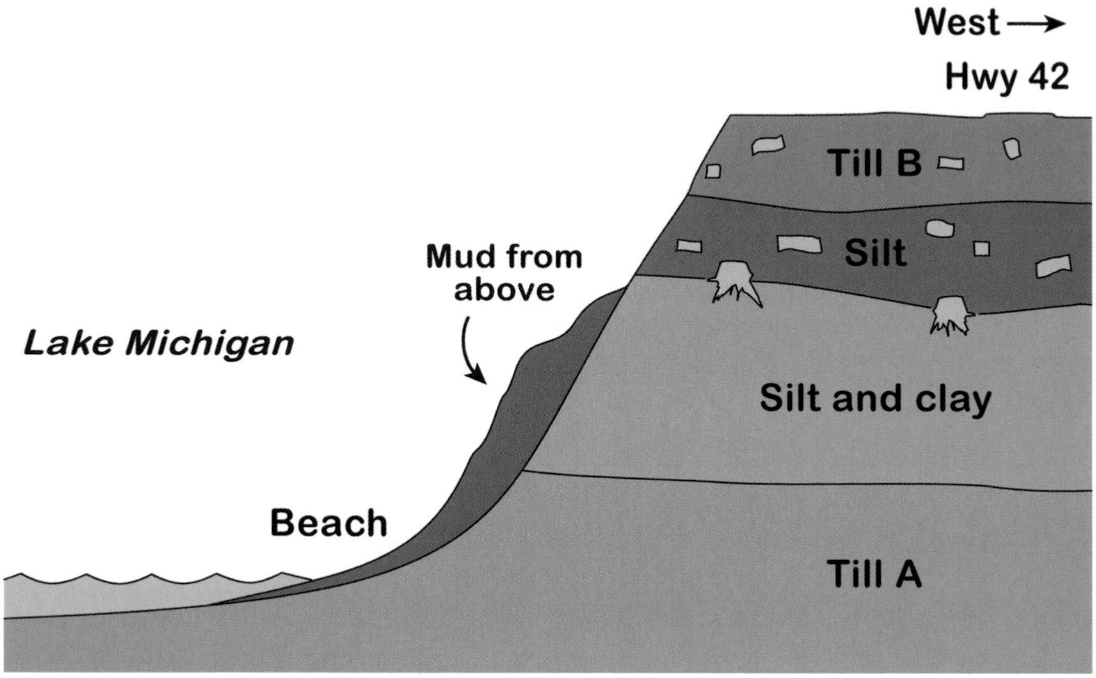

West →

Hwy 42

Till B

Mud from above

Silt

Lake Michigan

Silt and clay

Beach

Till A

Figure 77. Diagrammatic cross section of the bluff at Two Creeks Buried Forest State Natural Area showing arrangement of layers and location of Two Creeks wood. Bluff is about 35 feet high. "Till B" indicates the Two Rivers till that extends to the Two Rivers Moraine shown in figure 75. (Drafted by Mary Diman.)

Figure 78. Spruce tree stump in growth position on Two Creeks Buried Forest Bed, about 10 feet below the top of the bluff. Stump is about 10 inches across.

far enough north to allow the lake level to drop. When the water level dropped below the present bluff level, plants began to grow in the rich sediment. Based on revegetation histories in southeast Alaska, 50 years or more may have passed before spruce trees became the dominant species. The forest probably grew for almost 1,000 years, and the oldest tree found so far has 234 annual rings. The trees on this bluff died when the lake level rose, but at several sites between Green Bay and Denmark, Wisconsin, there is no lake sediment above the forest bed, indicating that the trees there perished under the advancing ice. In many of the logs we find in this layer, the tree rings become thinner near the outer bark, indicating that the trees were stressed by some combination of rising water and cold winds from the ice sheet.

Wood from the Two Creeks Forest Bed was the first geologic sample dated with the radiocarbon method by Libby in 1949 (Libby 1952). Until that time, people had little idea when the glaciers had retreated from North America. Many samples have been dated since then, and the uncalibrated radiocarbon dates indicate that this forest lived about 11,850 radiocarbon years ago, give or take 200 years. This is about 13,800 calendar years ago on the timescale that many archaeologists and geologists use today (SB 18).

The Two Creeks deposit is almost unique in the central United States, preserving a record of vegetation from a time right at the end of the last glaciation. During this time, it was warm enough for spruce forests to grow, whereas the scant evidence that there is for earlier times hints at cold, permafrost conditions with no trees, even though the glacier had at times retreated from the area. Figure 79 shows a plot of the relative temperature changes preserved in the Greenland Ice Sheet during deglaciation. Superimposed on that are colored bands indicating time ranges for which we have radiocarbon dates in northeastern Wisconsin. The gold and blue time bands are glacier-free times for which sparse remains of tundra plants have been found under till. Only for the pink range, the Two Creeks warm interval, has wood been found under till. The peak warming (in Greenland and probably globally) took place at about 14,300 cal. years BP. Tree growth at Two Creeks began sometime shortly after that in response to this warming. Temperature continued to decline until the Younger Dryas, about 12,500 cal. years ago (fig. 79), and it was this cooling that led to the glacial advance that covered the Two Creeks Forest Bed. The Younger Dryas cooling is also recorded in the pollen record from Devil's Lake, which is described in the chapter on the southern Green Bay Lobe.

## 6. East Twin River Segment
*Rockledge Rd. (0.4 miles)*

This is a short trail segment. To the east is red clay till (SB 5) shaped into drumlins (SB 14) by ice flowing toward the southwest (fig. 80). The ice advance that deposited the till and

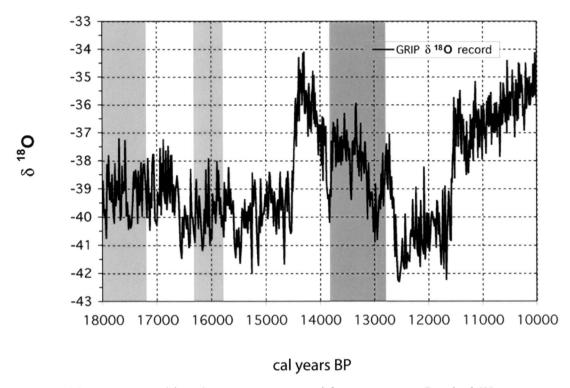

**cal years BP**

Figure 79. Temperature record (ratio between oxygen isotopes) from an ice core in Greenland. Warmer temperatures are shown toward the top and cooler toward the bottom. Pink shaded bar, between 13,800 and 12,800 cal. years ago, indicates the range of ages of wood (trees) from this site. Other colored bands indicate the age ranges of radiocarbon dates on tundra vegetation (no trees) under till in northeastern Wisconsin.

created the drumlins preceded the ice advance that built the Two Rivers Moraine, which lies about 1.5 miles to the east. The trail is entirely in the East Twin River valley on a low terrace and the modern floodplain. This was a south-flowing outwash stream (SB 8) when the ice was at the Two Rivers Moraine (see the Tisch Mills Segment).

## 7. Mishicot Segment

*Pit Rd. to CTH V (1.6 miles)*

Like the Tisch Mills and East Twin River segments, the Mishicot Segment (fig. 81) is in the East Twin River valley, an outwash stream (SB 8) that flowed more or less along the front of the Two Rivers Moraine. The Pit Rd. trailhead is on sand and gravel that may have been deposited as an esker in a tunnel beneath the ice or on ice as it was melting away during an older ice advance. When ice sat at the Two Rivers Moraine, outwash was deposited in the whole valley that the trail crosses. The CTH V trailhead is on this outwash plain.

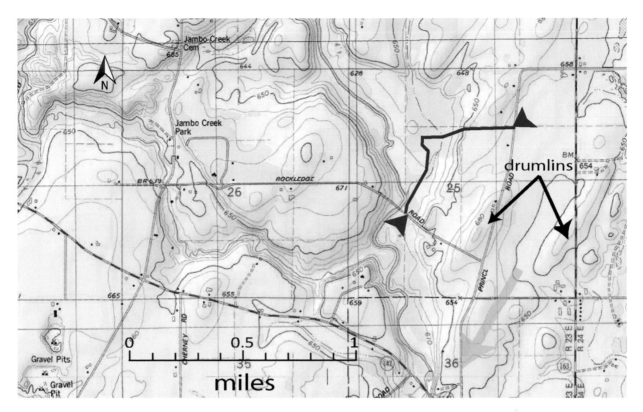

Figure 80.  Topography of the East Twin River Segment. Blue arrow shows ice-flow direction. (Map is part of the Larrabee USGS Quadrangle and was created with TOPO! © 2011 National Geographic Maps.)

## 8. Point Beach Segment

*Lake Shore Rd. to Park Rd. (10.1 miles)*

This segment of the trail crosses a much younger landscape than most of the IAT in Wisconsin. The unusual pattern of ridges and swales that the trail crosses in some places and to which it runs parallel in others was built out into the lake by waves and then later modified by wind during the last 6,000 years (fig. 82). Before discussing the path of the trail itself, let's look generally at the origin of Point Beach.

Between 12,000 and 11,000 cal. years ago, the level of Lake Michigan was as much as 300 feet below its present level, and the shoreline was far to the east (SB 20). The lake rose until it reached its present level about 7,000 cal. years ago, and continued to rise until about 6,000 cal. years ago, when it reached the Nipissing level (SB 20). At that time the shoreline was right at the Lakeshore Road trail access at the east end of the segment. The abandoned Nipissing shoreline extends south from here to the city of Two Rivers (fig. 83). Waves and

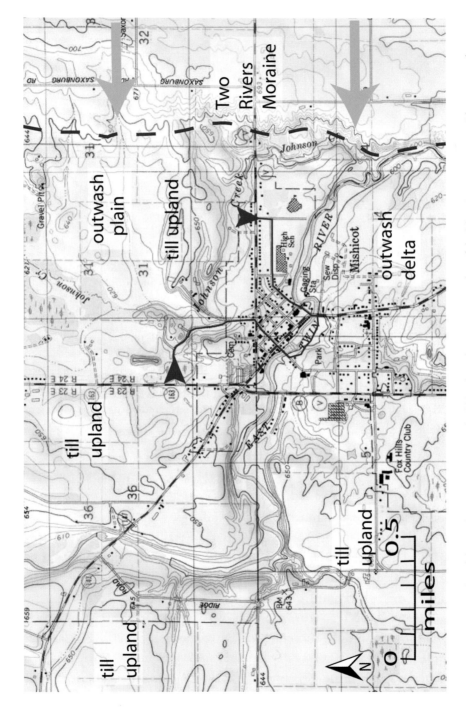

Figure 81. Topography of the Mishicot Segment. Dashed blue line shows western edge of the Two Rivers Moraine. Blue arrows show ice-flow direction. (Map is part of the Mishicot USGS Quadrangle and was created with TOPO! © 2011 National Geographic Maps.)

**Lake Michigan**

**lighthouse**

**Red Pine parking lot**

entrance road

County Hwy O

Figure 82. Low oblique aerial view looking southward across Point Beach State Forest. Note swales without trees alternating with tree-covered beach ridges. The ridge beneath CTH O is a spit built about 6,000 years ago, when the lake was at a higher level. Distance between CTH O and the lighthouse at Rawley Point is about a half mile.

currents carried sand to this part of the shoreline from eroding shore bluffs to the north and the south. Because of converging currents at this location, sand spits grew out into the lake. One of these spits grew northward from approximately the location where Molash Creek crosses the IAT (fig. 83). Another extended southward from a point about a mile north of the Lakeshore Road trailhead. Waves continued to extend these fingers of sand until they met, probably somewhere near the present location of the road intersection where the IAT and park entrance road head eastward to Rawley Point. At the same time, sand blew off the beach and piled up into dunes on the spit. A lagoon formed behind the spit and slowly filled

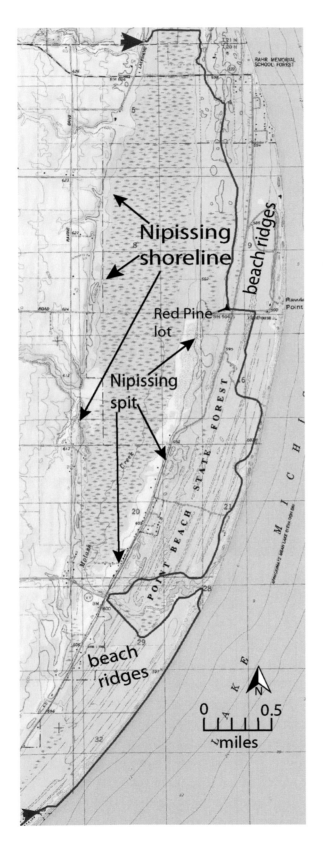

Figure 83. Topography of the Point Beach
Segment. Geologic features are discussed in
text. Rawley Point is just above center right.
(Map is part of the Two Rivers USGS
Quadrangle and was created with TOPO!
© 2011 National Geographic Maps.)

with peat. That is the present large wetland shown in figure 83 between the Nipissing shore-line and the Nipissing spit.

As lake level began to fall slowly from the Nipissing high to its present level, a series of beach ridges formed parallel to the shore. Figure 84 shows a cross section extending from approximately the location of CTH O eastward along the park entrance road to Lake Michigan. Note the profile of dunes and beach ridges with intervening swales. The highest dunes are on the Nipissing spit. Also shown are radiocarbon ages that date the development of the beach ridges as the lake fell to its present level. The oldest radiocarbon age is actually in the bottom of the diagram and dates organic matter deposited as the lake rose to the Nipissing level (SB 20). Other radiocarbon ages are from the lowest organic sediment at the bottom of several swales. Note that although the ridges are progressively younger moving toward the present shore, their formation has been irregular. The ridges probably represent times when the lake's level was low long enough to allow the buildup of dunes at the top of the beach. These dunes stabilized with vegetation, and a wide beach was built lakeward before the next high-water period.

From the Lakeshore Rd. trailhead, you will cross a wetland that is less than a mile wide before the trail rises onto the higher dunes to the east. The trail then winds southward through sand dunes on the spit before dropping off the spit's front into the beach-ridge complex and the Red Pine trailhead parking lot. The IAT then crosses small beach ridges and a large swale as it parallels the park entrance road. It turns southward and follows the crests of other beach ridges, cutting across some, to where it circles Molash Creek. From there nearly to the end of the segment at Park Rd., the IAT is on the modern beach.

## 9. City of Two Rivers Segment
*Park Rd. to Columbus St. (2.7 miles)*

This segment lies entirely on sandy lake sediment deposited during and after the Nipissing high stand (SB 20). At that time the whole Twin Rivers lowland was an embayment filled with shallow water. Note that there is very little topography except for the Two Rivers Moraine, which is a ridge with a noticeable water tower and hospital (fig. 85). This moraine lies east of the East Twin River, which formed during the maximum extent of the ice advance covering the Two Creeks Forest about 13,400 cal. years ago. From here the glacier margin went across Lake Michigan to what is now the state of Michigan. The glacier edge in the lake would have had a near-vertical face over 200 feet high with icebergs crashing off the cliff. Figure 86 shows a similar ice margin in southeast Alaska.

The West Twin River acted as an outlet for glacial Lake Oshkosh (SB 16) for a period of time during the early stages of ice retreat. The delta associated with that drainage event is farther to the west than the area shown in figure 85. The East Twin River, which flowed

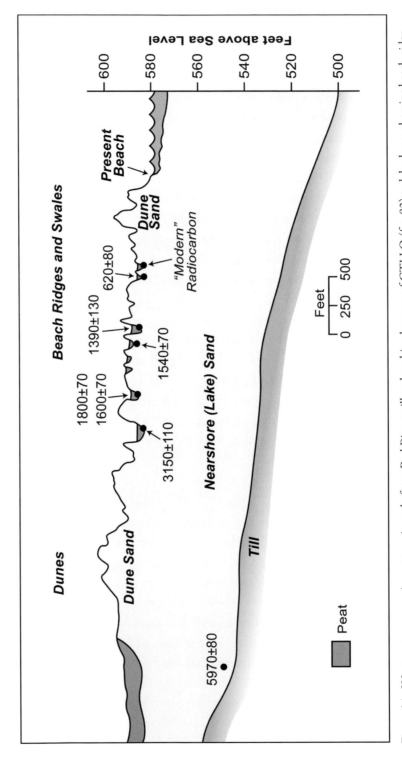

Figure 84. West-east cross section, approximately from Red Pine trailhead parking lot west of CTH O (fig. 83) to lakeshore, showing beach-ridge topography and intervening swales. Radiocarbon dates indicate deposition of organic matter at that time. All are radiocarbon years BP, not cal. years (SB 18). (Modified from Dott and Mickelson 1995; drafted by Mary Diman.)

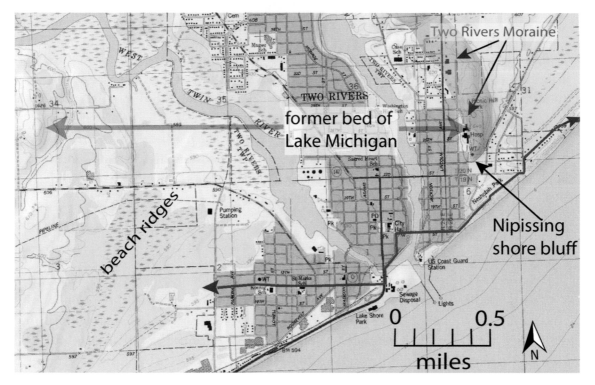

Figure 85. Topography of the City of Two Rivers Segment. Nearly all of this area was under water during the Nipissing high stand of Lake Michigan. Gray arrow shows extent of lake bottom west of the Two Rivers Moraine when ice sat at the moraine. The area east of the moraine was also under the lake as the glacier retreated from the moraine. (Map is part of the Two Rivers USGS Quadrangle and was created with TOPO! © 2011 National Geographic Maps.)

Figure 86. Calving McBride Glacier front in southeast Alaska. The glacier margin looked like this where it extended eastward across Lake Michigan from the Two Rivers Moraine at what is now Two Rivers. The face is over 200 feet high. The small dark feature on the water in front of the glacier is a 21-foot-long boat with three people in it.

southward parallel to the Two Rivers ice front, flowed into Lake Michigan farther upstream than the area shown in figure 85 because the lake level was higher than today and much of the surface sand now in the lowland had not yet been deposited. When the lake rose after the Chippewa low stand, about 6,000 cal. years ago, all of the area in figure 85 between the Two Rivers Moraine and the upland near the west edge of the map was inundated. At that time, waves eroded a bluff at the south end of the Two Rivers Moraine, which now forms an amphitheater-shaped steep slope near the baseball diamond at Neshotah Park. Both rivers downcut through the former lakebed sand as the lake level dropped (SB 16, 20).

## 10. Dunes Segment
*Columbus St. to Taylor St. (2.7 miles)*

This segment of the IAT is almost entirely in the Woodland Dunes State Natural Area (fig. 87). The Woodland Dunes Nature Center is located about 1 mile north of the Columbus Street trail access. The trail follows sandy beach ridges and wetland-filled swales similar to those along the Point Beach Segment. The ridges here also formed as the lake level slowly dropped after the Nipissing high stand (SB 20). All of the area crossed by the trail was under water 6,000 cal. years ago. Figure 88 shows a cross section of these lake deposits and the radiocarbon dates associated with them. West of the dashed line in figure 87 the landscape is blanketed with reddish-brown clayey till called the Valders till. This till is older than the Two Creeks Forest Bed. For a complete discussion of the geologic history of this area, see the discussions of that site and of the Point Beach and City of Two Rivers segments.

## 11. City of Manitowoc Segment
*Taylor St. to Rapids Rd. (STH 42) (7.3 miles)*

From Taylor St. westward, the trail follows the lakeshore (fig. 89). The lake covered most of this surface during the Nipissing stage (SB 20). Sand deposited in the shallow water and the falling lake level since the Nipissing phase have made it dry land. At the intersection of Waldo Blvd. with Memorial Dr./Maritime Dr., look straight west up Waldo Blvd. to see the Nipissing level—the lowest lake surface—across the river mouth. The land rises behind to the Calumet level, where there is a flat terrace, then farther west, near the intersection with 18th St., it rises to the Glenwood level. A gravelly Glenwood level beach (now covered with houses) sits at the top of that surface. The trail itself remains on the Nipissing surface to the edge of the Manitowoc River then turns north. Although there is a slight break in the slope, the Calumet-level surface near the intersection with State St. is not very obvious. The trail remains on a wave-washed till surface at the Glenwood level to south of Evergreen Cemetery. The remainder of the trail follows the floodplain and the Nipissing terrace of the Manitowoc River. The Manitowoc River is in a relatively wide valley compared to the

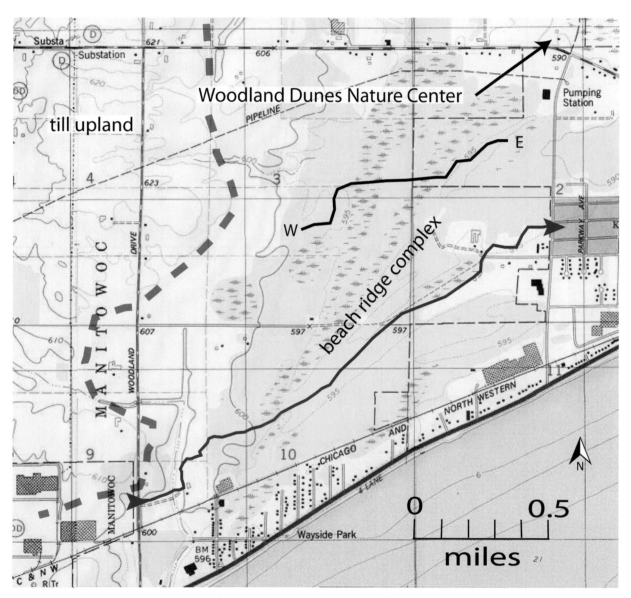

Figure 87. Topography of the Dunes Segment. Dashed gray line is the upper limit of Lake Michigan during the Nipissing high stand. Line labeled E–W shows the location of the cross section shown in figure 88. (Map is part of the Two Rivers USGS Quadrangle and was created with TOPO! © 2011 National Geographic Maps.)

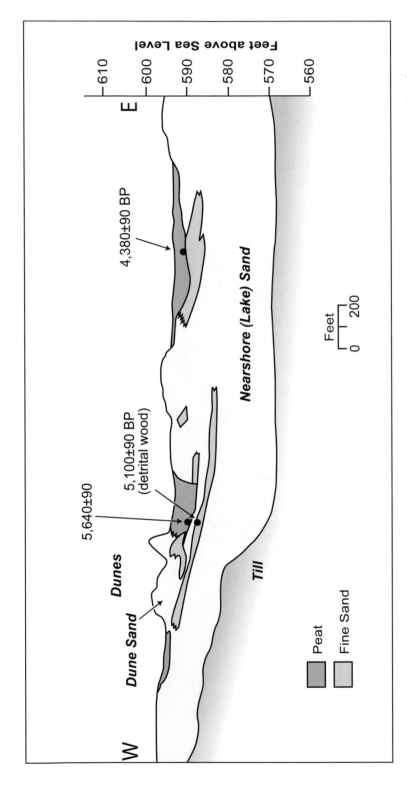

Figure 88. West-east cross section showing distribution of lake deposits and dunes in the Woodland Dunes area, specifically the E–W line shown in figure 87. Radiocarbon dates indicate deposition of organic matter at those times. All are radiocarbon years BP, not cal. years (SB 18). (Modified from Dott and Mickelson 1995; drafted by Mary Diman.)

size of the channel. This is especially true upstream of Manitowoc. At times in the past, the Manitowoc River was an outlet from glacial Lake Oshkosh into high levels of Lake Michigan (SB16; fig. 64). In this lower part of the valley, the river downcut well below its present floodplain level during the Chippewa low stand, and the valley slowly filled with sediment as the lake rose to the Nipissing level. At the time of the Nipissing high stand, the lake extended into the valley about a mile upstream of the Rapids Rd. trailhead.

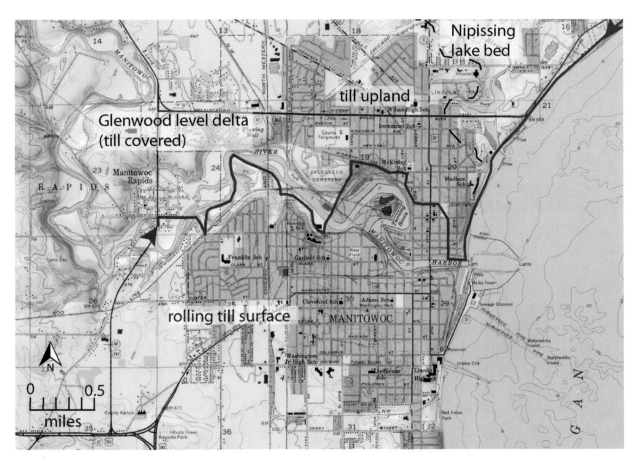

Figure 89. Topography of the City of Manitowoc Segment. Dashed blue line is approximate boundary between till surface to the west and lakebed surface to the east. During the Nipissing high stand of Lake Michigan, lake water extended up the Manitowoc River valley to the northwest corner of the map. (Map is part of the Manitowoc USGS Quadrangle and was created with TOPO! © 2011 National Geographic Maps.)

# Northern Kettle Moraine
# Ice Age Trail Segments

THE NORTHERN KETTLE MORAINE SEGMENTS of the IAT are located in eastern Wisconsin (fig. 3). The Kettle Moraine really isn't a moraine at all by most definitions. Its hills are composed mostly of stream-deposited sand and gravel instead of mostly till (SB 5) like other moraines. But it is a geographic feature that has had this name for over a hundred years and will continue to be called the Kettle Moraine for years to come! It formed between the Lake Michigan and Green Bay lobes over a period of several thousand years as the glacier retreated from its maximum extent position (fig. 1). It contains various landforms, including eskers (SB 13), hummocky sand and gravel (SB 11), moulin kames (SB 12), and kettles (SB9). Individual landforms are detailed in the trail-segment descriptions. In some places it is a single ridge and in others it bifurcates, leaving one ridge on the Lake Michigan Lobe side and another on the Green Bay Lobe side, with a low area in between. To the west and east of the Kettle Moraine, drumlins show the ice-flow direction (fig. 90).

Although there are only four trail segments in the northern Kettle Moraine, they total more than 30 miles in length and exhibit spectacular glacial features. Because much of the surrounding area is state forest, much of the trail is on public land. Numerous other trails and the Henry S. Reuss Ice Age Visitor Center near Dundee offer great opportunities for recreation and learning. The easternmost segment of the IAT is La Budde Creek. It follows a glacial drainageway along the Lake Michigan Lobe edge of the Kettle Moraine. The Greenbush Segment (fig. 90) winds mostly through high-relief hummocky topography (SB 11) on the Green Bay Lobe side of the Kettle Moraine. The southernmost half-mile crosses open land that is pitted outwash (SB 10) deposited when all but isolated masses of buried ice had disappeared from the low, and now central, Kettle Moraine area.

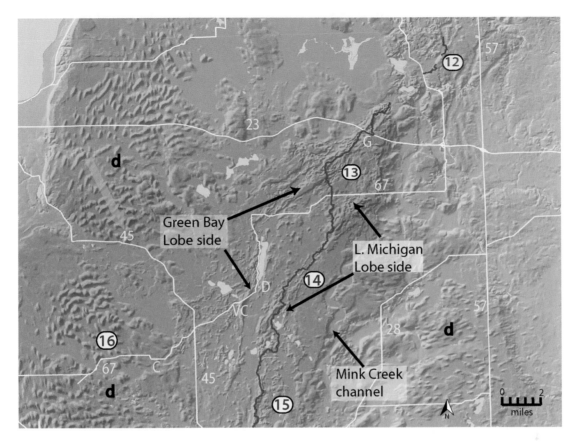

Figure 90. Shaded relief of the northern Kettle Moraine IAT segments (red): (12) La Budde Creek, (13) Greenbush, (14) Parnell, (15) Milwaukee River Segment in Fond du Lac County. Green Bay Lobe (western) and Lake Michigan Lobe (eastern) sides of the Kettle Moraine and meltwater channel are shown with black arrows. Blue arrows show ice-flow direction. The letter *d* on the right indicates Lake Michigan Lobe drumlins. The letters *d* on the left indicate Green Bay Lobe drumlins. The Campbellsport Drumlins Unit of the Ice Age National Scientific Reserve (16) is in the middle group of Green Bay Lobe drumlins. Note that at this scale the individual drumlins appear to cluster in bands perpendicular to ice flow. Yellow lines and numbers indicate highways. Cities shown (yellow): (C) Campbellsport, (D) Dundee, (G) Greenbush, (K) Kewaskum. VC refers to the Henry S. Reuss Ice Age Visitor Center. (Base map constructed from USGS National Elevation Dataset and modified by WGNHS.)

The Parnell Segment crosses more pitted outwash and some till surface before climbing into the Lake Michigan side of the Kettle Moraine about a half mile before the junction with the Parnell Tower trail. The remainder of this segment and the Milwaukee River Segment are in the Lake Michigan Lobe side of the Kettle Moraine.

The northern Kettle Moraine is mostly a double ridge, and the central and southern Kettle Moraine is mostly a single broad ridge. Figure 91 shows cross sections that illustrate how we believe the northern Kettle Moraine formed. Throughout the period of glaciation this was a low area on the ice surface (fig. 91a). Meltwater from both lobes drained into this low trough and then southwestward toward Walworth County. Sand and gravel accumulated on this riverbed on the ice surface. Eventually the sand and gravel was thick enough that it insulated the ice below. Off to either side, where clean ice or ice with a very thin debris layer could absorb more radiation from the sun, melting was more rapid. Eventually two troughs formed, with the ridge of debris-covered ice between them (figs. 91b, 91c). As these

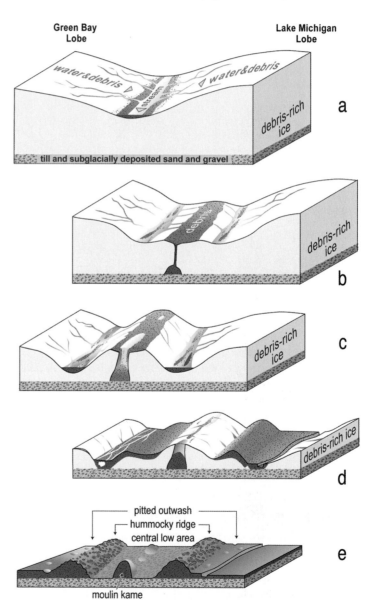

Figure 91. Three-dimensional diagrams of probable sequence in the formation of the double ridge of the northern Kettle Moraine: (a) as ice begins to thin, meltwater carrying sediment flows to the low area between the Green Bay Lobe and the Lake Michigan Lobe; (b) sand and gravel thicken and begin to insulate ice beneath, while cleaner ice on either side melts more rapidly and two streams develop on either side of the high debris-covered ice ridge; (c) the debris-covered central ridge sheds sediment to the streams on either side as well as into moulins; (d) accumulating sediment insulates ice on either side of the central ridge, which begins to melt down more rapidly than ridges on either side; (e) ice finally melts from beneath all of the Kettle Moraine zone. (Modified from Carlson et al. 2004; drafted by Mary Diman.)

channels deepened, more and more sand and gravel accumulated in them. It is this sand and gravel that eventually collapsed to form the hummocky ridges that now form the outer edges of the northern Kettle Moraine (figs. 91d, 91e). The low area between the hummocky ridges represents less accumulation of sediment from the ice surface. Part of the flat surface is till, and part is outwash deposited in the very late stages of the formation of the Kettle Moraine.

Spectacular moulin kames (SB 12) are located in the central area, west of the trail (fig. 92). These are "world class" features that are unique in size and number to the Kettle Moraine. They probably formed as the central ridge of debris-covered ice slowly decayed. Eskers (SB 13) were deposited by streams flowing in tunnels in this central ice mass. Figure 93 shows one that crosses STH 67 about 2 miles west of the Ice Age Visitor Center. There are several other unnamed eskers that the IAT follows in places (see segment descriptions).

Figure 92. Oblique aerial view, looking west, of McMullen Hill, a large moulin kame in the northern Kettle Moraine. Note farm buildings for scale. Long Lake is at the top of the photo.

It runs along the crest of the Parnell Esker, and explanatory signs are located at Butler Lake. Probably the last features to form, as buried ice slowly melted, were the thousands of kettles (SB 9). One water-filled kettle is Spruce Lake Bog (fig. 32), a National Natural Landmark and a State Natural Area. This kettle, a little over a mile west of Long Lake, has been slowly filling with peat since the buried ice melted out. Sphagnum and other plants also have formed a floating mat with many unusual plants. You can see this kettle with a floating bog from a boardwalk accessible from Airport Rd.

## 12. La Budde Creek Segment

*CTH FF to STH 67 (3.1 miles)*

The La Budde Creek Segment parallels the east side of the northern Kettle Moraine (fig. 94). High-relief hummocky topography (SB 11) lies to the west and smoother pitted outwash (SB 10) and till surfaces (SB 5) lie to the east. The northern trailhead is in the La Budde Creek valley, and the creek is an underfit stream (SB 8). This valley was cut by glacial

Figure 93. Oblique aerial view, looking east, of unnamed esker in the Green Bay Lobe (west) side of the Kettle Moraine. Highway 67 is to the left of the area shown.

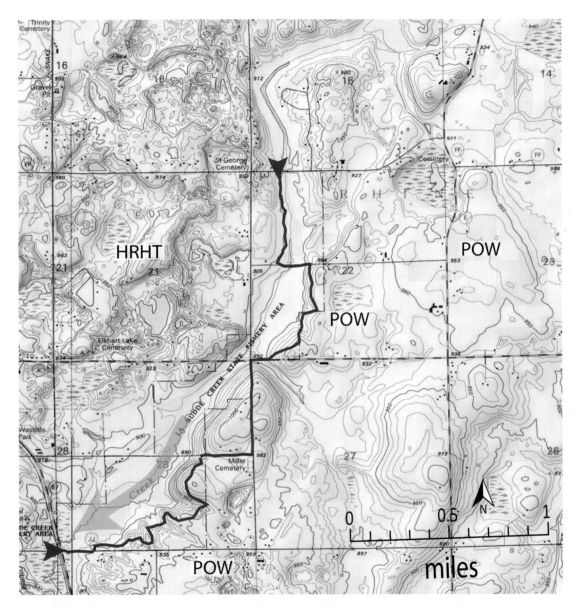

Figure 94. Topography of the La Budde Creek Segment along the east side of the northern Kettle Moraine. HRHT: high-relief hummocky topography (SB 11); POW: pitted outwash (SB 10). (Map is part of the Elkhart Lake and Franklin USGS Quadrangles and was created with TOPO! © 2011 National Geographic Maps.)

meltwater flowing from the northeast toward the southwest as the ice sheet retreated away from the Kettle Moraine. It was also fed by debris-covered ice melting out to produce the high-relief hummocky topography to the west. From where the trail rises onto the east bank of the channel it is on pitted outwash almost to CTH A. The hilltops here, including the one south of CTH A, appear to contain Lake Michigan Lobe till. There is a large, deep kettle south of the intersection of Badger Rd. and Little Elkhart Lake Rd. South of Badger Rd. the trail follows the rim of the deep La Budde Creek channel to Garton Rd.

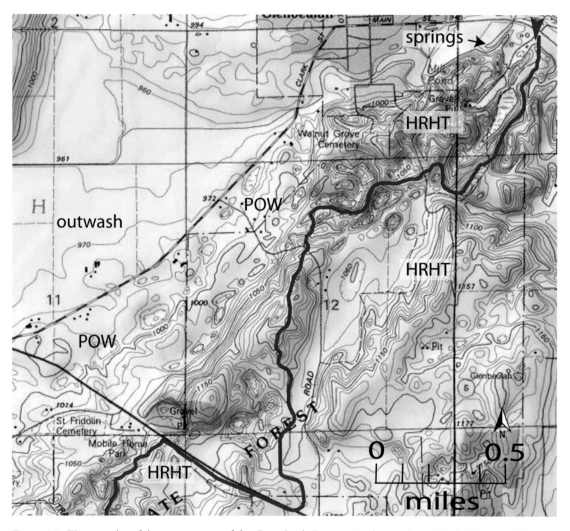

Figure 95. Topography of the eastern part of the Greenbush Segment in the northern Kettle Moraine. Note large gravel pit on Highway 23, shown in southern part of the map. HRHT: high-relief hummocky topography (SB 11); POW: pitted outwash (SB 10). (Map is part of the Elkhart Lake USGS Quadrangle and was created with TOPO! © 2011 National Geographic Maps.)

## 13. Greenbush Segment

*CTH P to STH 67 (9 miles)*

The Greenbush Segment traverses high-relief hummocky topography (SB 11) from its northern terminus to STH 23 (fig. 95). At low points in the landscape, springs flow from the sand and gravel hummock bases. One such spring feeds the small creek that crosses CTH P just west of the trail terminus. The trail winds around deep kettles (SB 9) and over hummocks that form the Green Bay Lobe side of the Kettle Moraine. There are a few views of flat outwash (SB 8) immediately to the west, stretching to drumlins (SB 14) on the horizon. The huge gravel pit on STH 23 shows the coarse sand and gravel that underlie this terrain. Be sure to also look at the *Ice Age Trail Companion Guide 2011* for information on historic sites, spur trails, and features of interest in this area.

West of STH 23 (fig. 96), the trail continues through the hummocky topography of the Green Bay Lobe side of the Kettle Moraine. The Greenbush kettle is only one among thousands here, but someone decided to distinguish it with a sign on Kettle Moraine Dr. From this point to STH 67, the trail crosses outwash with a fairly flat surface and only a few kettles. Note the difference in vegetation as you stride across the outwash. The surface is relatively good for farming because it is fairly flat, but the high-relief hummocky topography is poor by comparison because of its steep slopes.

## 14. Parnell Segment

*STH 67 to Mauthe Lake Recreation Area (13 miles)*

The eastern end of the Parnell Segment is on pitted outwash (SB 10) in the central part of the Kettle Moraine (fig. 97). Look east to see the Lake Michigan Lobe side of the Kettle Moraine, and west to see the Green Bay Lobe side. Kettles interrupt the outwash surface of this segment, but they aren't as deep or as numerous as the kettles on either side of the Kettle Moraine (fig. 97). Note the many other trails crisscrossing the landscape; you can find trail maps and descriptions for these from the Henry S. Reuss Ice Age Visitor Center on STH 67, west of the trail.

About 1 mile south of STH 67, you'll enter the higher-relief hummocky topography (SB 11) of the Lake Michigan Lobe side of the Kettle Moraine. A spur trail goes eastward to the Parnell Observation tower. From the tower there is a great view of the large moulin kames a little over a mile to the west. The hills on the western horizon are Green Bay Lobe drumlins. Westward from the spur trail the IAT follows the Lake Michigan Lobe side of the Kettle Moraine past kettles and hummocks, but you're in for a less strenuous hike, because the relief is lower than on the Greenbush Segment.

Figure 98 shows the western part of the Parnell Segment. You can see the Parnell Esker to the west of the trail north of CTH V. Then follow the IAT as it climbs onto the esker

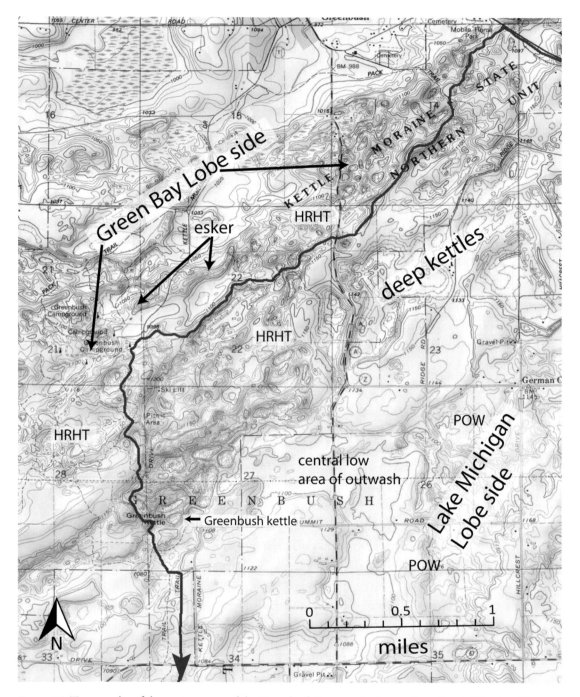

Figure 96. Topography of the western part of the Greenbush Segment in the northern Kettle Moraine. Note transition in landscape from the Green Bay Lobe hummocky topography to flatter outwash in the central Kettle Moraine area. This is noticeable on the trail south of Greenbush Kettle. HRHT: high-relief hummocky topography (SB 11); POW: pitted outwash (SB 10). (Map is part of the Elkhart Lake and Cascade USGS Quadrangles and was created with TOPO! © 2011 National Geographic Maps.)

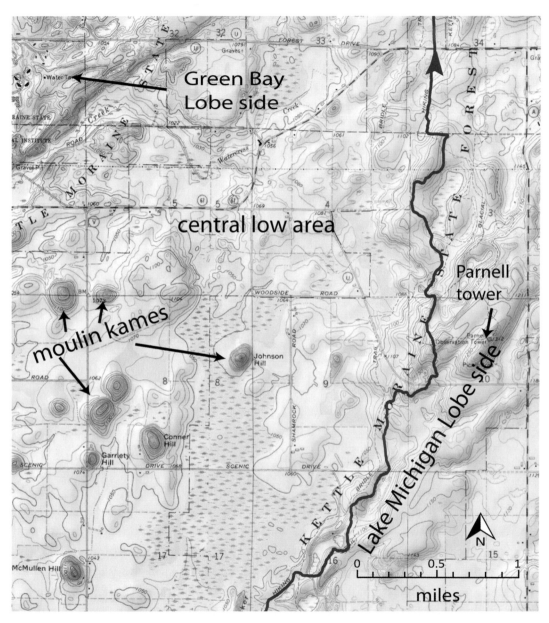

Figure 97. Topography of the eastern part of the Parnell Segment. Map shows the Lake Michigan Lobe side of the Kettle Moraine, the Green Bay Lobe side of the Kettle Moraine, and the low area between. Large moulin kames are present west of the trail. (Map is part of the Cascade USGS Quadrangle and was created with TOPO! © 2011 National Geographic Maps.)

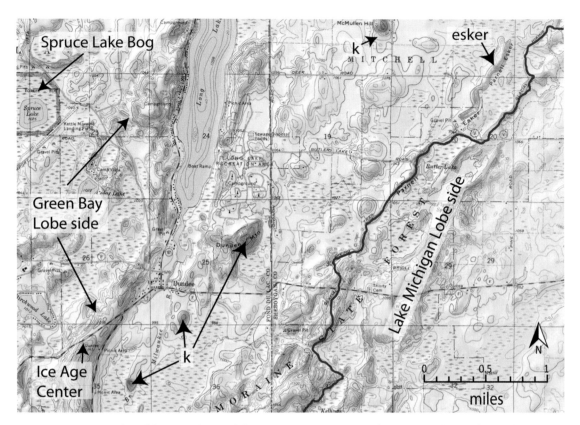

Figure 98.  Topography of the central part of the Parnell Segment. Note the Parnell Esker in the northern part of the map. The trail follows the crest of the esker south of Butler Lake. Some moulin kames are labeled *k*. The Henry S. Reuss Ice Age Visitor Center lies on the Green Bay Lobe side of the Kettle Moraine. Spruce Lake Bog is shown in figure 32. (Map is part of the Cascade and Dundee USGS Quadrangles and was created with TOPO! © 2011 National Geographic Maps.)

(SB 13) at Butler Lake parking lot. Butler Lake, part of the Butler Lake Flynn's Spring State Natural Area, is spring fed and sits in a large kettle mostly bounded by peat bog. The spring itself is 0.3 miles southeast of the lake. The IAT follows the crest of the Parnell Esker along the west side of Butler Lake (fig. 99). Several breaks in the esker indicate areas where there was no deposition of sand and gravel in the esker tube. Humans probably cut the break closest to Butler Lake sometime in the past, but the rest appear to be natural breaks. Dundee Mountain is a large moulin kame about a mile west of the IAT.

About a quarter mile north of CTH F, the trail turns to the southeast and crosses parallel ridges that are probably crevasse fillings (SB 13) and other collapse features that formed as the last of the ice buried in the Lake Michigan Lobe side of the Kettle Moraine melted. South of CTH F, the IAT continues in this complex of short elongate ridges and kettles that

Figure 99. View of the Ice Age Trail on the crest of the Parnell Esker near Butler Lake.

make up this side of the Kettle Moraine for about a half mile. It then follows the edge of large interconnected kettles westward to Crooked Lake and CTH SS (fig. 100). A large, symmetrical moulin kame lies just south of the highway. From there the IAT follows the crest of a small esker before climbing onto a sand and gravel hummock and continuing on hummocky sand and gravel to its terminus at Mauthe Lake Recreation Area.

While you're in the area, be sure to visit the Ice Age Visitor Center, which is located on STH 67 2.5 miles west of where the trail crosses the highway (fig. 98). Exhibits, presentations, and literature take you back into the frozen history of the Kettle Moraine area and the Ice Age in general, and you can stay warm (or cool) watching a film about glaciation and glacial features in Wisconsin. A stop here is highly recommended!

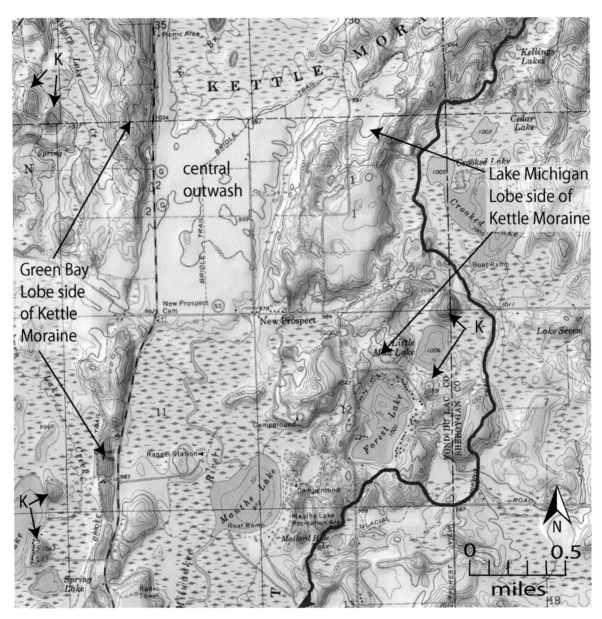

Figure 100. Topography of the western part of the Parnell Segment. Both sides of the Kettle Moraine and moulin kames (K) are labeled. The Green Bay Lobe side of the Kettle Moraine west of Mauthe Lake is likely an esker. (Map is part of the Dundee and Kewaskum USGS Quadrangles and was created with TOPO! © 2011 National Geographic Maps.)

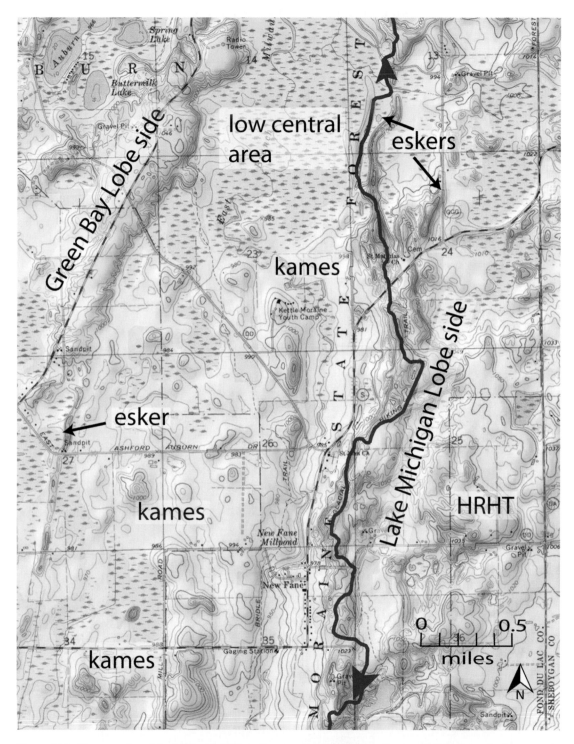

Figure 101. Topography of the Milwaukee River Segment in Fond du Lac County. Selected moulin kames and eskers are labeled. HRHT: high-relief hummocky topography (SB 11). (Map is part of the Kewaskum USGS Quadrangle and was created with TOPO! © 2011 National Geographic Maps.)

## 15. Milwaukee River Segment in Fond du Lac County

*Mauthe Lake Recreation Area to Kettle Moraine Dr. (4.3 miles)*

Westward from Mauthe Lake Rd. the trail follows the edge of a low esker (SB 13) for a couple of hundred yards before turning westward onto another small esker. A short distance to the east, you can see a large esker that runs along the trail for about a half mile (fig. 101). The trail then rises onto an irregular hummocky (SB 11) sand and gravel ridge that is the Lake Michigan Lobe side of the Kettle Moraine. There are good views of numerous moulin kames (SB 12) in the lowland to the west. Drumlins (SB 14) northwest of Kewaskum can be seen on the skyline to the west. This IAT segment remains in this ridge complex to its western terminus just north of the county line (fig. 101).

Figure 102. Contour plowing on drumlins in the Campbellsport drumlin field.

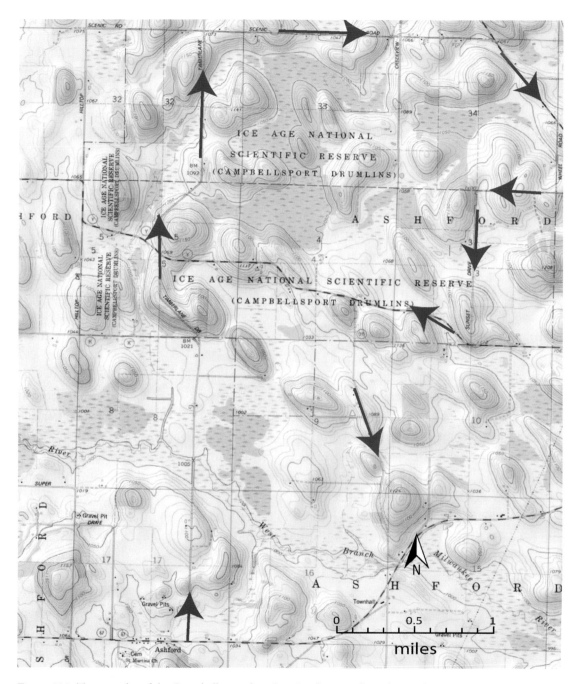

Figure 103. Topography of the Campbellsport drumlins. Ice flow was from the northwest. Red arrows indicate suggested auto- or bike-tour route described in text. (Map is part of the Campbellsport USGS Quadrangle and was created with TOPO! © 2011 National Geographic Maps.)

## 16. Campbellsport Drumlins Unit of the
## Ice Age National Scientific Reserve

This small area west of Campbellsport exhibits some of the highest drumlins (SB 14) in Wisconsin. They formed beneath the eastern side of the Green Bay Lobe when the edge of the lobe was in the area of the Kettle Moraine and was bumping against ice of the Lake Michigan Lobe. Ice flowed toward the southeast or south-southeast, and several of the drumlins are elongate in that direction. Quite a few of these drumlins, however, are actually more or less equidimensional (figs. 102, 103).

The drumlins here are distinct from those in other parts of Wisconsin: they are higher, have steeper sides, are more equally dimensional, and are more tightly clustered. They have these characteristics probably because ice flowed uphill from the Lake Winnebago lowland and also because the Lake Michigan Lobe obstructed the flow of the ice.

From Campbellsport the landscape slopes gently toward Lake Michigan. Although the drumlins are west of the Kettle Moraine, the lowlands between the drumlins are the headwaters for the Milwaukee River. We cannot see the material that makes up the drumlins, but we have clues as to their composition: wells here go down about 100 feet before hitting the Silurian dolomite bedrock, so it appears that the drumlin hills are made up almost entirely of till or sand and gravel.

No hiking trails or public lands exist within the Campbellsport drumlins at this time. However, you can find spectacular views of the drumlins on a short driving loop north on CTH K from STH 67 (fig. 103). To do the loop, from CTH K turn right on Timberlane Dr. and continue north across several drumlins. Note the high drumlins to the west and east. Turn right on Scenic Rd., and right again on CTH V. Turn right on Wheel Rd., then right on Sunset Dr. Turn right for a final time on Campbell Dr. (CTH Y). Take a left on Rolling Dr. to return to STH 67, about 2 miles south.

Similar high drumlins can be seen in the whole area south of STH 67, north of STH 28, and between USH 45 on the east and USH 41 on the west.

# Middle Kettle Moraine
# Ice Age Trail Segments

THE MIDDLE KETTLE MORAINE, as used in this book, extends from the Fond du Lac—Washington County line southward to Interstate 94 (I-94) near Delafield (fig. 104). Unlike the northern and southern parts of the Kettle Moraine, much of the land here is under private ownership. Nonetheless, there are 13 trail segments that exhibit glacial features typical of the Kettle Moraine: eskers (SB 13), moulin kames (SB 12), and kettles (SB 9) dominate the landscape. In places, pitted outwash (SB 10) is extensive, and in the southern part of this area there are a number of large meltwater channels. As in the Kettle Moraine north and south of here, nearly all of the sediment is sand or a combination of sand and gravel that was deposited by meltwater. There are few examples of till exposed anywhere in this part of the Kettle Moraine. Thus, the Kettle Moraine is not a moraine by most definitions but instead is an interlobate zone where supraglacial and subglacial streams deposited sand and gravel.

Much of the northern Kettle Moraine consists of two distinct ridges separated by a low area of outwash or till. The middle Kettle Moraine is more complex. In places there are two clearly defined ridges separated by a low area, but throughout most of this section, the ridges merge into a single broad mass of sand and gravel. North of the Slinger or Cedar Lakes Segment (fig. 104), meltwater from the retreating Lake Michigan Lobe drained away from the Kettle Moraine and toward Lake Michigan. From there southward the Lake Michigan Lobe meltwater flowed into and sometimes across the Kettle Moraine, leaving behind broad channels that are now occupied only by small streams. Mickelson and Syverson (1997) interpreted the formation of the single-ridge Kettle Moraine in Washington County to be similar to that shown in figure 105.

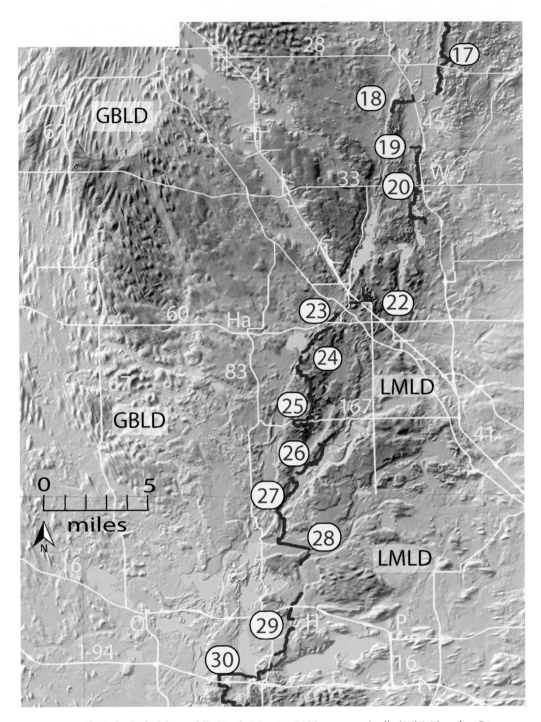

Figure 104. Shaded relief of the middle Kettle Moraine IAT segments (red): (17) Milwaukee River Segment in Washington County, (18) Kewaskum, (19) Southern Kewaskum, (20) West Bend, (22) Cedar Lakes, (23) Slinger, (24) Pike Lake, (25) Holy Hill, (26) Loew Lake, (27) Monches, (28) Merton, (29) Hartland, (30) Delafield. Blue arrows show ice-flow direction. Yellow lines and numbers indicate highways. Cities shown (yellow): (H) Hartland, (Ha) Hartford, (K) Kewaskum, (O) Oconomowoc, (P) Pewaukee, (T) Theresa, (W) West Bend. GBLD: Green Bay Lobe drumlins; LMLD: Lake Michigan Lobe drumlins. (Base map constructed from USGS National Elevation Dataset and modified by WGNHS.)

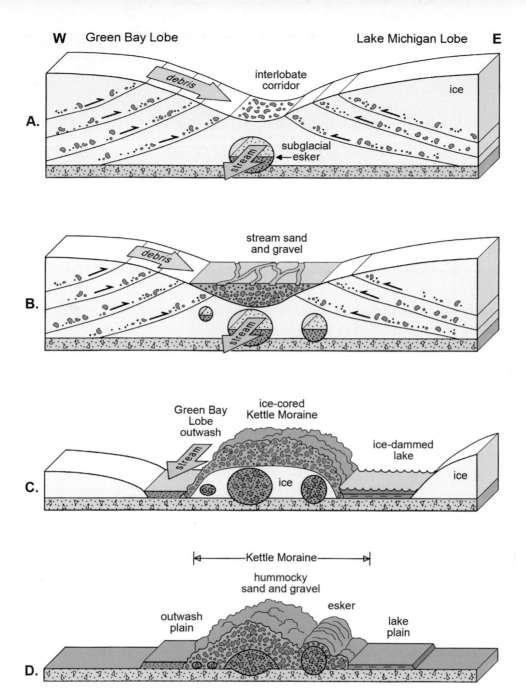

Figure 105. History of single-ridge Kettle Moraine formation in Washington County. (A) Advancing Green Bay and Lake Michigan lobes meet in the interlobate region. The upward flow of ice brings debris from base of ice to surface. Subglacial streams flowing in tunnels deposit sand and gravel. (B) The ice thins. Meltwater streams at surface transport and deposit sand and gravel in interlobate corridor. More subglacial tunnels form in interlobate region. Meltwater finds its way to base of ice through moulins. (C) Glacier lobes separate from the interlobate region. Thick debris insulates underlying ice. Hummocky topography develops as underlying ice slowly melts. (D) Hummocky sand and gravel lowers as underlying ice continues to melt, forming most high areas in Kettle Moraine. Eskers remain as prominent ridges in the Kettle Moraine. (Modified from Mickelson and Syverson 1997; drafted by Mary Diman.)

## 17. Milwaukee River Segment in Washington County
*Kettle Moraine Dr. to CTH H (4.7 miles)*

This segment of the IAT follows the high-relief hummocky topography (SB 11) of the Lake Michigan Lobe side of the Kettle Moraine. You will find deep kettles throughout (fig. 106). There are some good views through the trees of individual moulin kames (SB 12) in the low, central part of the Kettle Moraine to the west. The low flat areas are outwash (SB 8), but in places they have thin peat on the surface. About 1 mile west of where the trail crosses STH 28, a small esker (SB 13), partly mined away, crosses the highway. You'll find a cluster of small moulin kames (unlabeled) just west of the trail (north of STH 28) and a very large kame (unlabeled) south of the highway. The trail continues on the Lake Michigan Lobe side of the Kettle Moraine to CTH H across high-relief hummocky topography.

## 18. Kewaskum Segment
*Eisenbahn State Trail to Ridge Rd. (2.2 miles)*

At the eastern trailhead of the Kewaskum Segment, the Eisenbahn Trail follows an old railroad grade that crosses CTH H on outwash (SB 8) in the low, central part of the Kettle Moraine just east of STH 45 (fig. 107). From STH 45 the IAT runs westward across more outwash before climbing into the high-relief hummocky topography (SB 11) of the Green Bay Lobe side of the Kettle Moraine near Prospect Drive. The main Sunburst Ski Area hill is a high gravel hill, one of many hummocks (SB 11) in this ridge. High drumlins (SB 14) of the Green Bay Lobe form most of the skyline to the west. These drumlins are among the highest in Wisconsin and are similar to the Campbellsport drumlins. You can see the low, central Kettle Moraine immediately to the east, and the Lake Michigan Lobe side of the Kettle Moraine just beyond it. The Milwaukee River lies immediately to the east of that ridge (fig. 107). All of the hills along this segment are hummocks of sand and gravel, but it seems unlikely that most are moulin kames (SB 12), like the conical hills in the central Kettle Moraine area, because they do not appear to have formed in individual moulins.

## 19. Southern Kewaskum Segment
*Wildwood Rd. to CTH D (1.3 miles)*

This short segment is in hummocky terrain (SB 11) of the Lake Michigan Lobe side of the Kettle Moraine (fig. 108). This side is more ridgelike than it is farther to the north and it seems to be made up mostly of hummocky sand and gravel and short eskers (SB 13) oriented parallel to the Kettle Moraine.

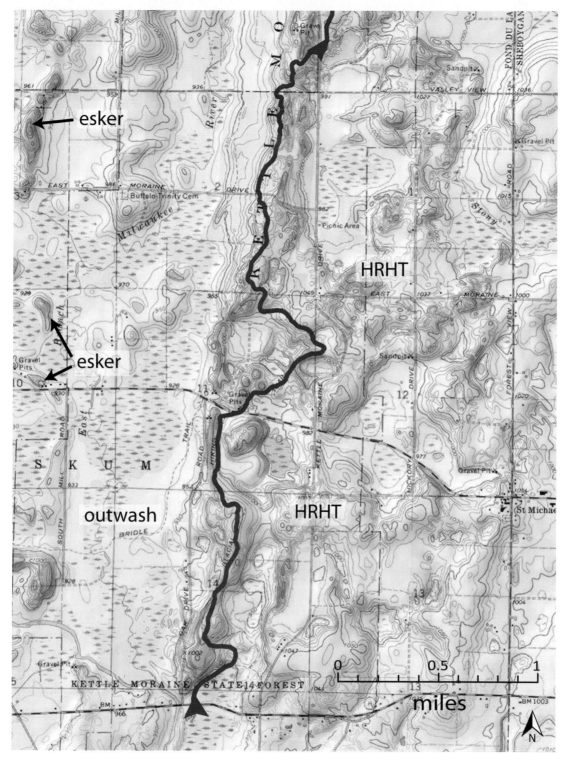

Figure 106. Topography of the Milwaukee River Segment in Washington County. Numerous moulin kames have a "bull's-eye" pattern to their contours. HRHT: high-relief hummocky topography (SB 11). (Map is part of the Kewaskum and West Bend USGS Quadrangles and was created with TOPO! © 2011 National Geographic Maps.)

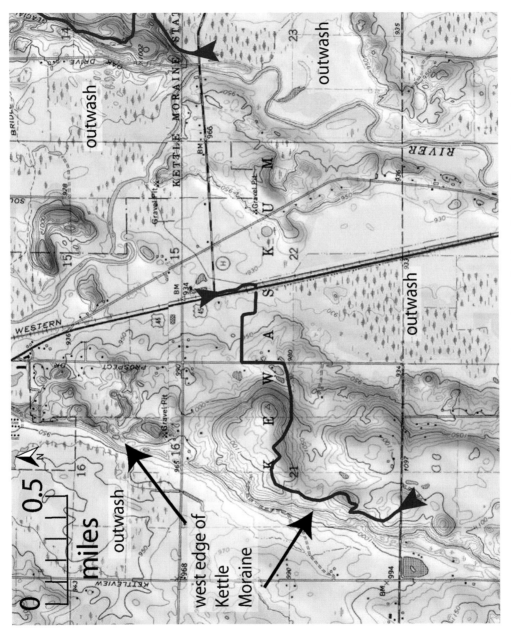

Figure 107. Topography of the Kewaskum Segment. The southern end of the Milwaukee River Segment is in the upper right corner. Entire segment is on high-relief hummocky topography of the Green Bay Lobe side of Kettle Moraine. Moulin kames have a "bull's-eye" pattern to their contours. Thin peat covers outwash (SB 8) in many places. (Map is part of the West Bend USGS Quadrangle and was created with TOPO! © 2011 National Geographic Maps.)

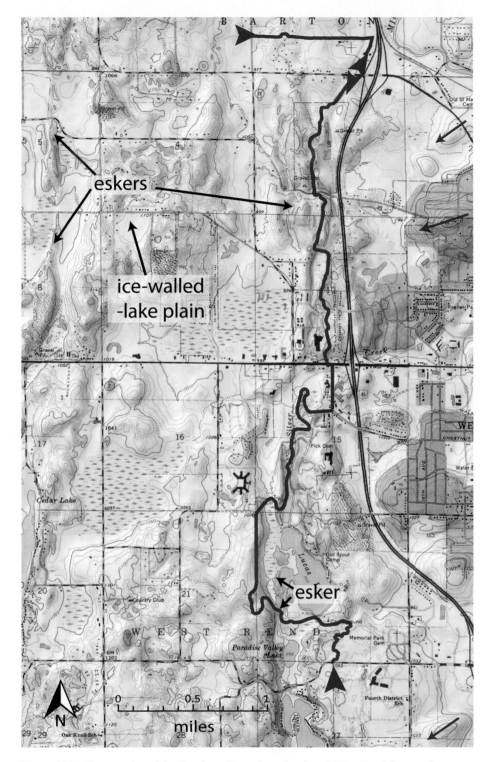

Figure 108. Topography of the Southern Kewaskum (top) and West Bend (bottom) segments. Trail follows Lake Michigan Lobe edge of Kettle Moraine. Thin peat covers many low outwash areas. Drumlins of Lake Michigan Lobe are indicated by blue arrows. (Map is part of the West Bend USGS Quadrangle and was created with TOPO! © 2011 National Geographic Maps.)

## 20. West Bend Segment
*CTH D to Paradise Dr. (7 miles)*

The West Bend Segment continues in the east side of the Kettle Moraine. About two miles to the west, Beaver Dam Rd. skirts part of an esker (SB 13) that you can easily see (fig. 108). This esker continues south and crosses Washington St. (STH 33) at the West Bend Sand and Gravel pit. South of Beaver Dam Rd., the IAT runs along the crest of a small esker, parallels the ridge to the west, then skirts a large kettle (SB 9). From there southward the trail drops into kettles and over ridges that may be in part subglacial. A little more than a mile west of the IAT on Schuster Drive is an ice-walled-lake plain (SB 15), a relatively rare landform in this part of the Kettle Moraine. South of West Washington St., the trail goes through more high-relief hummocky topography (SB 11). Lucas Lake is a large kettle; parts of a small esker form its southwest edge. The segment ends at West Paradise Dr. on pitted outwash (SB 10). The hill about a half mile southeast of the Paradise Drive trailhead is a Lake Michigan Lobe drumlin (SB 9), as are many of the hills in West Bend itself. All of the lakes in the vicinity are kettles. About 22 miles west of West Bend on STH 33 is Horicon Marsh, an extinct glacial lake that is now a unit of the Ice Age National Scientific Reserve.

## 21. Horicon Marsh Unit of the Ice Age National Scientific Reserve

Horicon Marsh is a wetland about 15 miles long and up to 5 miles wide in some places (figs. 109, 110). The northern half is administered as a national wildlife refuge and the southern half as a state wildlife area, so the public has access to most of the wetland. Horicon Marsh lies in the lowland gouged out as the Green Bay Lobe flowed toward the southwest. Green Bay, Lake Winnebago, and Lake Sinissippi (originally a marsh) all occupy this lowland. The divide between the south-flowing Rock River and the north-flowing Fox River drainage is a few miles to the north, just off the map in figure 110.

The Niagara Escarpment (SB 19), composed of Silurian dolomite, bounds the marsh to the east. Beneath this dolomite is the Maquoketa Formation, made of shale that erodes easily. To the west of the marsh, thin till over a bedrock layer of Ordovician dolomite extends for miles. A generalized cross section of the escarpment is shown in figure 58.

To understand how and why Horicon Marsh exists, we have to look back nearly 20,000 years, when the ice began to retreat and expose the barren land. In the Horicon area, the Green Bay Lobe slowly melted, forming a small, discontinuous end moraine (SB 6), called the Green Lake Moraine, across this broad lowland. As the ice continued to retreat, a large lake formed between the Green Lake deposits (which in Horicon are composed of hummocky sand and gravel) and the retreating ice margin. The lake was deepest on its eastern side. Under the marsh surface lies more than 50 feet of sand, silt, and clay, which was deposited in this lake since deglaciation.

Figure 109.  Low oblique aerial view looking toward the north of Horicon Marsh. City of Horicon is in foreground.

Over time the lake basin shallowed as it filled with lake sediment and organic matter, creating what we recognize as a marsh, or wetland. The marsh expanded westward in a cyclical manner: organic sediment slowed water flow, allowing more peat to grow in the stagnant, warm waters; the additional organic matter, in turn, caught and slowed more water. At the time of European settlement, thick peat covered much of the area. During the second half of the nineteenth and the early twentieth centuries, people tried many times to drain the marsh. Although never successful, their attempts lowered the water table and exposed the peat to the open air. The peat dried out and oxidized (the same process as burning, but without the flame). Over the decades, wind carried much of the dried peat away, leaving only a thin (less than 5 feet) layer above the glacial lake sediment. Although no one has reported finding large mammal remains in Horicon Marsh, other mixed glacial-organic deposits like this

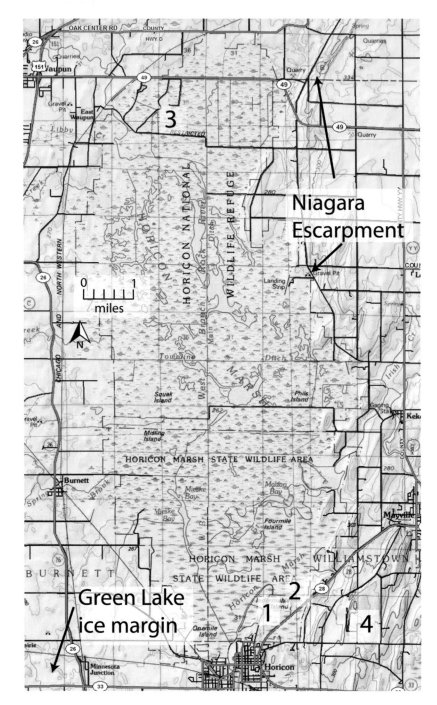

Figure 110. Topography of the Horicon Marsh and part of the Niagara Escarpment. The remnants of the Green Lake ice margin (not a significant moraine here) naturally dam the Fox River at the south end of the marsh. There is now a human-made dam that controls water levels. Suggested viewing points (keyed to numbers on map): (1) the Wisconsin DNR office at the north end of North Palmatory St. in Horicon (note the parking lot and viewing area are on the top of a drumlin); (2) the visitor center on Clinic Drive off STH 28 between Horicon and Mayville; (3) several places along STH 49, which provide good views of the north marsh; and (4) from Horicon Ledge County Park, which offers an excellent view from a higher perspective. (Map is part of the 1:100,000 Oconomowoc and Fond du Lac USGS Quadrangles and was created with TOPO! © 2011 National Geographic Maps.)

contain a record of plants and animals since the time of deglaciation. In fact, throughout southern Wisconsin the remains of mastodon, mammoth, giant beaver, and other extinct animal species have been found.

Although the number of birds using the marsh has decreased dramatically since European settlement, many migrating geese and other waterfowl still visit the marsh. You can view the marsh—and hopefully some of these birds—from the locations suggested in figure 110.

## 22. Cedar Lakes Segment
*CTH NN to Kettle Moraine Dr. (2.8 miles)*

In this area the Kettle Moraine has two fairly distinct hummocky ridges (SB 11) with a narrow, low, central zone of slightly pitted outwash (SB 10; fig. 111). North and south of here, the Green Bay Lobe and Lake Michigan Lobe sides of the Kettle Moraine are less distinct. The Cedar Lakes Segment of the IAT is on the Lake Michigan Lobe side of the Kettle Moraine (fig. 111). The CTH NN terminus of the trail is on pitted outwash. However, the dominant landform is the moulin kame (SB 12), and this group of kames is known as the Polk kames. They are excellent examples of this landform and are among the highest and most abundant anywhere in this part of the Kettle Moraine. The trail passes between and around them. The trail continues south to Cedar Creek Rd., then follows it about a mile west to the intersection with Kettle Moraine Drive.

## 23. Slinger Segment
*Kettle Moraine Dr. to STH 60 (1.5 miles)*

Large kames (SB 12) are abundant in this area and include the ski slope southeast of the Kettle Moraine Dr. trailhead (fig. 111). The flatter surfaces in Slinger and east of Howard Ave. are pitted outwash (SB 10). The ridge west of Howard Ave. may have formed in a subglacial tunnel or a collapsed trench open to the sky. There is not enough evidence to say for sure that it is an esker (SB 13). The nearly flat area west of the Kettle Moraine is the bed of a former lake dammed between the Kettle Moraine and the ice of the Green Bay Lobe.

## 24. Pike Lake Segment
*STH 60 to CTH E (3.2 miles)*

From the STH 60 trailhead, the IAT traverses the high-relief hummocky topography (SB 11) of the Green Bay Lobe side of the Kettle Moraine, exiting the Kettle Moraine at the parking lot near Pike Lake (fig. 112). A spur trail goes to the top of Powder Hill, a large moulin kame (SB 12) that has an observation tower at the top and great views in all directions (fig. 113). Pike Lake is a kettle (SB 9) in pitted outwash (SB 10) of the Green Bay Lobe. The hills immediately west of Pike Lake are drumlins (SB 14) of the Green Bay Lobe

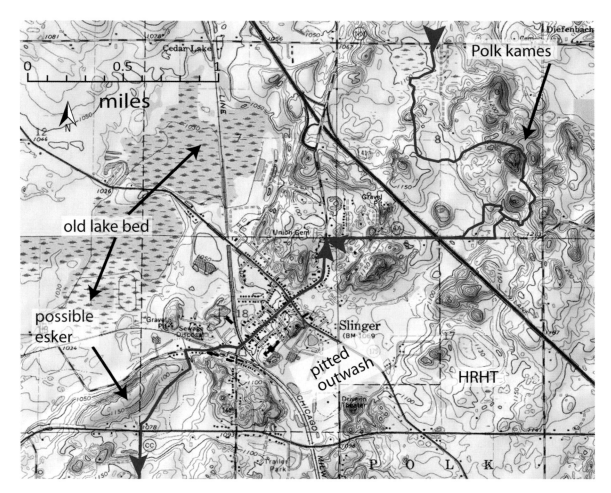

Figure 111. Topography of the Cedar Lakes (top) and Slinger (bottom) segments. Trail follows Lake Michigan Lobe side of Kettle Moraine in vicinity of the Polk kames. The Slinger Segment is on the Green Bay Lobe side of Kettle Moraine. Moulin kames have a "bull's-eye" pattern to their contours. HRHT: high-relief hummocky topography (SB 11). (Map is part of the Hartford East USGS Quadrangle and was created with TOPO! © 2011 National Geographic Maps.)

(fig. 112). Where the trail follows Glassgo Dr., it climbs back into the high-relief hummocky topography of the Green Bay Lobe side of the Kettle Moraine and remains in it to its terminus at CTH E. The broad, nearly flat open area to the west is pitted outwash.

## 25. Holy Hill Segment
*CTH E to Donegal Rd. (7 miles)*

The IAT follows Glassgo Dr. in the hummocky (SB 11) Green Bay Lobe side of the Kettle Moraine (fig. 114). Where the trail leaves Glassgo Dr., it skirts the base of a high ridge that

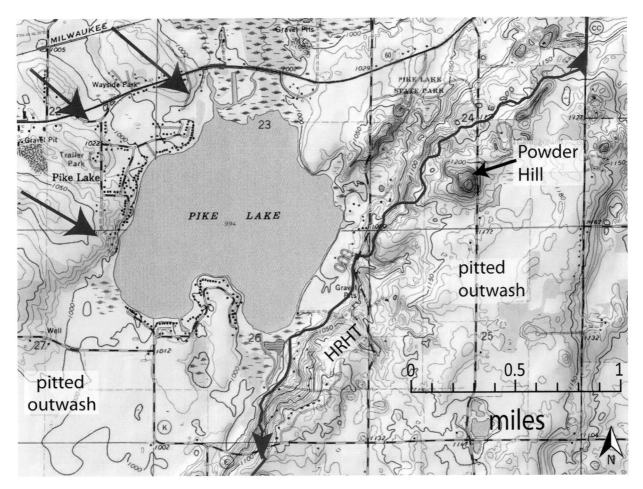

Figure 112. Topography of the Pike Lake Segment. Powder Hill summit is on a spur trail. Blue arrows are on drumlins and show ice-flow direction of Green Bay Lobe. HRHT: high-relief hummocky topography (SB 11). (Map is part of the Hartford East USGS Quadrangle and was created with TOPO! © 2011 National Geographic Maps.)

may have formed in a subglacial tunnel and is therefore an esker (SB 13). It could also be a crevasse filling that formed in a channel open to the sky. The flat landscape in the west is outwash (SB 8), some of which has a thin peat cover. The ridge west of CTH K (fig. 114) is an esker. The IAT then crosses pitted outwash (SB 10) and hummocky terrain to Waterford Rd. This is in the central, low part of the Kettle Moraine. South of Waterford Rd. the trail crosses a wetland that occupies a long, shallow depression underlain by silt and sand. This was the site of an ice-walled lake during deglaciation (fig. 114), but it is not very noticeable on the map. The IAT then rises into a complex of gravel hummocks. These appear to be moulin kames that have been surrounded by pitted outwash.

Figure 113. View from the top of the tower at Powder Hill looking south along the Kettle Moraine. Holy Hill (with church) is one of several tall moulin kames in view.

Between Pleasant Hill Rd. (fig. 114) and Shannon Rd. (fig. 115) the IAT is on pitted outwash. Just to the east as you approach STH 67 is high-relief hummocky topography (SB 11) that forms the axis of the Kettle Moraine in this area (fig. 115). There is no low, central area between the Green Bay Lobe and Lake Michigan Lobe sides of the Kettle Moraine as there is father north. The high moulin kames (SB 12) visible to the east and south are among the highest anywhere in the Kettle Moraine. Holy Hill and the unnamed moulin kame just south of STH 167 and north of Holy Hill both have summits at about 1,330 feet above sea level—750 feet above Lake Michigan!

Erratics (SB 5) are present here and there along the IAT in most places in the Kettle Moraine. A spur trail climbs to a large erratic on a kame north of Holy Hill. Eskers are present north and south of STH 167 west of where the IAT crosses the highway (fig. 115).

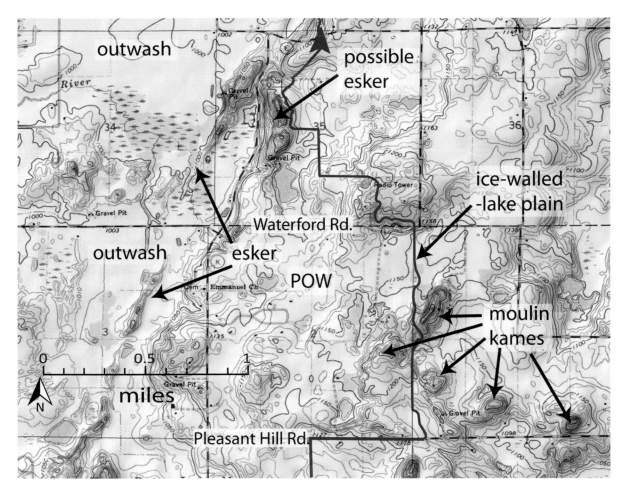

Figure 114. Topography of the eastern Holy Hill Segment. POW: pitted outwash (SB 10). (Map is part of the Hartford East USGS Quadrangle and was created with TOPO! © 2011 National Geographic Maps.)

East of Holy Hill there are good views of Flynn Creek valley. This is a good example of an underfit stream (SB 8), where the valley is much larger than could be cut by the present creek. The valley heads only a mile to the north and does not have tributaries. This channel was eroded late in the formation of the Kettle Moraine. Note that there are no kettles in the bottom of the channel (fig. 115), indicating that there was no buried ice under the channel when it was last occupied by a major stream. In all probability a major source of the water was the melting of buried ice in the complex of hummocky topography to the west, but it may have also had drainage from a small ice-marginal lake that formed in front of the retreating Lake Michigan Lobe. The next channel to the east, the Friese Lake channel, is discussed with the Loew Lake Segment.

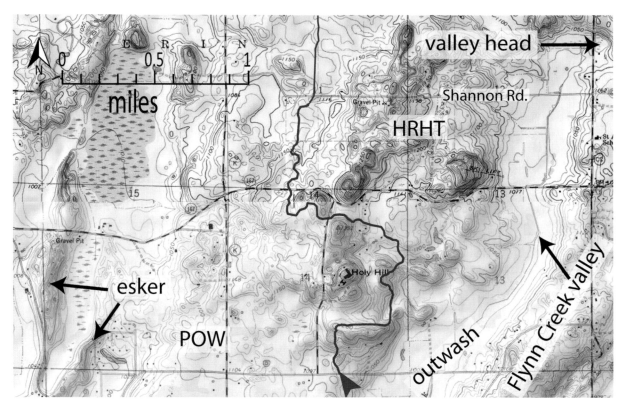

Figure 115. Topography of the western Holy Hill Segment. Holy Hill is large moulin kame, as are conical hills north. Flynn Creek valley is a broad meltwater channel that drained into the Friess Lake channel. HRHT: high-relief hummocky topography (SB 11); POW: pitted outwash (SB 10). (Map is part of the Hartford East and Merton USGS Quadrangles and was created with TOPO! © 2011 National Geographic Maps.)

## 26. Loew Lake Segment
*Emerald Dr. to CTH K at CTH Q (5 miles)*

This trail segment starts on a high, thin-till-covered (SB 5) gravel terrace (SB 8) that at one time was continuous with the more or less flat surface directly to the west across Flynn Creek valley (fig. 116). The same surface makes up the skyline to the east and south of here (southeast part of fig. 116, till-covered upland). This terrace is close to the north end of a huge sand-and-gravel deposit that apparently formed as a Lake Michigan Lobe outwash plain in an early stage of the last glaciation. It extends southward through eastern Waukesha County into eastern Walworth and western Racine counties. It was covered by till from the last major ice advance, the one that went to the Kettle Moraine and that formed nearly all of the glacial landforms in eastern Wisconsin. In some places, this extensive outwash plain was shaped at the bottom of the glacier to form drumlins that have a gravel core. These are common east of the Kettle Moraine in southern Washington, Waukesha, and Walworth counties. Rivers flowing beneath and away from the glacier as it retreated eroded channels with steep walls and broad flat floors. The channel in which Loew Lake has formed is one of these channels. It is been called the Friess Lake channel because it is occupied by Friess Lake, which you will find a short distance northeast of this trail segment. The channel occupied by Flynn Creek is also a meltwater channel, as is the channel occupied by the Oconomowoc River east of Loew Lake.

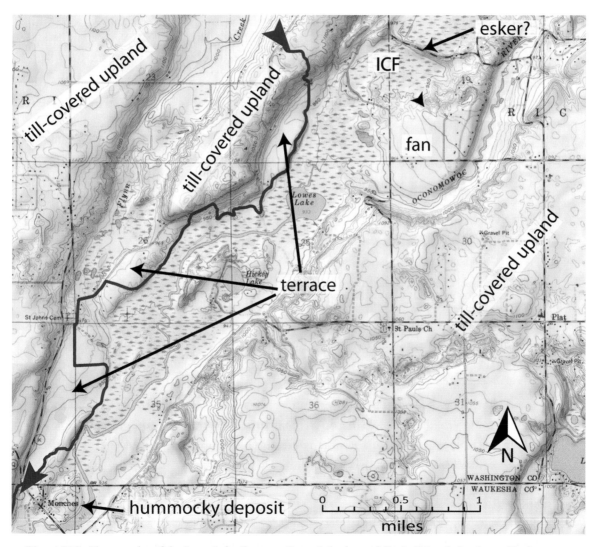

Figure 116. Topography of the Loew Lake Segment. Loew Lake (center) lies in Friess Lake channel bed, a former drainage way. Water flowed southwest. ICF: ice-contact face of outwash fan (SB 10). (Map is part of the Merton USGS Quadrangle and was created with TOPO! © 2011 National Geographic Maps.)

From the Emerald Dr. trailhead the trail crosses the till-covered upland for 0.25 miles before descending into the Friess Lake channel. This channel clearly had some ice in and under it when meltwater was flowing toward the southwest. Several small ridges had ice walls and may be eskers, although they could have been channels open to the sky (fig. 116). The large, gently sloping surface across the channel from the IAT is an outwash fan that was built by a river flowing off an ice margin sitting across the channel at its north end. The ice-contact face (ICF in fig. 116) is highly collapsed and has many small kettles. The trail itself goes along the edge of a terrace that represents a former streambed that was active before the stream cut even lower, to the level now occupied by the wetland. This terrace is discontinuous but the trail follows it for much of the way to CTH Q. The wide Friess Lake channel ends about where the trail meets CTH Q. At that point hummocky sand and gravel (SB 11) fills most of the channel except where the Oconomowoc River has cut through the deposit. Note that the river valley is much smaller southward from here (see the Monches Segment description). This hummocky deposit, along with the numerous small gravel deposits that are now in the channel bed, indicates that glacier ice was present beneath parts of the Friess Lake channel when a braided river was flowing in it. If that were not so, the river would have eroded or buried these small features.

## 27. Monches Segment
*CTH K at CTH Q to Funk Rd. (3.1 miles)*

The IAT climbs out of the Oconomowoc River valley into high-relief hummocky topography (SB 11) a short distance south of CTH Q (fig. 117). It then drops back into the valley bottom and across the river onto a low terrace (SB 8) and remains on that terrace to Funk Rd. Note that the river valley is narrower throughout this area than it is to the north of CTH Q, probably because some meltwater flowed over ice and did not contribute to meltwater channel erosion. The high terrace west of the river in the half mile north of Funk Rd. is at roughly the same elevation as the high terrace east of the trail on the east side of the river. The elongate hills west of the river appear to be erosional remnants on the former riverbed. Although they have the teardrop shape of drumlins, they are not. A small esker (SB 13) on private land is present in the northwest part of figure 117. All of the depressions except the river valley are kettles (SB 9).

## 28. Merton Segment
*Funk Rd. to CTH K at Centennial Park (5.8 miles)*

For the first 2 miles, the trail crosses an outwash terrace (SB 8; fig. 118). The river that deposited the sand and gravel in this terrace came from the Friess Lake channel to the north (see the Loew Lake Segment description). Where the trail turns directly southward, it

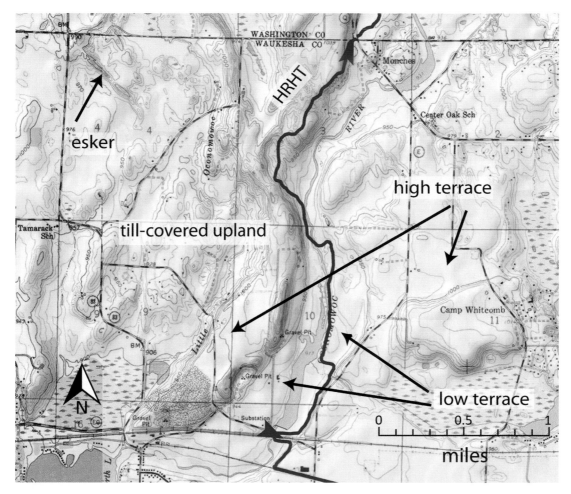

Figure 117. Topography of the Monches Segment. All depressions except the Oconomowoc River valley are kettle holes. This is a continuation of the Friess Lake channel discussed with the Loew Lake Segment. HRHT: high-relief hummocky topography (SB 11). (Map is part of the Merton USGS Quadrangle and was created with TOPO! © 2011 National Geographic Maps.)

crosses a narrow upland that is till-covered sand and gravel. This sand and gravel is the remnant of an extensive sheet of outwash that was deposited before the last glacial advance, which was in turn covered by thin, light-brown, sandy till (SB 5) deposited by the most recent advance. In places, the Lake Michigan Lobe made drumlins (SB 14) on this surface. The IAT then descends from the upland and follows the floodplain of the Bark River. This is another underfit stream (SB 8), a very large stream channel that today is occupied by a very small stream. This Bark channel carried a huge flow of meltwater from the retreating ice edge to the northeast. The IAT remains in the Bark River floodplain for 0.4 miles before climbing onto another terrace. About a third mile south of CTH EF the trail climbs onto another remnant of the till upland for a short distance. This surface has small drumlins (SB 16; fig. 118). From there it continues on outwash to the south trailhead at Centennial Park on CTH K. Beaver Lake, to the west, is a kettle (SB 9).

## 29. Hartland Segment
*CTH K at Centennial Park to STH 83 (6.9 miles)*

From Centennial Park through Hartland, the trail follows the floodplain of the Bark River (fig. 119). The till-covered upland just west of Hartland is an eroded remnant of a much

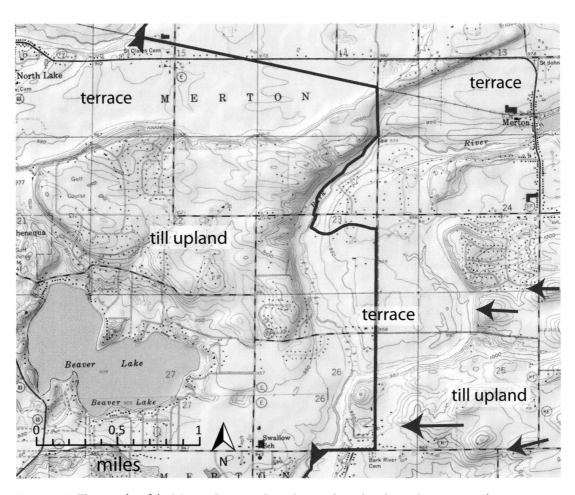

Figure 118. Topography of the Merton Segment. Several outwash sand-and-gravel terraces are shown separated by till uplands that have a sand-and-gravel core. Small drumlins formed by the Lake Michigan Lobe are indicated with blue arrows. (Map is part of the Merton and Hartland USGS Quadrangles and was created with TOPO! © 2011 National Geographic Maps.)

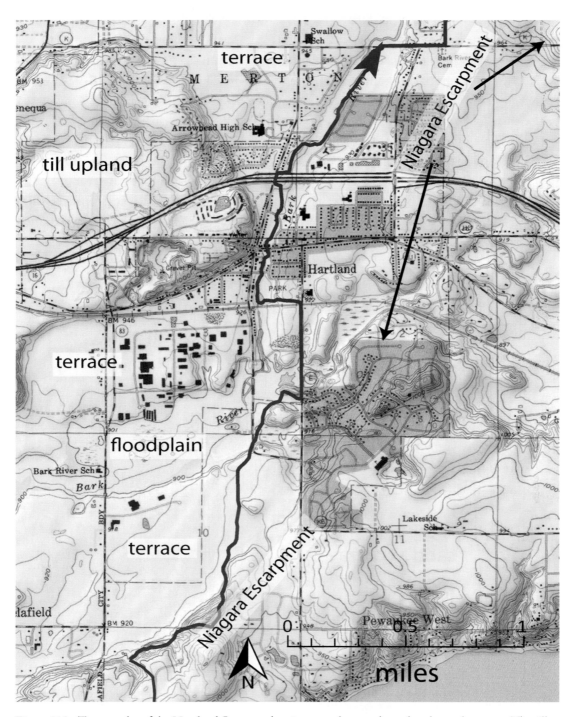

Figure 119. Topography of the Hartland Segment showing several outwash sand-and-gravel terraces. The till upland west of Hartland has a sand-and-gravel core. The upland west of the trail has Silurian dolomite close to the surface. The best expression of the Niagara Escarpment is along the western half-mile of the segment. West end of Pewaukee Lake is in lower right corner. (Map is part of the Hartland USGS Quadrangle and was created with TOPO! © 2011 National Geographic Maps.)

more extensive sand-and-gravel surface deposited before the last advance of late Wisconsin ice. South of Hartland, CTH E is on a sand-and-gravel outwash terrace (SB 10) interrupted by kettle holes (SB 13). Lake Michigan Lobe drumlins (SB 14) lie only a mile or so to the east. The upland east of the trail is underlain by Silurian-age dolomite. The western edge of this area of shallow dolomite is the Niagara Escarpment (SB 19). It is not very high here, and it underlies till in the upland east of this part of the IAT (figs. 119, 120). Pewaukee Lake lies in a valley that cuts across the escarpment just south of this trail segment. The lake is not a kettle, but it is in a valley in the dolomite bedrock that was probably first cut by a stream then carved wider and deeper by flowing glacier ice before being dammed by glacial deposits.

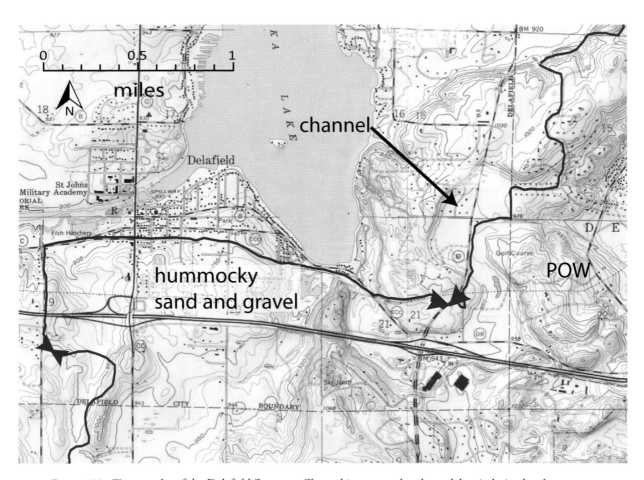

Figure 120. Topography of the Delafield Segment. Channel is a now a dry channel, but it drained meltwater from the Pewaukee Lake depression. Lake shown is south end of Nagawicka Lake. POW: pitted outwash (SB 10). (Map is part of the Hartland and Oconomowoc East USGS Quadrangles and was created with TOPO! © 2011 National Geographic Maps.)

## 30. Delafield Segment

*STH 83 to Cushing Park Rd. (3.5 miles)*

The STH 83 terminus of the IAT is on the floor of an abandoned river channel (fig. 120). This mostly dry channel was the outlet for a high level of Pewaukee Lake when ice dammed the east end of the lake. Water flowed to the south through the channel, which crosses I-94 at the intersection with STH 83. Much of the channel at this location is occupied by a shopping center. This was called the Ethan Allen Channel by Clayton (2001). The trail then climbs out of the channel and crosses a hummocky gravel (SB 11) upland. Numerous shallow kettles (SB 9) are present. This upland is the Kettle Moraine, and here it sits right on top of the Niagara Escarpment (SB 19). Green Bay Lobe drumlins (SB 14) can be seen on the horizon to the west. All of the lakes to the west and north are kettles in pitted outwash (SB 10). Through the city of Delafield the trail crosses outwash that is collapsed in places. You may see several low Green Bay Lobe drumlins just west of Cushing Park Rd. south of I-94.

# Southern Kettle Moraine Ice Age Trail Segments

THE ICE AGE NATIONAL SCENIC TRAIL winds through forest and old fields of the southern Kettle Moraine throughout most of its length, although the trail still has road connectors in places. What follows in this chapter is a description of trail segments from I-94 near Delafield to the west end of Kettle Moraine near Whitewater Lake (fig. 121).

Unlike the Kettle Moraine farther north, it appears that the southern part of the Kettle Moraine is mostly sand and gravel deposited by glacial meltwater in a trough between the Lake Michigan and Green Bay lobes. This low area sloped toward the southwest, and the water, sand, and gravel that it carried flowed in that direction (fig. 122). As the glacier margin retreated toward the north, the V-shaped opening between the lobes widened and extended toward the northeast.

Figure 122 shows an interpretation of the positions of the ice margin at the time the Ethan Allen Channel (see the Lapham Peak Segment description) was carrying meltwater. Much of the Kettle Moraine here appears to be a single, broad ridge with many kettles, but also with some broad, nearly flat outwash surfaces. The kettles themselves formed when glacier ice that was buried in the sand and gravel slowly melted, perhaps hundreds or thousands of years after water stopped flowing toward the southwest (SB 9). The high gravel surfaces were bounded (but not underlain) by glacial ice when water flowed on them. Glacier ice on both sides of the southwestward flowing river formed the walls of the channel, allowing the sand and gravel to be deposited well above what would become the surrounding land surface. Subsequently the surrounding ice melted out, leaving the former riverbed as the highest part of the present-day landscape (see the Scuppernong Segment description).

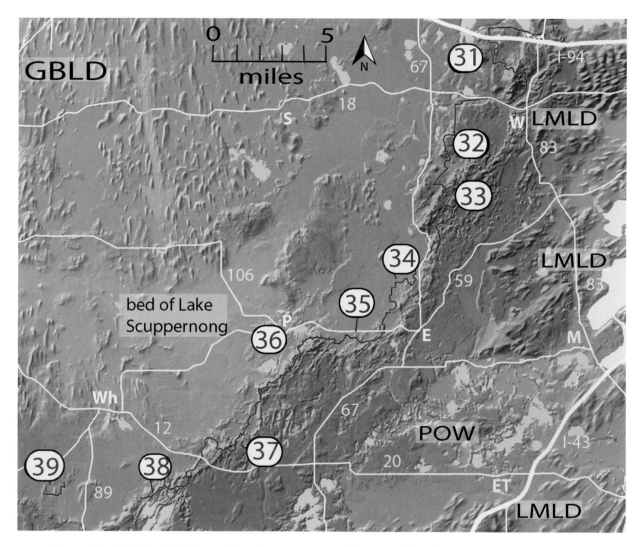

Figure 121. Shaded relief of the southern Kettle Moraine IAT segments (red): (31) Lapham Peak, (32) Waterville, (33) Scuppernong, (34) Eagle, (35) Stony Ridge, (36) Blue Spring Lake, (37) Blackhawk, (38) Whitewater Lake. The Clover Valley Segment (39) is not in the Kettle Moraine but is described with the southern Green Bay Lobe segments. Yellow lines and numbers indicate highways. Cities shown (yellow): (E) Eagle, (ET) East Troy, (P) Palmyra, (M) Muskego, (W) Wales, (Wh) Whitewater, (S) Sullivan. Blue arrows show ice-flow direction. All of the low area west of the Kettle Moraine was under glacial Lake Scuppernong as ice retreated. The drumlins were islands in the lake. LMLD: Lake Michigan Lobe drumlins; GBLD: Green Bay Lobe drumlins; POW: pitted outwash (SB 10). (Base map constructed from USGS National Elevation Dataset and modified by WGNHS.)

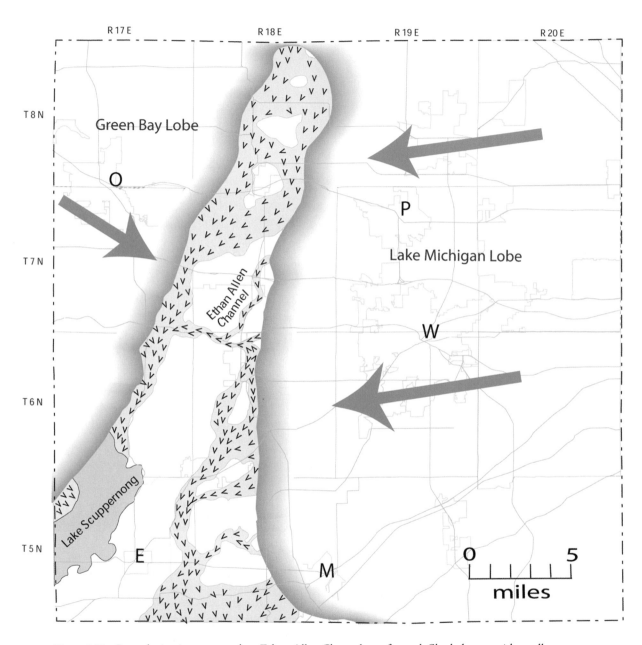

Figure 122. Stage during ice retreat when Ethan Allen Channel was formed. Shaded areas with small arrows indicate active outwash rivers at that time. Blue arrows show ice-flow direction. Cities shown: (E) Eagle, (M) Mukwango, (O) Oconomowoc, (P) Pewaukee, (W) Waukesha. (Modified from Clayton 2001.)

As the glacier continued to retreat, hills to the east (part of the Lake Michigan Lobe's Waukesha drumlin field) were exposed (fig. 121). To the west a large lake, glacial Lake Scuppernong (SB 16), formed in front of the retreating ice margin (fig. 122). Drumlins of the Green Bay Lobe drumlin field were islands in this huge shallow lake as the Green Bay Lobe retreated to the north (fig. 121). Along much of its length south of I-94, the Kettle Moraine more or less lies atop the Niagara Escarpment (SB 20).

## 31. Lapham Peak Segment
*Cushing Park Rd. to the UW–Waukesha Field Station (7.8 miles)*

Southwest of the Lapham Peak Segment trailhead there are Green Bay Lobe drumlins (SB 14) partly buried in outwash (fig. 123). A half mile to the west of the trailhead is Lower Nemahbin Lake, which lies in a large kettle in pitted outwash (fig. 123). The surface of the lake is about 100 feet lower than the elevation of the trailhead, which is at a higher elevation because it is on the Niagara Escarpment (SB 19). Silurian dolomite is close to the surface here, and the sediment cover on the bedrock is only about 20 feet thick. Note that the land between Cushing Park Rd. and CTH C has lower-relief topography with fewer deep kettles than the Kettle Moraine east of CTH C. Climb onto the Green Bay Lobe side of the Kettle Moraine and up to the tower for a broad view of this part of the Kettle Moraine. Lapham Peak is the highest point in Waukesha County! You can see excellent examples of the surrounding high-relief hummocky topography (SB 11). From the tower you can also look northward and see the middle Kettle Moraine with Holy Hill (fig. 124), which is easily identifiable because of the church on top of the large moulin kame. The nearly flat surface with lakes to the west is pitted outwash, and the Green Bay Lobe drumlin field forms the hills in the distance. To the southwest is the flat bed of glacial Lake Scuppernong. From the tower you can look eastward into the Ethan Allen meltwater channel and beyond it to the Waukesha drumlin field. As ice was melting away, the Ethan Allen Channel carried large volumes of glacial meltwater toward the south from the vicinity of Pewaukee Lake (see the Hartland Segment description). A small spring-fed stream, Scuppernong Creek, now flows in the channel. This is a great example of an underfit stream—one that is too small to have cut the channel through which it flows (SB 8).

From the base of the tower the trail drops into the Ethan Allen meltwater channel. It then follows a terrace (SB 8) along the west side of this channel. This terrace was the stream bottom at the time glacial meltwater flowed southwestward down this wide valley. Kettles interrupt the flat-topped terrace in places, indicating that there was still a small amount of glacial ice present beneath the channel bottom when the stream flowed at the terrace level. Along the side of the valley just south of the access road to the Ethan Allen facility, you can see kettles in this surface that have been modified by gravel mining at some time in the past.

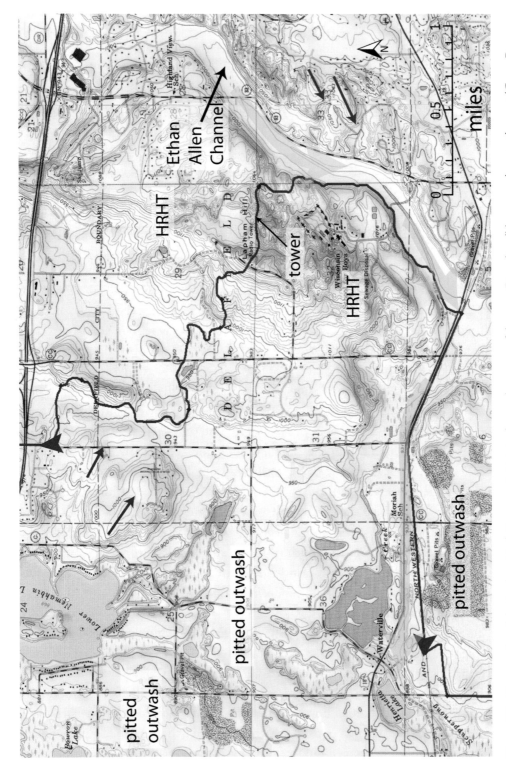

Figure 123. Topography of the Lapham Peak Segment. I-94 is in the northern part of the map. Drumlins (blue arrows) are to the west (Green Bay Lobe) and east (Lake Michigan Lobe) of the Kettle Moraine. The Lapham Peak tower is on a high point in the Green Bay Lobe side of the Kettle Moraine. HRHT: high-relief hummocky topography (SB 11). (Map is part of the Oconomowoc East USGS Quadrangle and was created with TOPO! © 2011 National Geographic Maps.)

Just north of USH 18, the IAT drops off the terrace onto the floodplain. South of USH 18, it follows the Scuppernong Creek valley westward along an abandoned railroad grade, which is now part of the Glacial Drumlin Trail and is primarily on pitted outwash (SB 10). After meltwater abandoned the channel, much of the valley bottom collapsed because of the meltout of glacier ice beneath, so we are not looking at the slope of the stream bed as it originally formed. However, it appears that the river flowed toward the west and was carrying water from the Ethan Allen Channel. Water that flowed down this channel probably went into Lake Scuppernong, a large glacial lake that formed in front of the retreating Green Bay Lobe. The valley was cut through a higher surface of pitted outwash and hummocky sand and gravel. Two large active gravel pits and several abandoned ones expose the coarse sand and gravel that underlies this higher surface south of the trail. Dutchman and Henrietta lakes are kettle lakes.

Figure 124. View looking north from the Lapham Peak tower along the axis of the Kettle Moraine. Highest hill in the distance is Holy Hill, a moulin kame. Nagawicka Lake is in the foreground. (Photo by Brad Singer.)

## 32. Waterville Segment

*Waterville Rd. to CTH D (3.6 miles)*

The Waterville Segment of the IAT begins on pitted outwash (SB 10) on Waterville Road (fig. 125). The connector from the Lapham Peak Segment is Waterville Road, which is on pitted outwash and till between the two segments. Unlike much of the IAT in this area, the Waterville Segment is west (Green Bay Lobe side) of the Kettle Moraine. As you walk, note that the landscape lacks the high-relief hummocky topography (SB 11) that is in the Kettle Moraine to the east. Instead, another geologic feature stands out: from Waterville Road, the trail rises across a little pitted outwash, then onto till, to the edge of the Niagara Escarpment (SB 19). There it follows near the crest of the more than 100-foot-high, steep slope. Look for small outcrops of dolomite near the top of this slope. Where the trail approaches CTH D, the escarpment is interrupted by a valley that may be a preglacial valley later reoccupied by a meltwater stream. Several moulin kames are present in this area (fig. 125).

## 33. Scuppernong Segment

*CTH C to STH 67 wayside (5.5 miles)*

From CTH C the trail crosses pitted outwash (SB 10) for 0.4 miles before rising into the Kettle Moraine (fig. 126). It then climbs onto a high pitted-outwash surface. The steep slopes on both sides of the high surface are called ice-contact slopes. What is now the high area was the low area when braided streams (SB 8) flowed southwestward between the Lake Michigan and Green Bay lobes (fig. 127). The streams deposited gravel on top of thin, discontinuous, glacier ice. When the ice melted, the gravel deposits ended up higher than the surrounding landscape. From the high, pitted-outwash surface, the trail drops into high-relief hummocky topography (SB 11) that makes up the landscape southward to the Mackie Group Picnic Area. Just northwest of the picnic area, there is a large, very symmetrical moulin kame (SB 12; fig. 126). The trail continues in high-relief hummocky topography south of the small area of outwash at CTH ZZ. Just north of Piper Rd., the IAT leaves the high-relief hummocky topography and drops onto pitted outwash. The low area to the west was inundated by glacial Lake Scuppernong (SB 16).

About a mile to the west of the Scuppernong Segment of the IAT is Ottawa Lake Recreation Area and Scuppernong Springs Nature Trail. Most of the area is on the bed of glacial Lake Scuppernong. In the early twentieth century a train took marl, a slippery gray limey mud that was mined here, to Dousman, where it was redistributed for use in agriculture and mortar for buildings. This marl had been deposited on the bottom of glacial Lake Scuppernong (SB 16) and the smaller post-glacial lake that remained after ice retreat. The marl is composed of calcium carbonate secreted by plants growing in the shallow water. The area

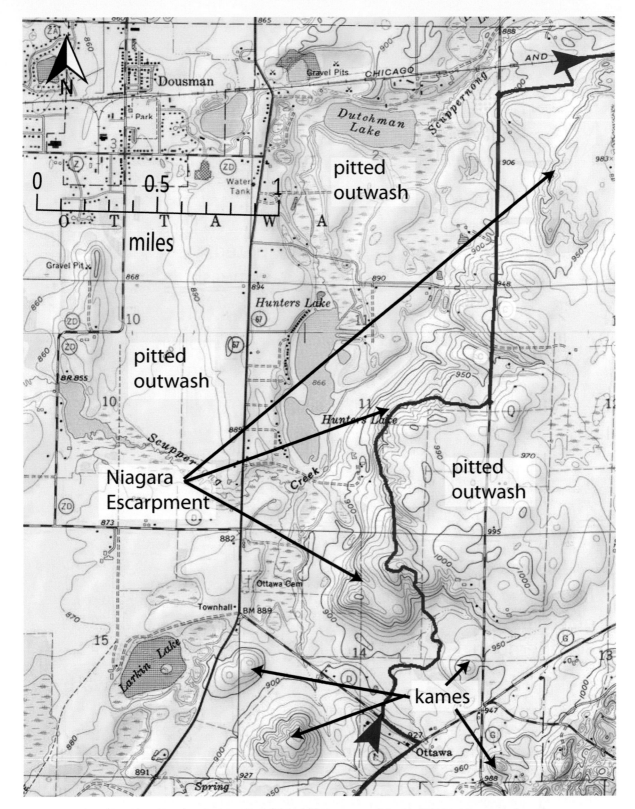

Figure 125. Topography of the Waterville Segment. (Map is part of the Eagle USGS Quadrangle and was created with TOPO! © 2011 National Geographic Maps.)

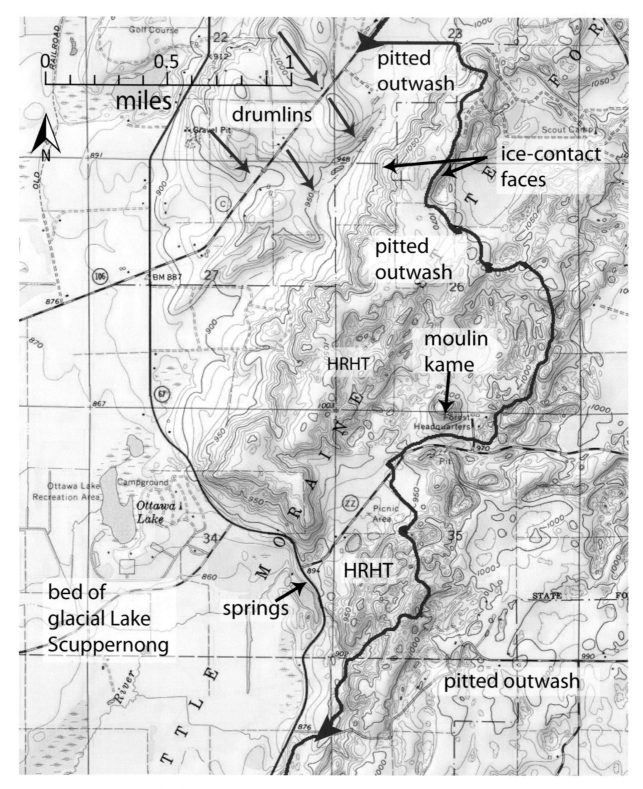

Figure 126. Topography of the Scuppernong Segment. Blue arrows show ice-flow direction. HRHT: high-relief hummocky topography (SB 11). (Map is part of the Eagle and Palmyra USGS Quadrangles and was created with TOPO! © 2011 National Geographic Maps.)

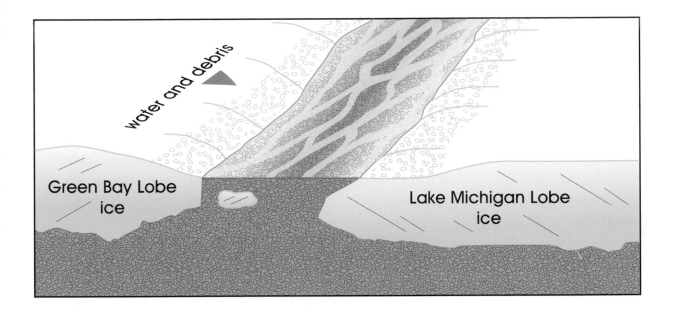

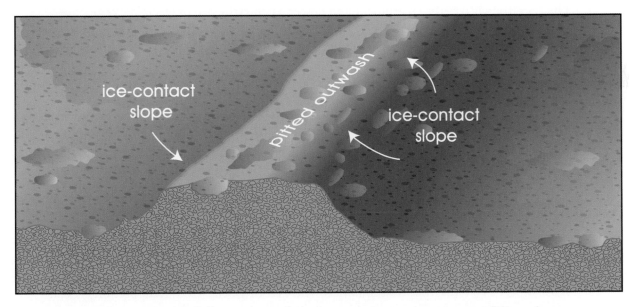

Figure 127. Formation of the high central area of pitted outwash that is common in parts of the southern Kettle Moraine. Steep slopes of edges of the deposit are called ice-contact slopes. (Drafted by Mary Diman.)

remains wet today because springs bubble out where groundwater reaches the surface. This water comes from rain infiltrating the gravelly Kettle Moraine just to the east and is typical of hundreds of springs along the edge of the Kettle Moraine. The temperature of the spring water is about 48°F, our mean annual temperature. The spring area was the site of a trout hatchery, cheese factory, sawmill, and hotel at various times in the late 1800s and early 1900s.

## 34. Eagle Segment

*STH 67 wayside to STH 59 (5.5 miles)*

From Piper Rd. the IAT crosses pitted outwash as it skirts the west edge of the Kettle Moraine (fig. 128). The flat area to the west is the bed of glacial Lake Scuppernong. To the east, the Kettle Moraine sits atop the Niagara Escarpment (SB 19). West of STH 67 the trail rises gently, then steeply, onto this escarpment. The high-relief hummocky part of the Kettle Moraine is to the east. There are good views of the bed of Lake Scuppernong from several points along the trail. A short side trail goes to Brady's Rocks, an outcrop of Silurian dolomite that was mined in the 1800s. Notice the difference between the angular limestone of the bedrock and the very rounded erratics left by the glacier. This is right on the edge of the Niagara Escarpment.

The IAT then leaves the escarpment and descends onto the bed of glacial Lake Scuppernong and remains on it almost to the CTH N crossing, where it rises onto pitted outwash (SB 10). Paradise Springs is located on CTH N about a half mile south of the trail crossing (fig. 129). This excellent example of a Kettle Moraine spring discharges about 500 gallons of water per minute at a temperature of 47°F all year long. In summer, note the trout that are attracted to the cold water where it exits the stone wall. The spring was once the site of a small resort hotel, and water from the spring was bottled for sale until the 1960s. Many springs occur along the edge of the Kettle Moraine because rainwater infiltrates the ground readily and flows to the lower-elevation land surrounding it. The gravelly soils and closed depressions (kettles, SB 9) combine to act as natural rain gardens. Water moves through the permeable sand and gravel of the moraine and discharges where less permeable sediments block further downward flow or where the water table intersects the ground surface. You can spot these natural springs by looking for water bubbling out of the ground. Across CTH N from the spring is the Henry Gotten Log Cabin, built in the early 1900s.

West of CTH N the trail descends onto the bed of glacial Lake Scuppernong. Near the southern edge of the lakebed, there is a small parking lot. The trail follows a dirt road between the parking lot and trailhead and skirts a complex of moulin kames (SB 12). Several gravel pits are present along the dirt road and on the north side of STH 59. You can see the moulin kames (which are on private land) from a distance.

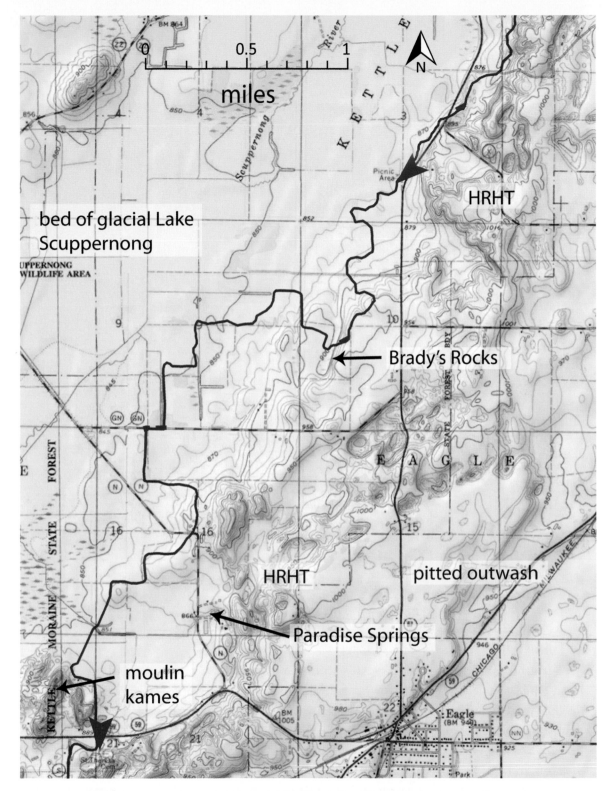

Figure 128.   Topography of the Eagle Segment. HRHT: high-relief hummocky topography (SB 11).
(Map is part of the Eagle and Palmyra USGS Quadrangles and was created with TOPO! © 2011 National
Geographic Maps.)

Figure 129.  Paradise Springs near the Eagle Segment (see fig. 128).

## 35. Stoney Ridge Segment
*STH 59 to CTH Z (3.1 miles)*

Just south of STH 59 the trail crosses a pitted-outwash surface (SB 10) and an old rail-road grade, then crosses a southwest-northeast-trending esker (SB 13; fig. 130). Water flow in the tunnel would have been to the southwest. After crossing the esker, the IAT enters high-relief hummocky topography (SB 11) of the Kettle Moraine. On a spur trail north of the IAT is the Department of Natural Resources Southern Kettle Moraine State Forest headquarters and visitor center. When you stop by, note that the steep-sided ridge south and southeast of the visitor center and just east of the water-filled kettle is an esker. Part of the Stoney Ridge Nature Trail runs along its crest. West of the spur trail, the IAT descends to the bed of glacial Lake Scuppernong and remains on this surface to CTH Z.

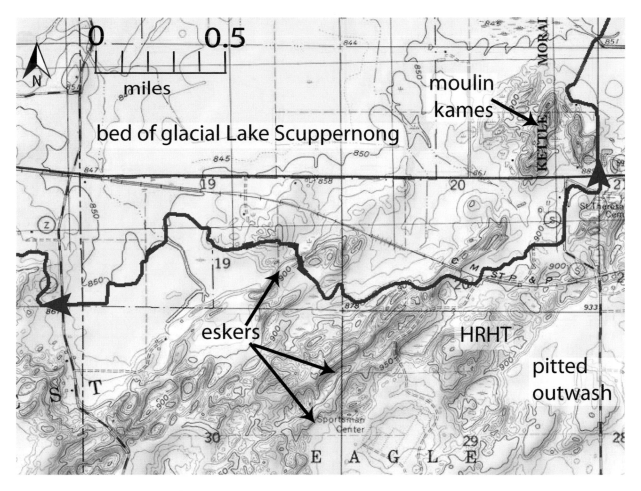

Figure 130. Topography of the Stoney Ridge Segment. HRHT: high-relief hummocky topography (SB 11). (Map is part of the Palmyra USGS Quadrangle and was created with TOPO! © 2011 National Geographic Maps.)

## 36. Blue Spring Lake Segment

*CTH Z to Young Rd. (7.1 miles)*

From the CTH Z trailhead the IAT passes through high-relief hummocky topography (SB 11) that makes up the Kettle Moraine in this area (fig. 131). There are no well-defined eskers as there are in the Stony Ridge Segment to the east. There are many excellent examples of kettles and hummocks, and you'll see that some of the surface was modified by gravel mining long ago. West of Little Prairie Rd. the trail follows a steep-sided ridge that could be an esker but it seems more likely a ridge left by melting of buried ice on either side. Just west of Tamarack Rd. the trail rises onto a high, isolated, pitted-outwash plain (SB 10). About halfway between Tamarack Rd. and Young Rd. stands the stone elephant, a large erratic

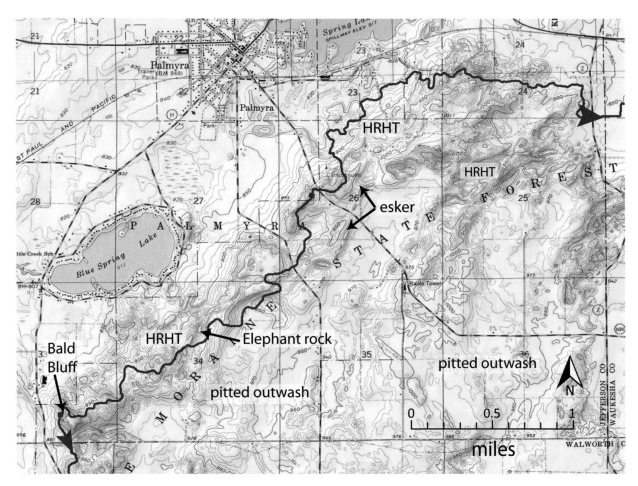

Figure 131.  Topography of the Blue Spring Lake Segment. HRHT: high-relief hummocky topography (SB 11). (Map is part of the Palmyra and Little Prairie USGS Quadrangles and was created with TOPO! © 2011 National Geographic Maps.)

boulder (SB 5) of granite. The pinkish grains are potassium feldspar (SB 21). Just north of Young Rd. the trail crosses Bald Bluff, the highest point in Jefferson County. This tall moulin kame (SB 12) offers a great view to the west across the bed of glacial Lake Scuppernong and southwestward along the edge of the Kettle Moraine (fig. 132).

## 37. Blackhawk Segment
*Young Rd. to STH 12 (7.0 miles)*

From the trailhead at Young Rd. the IAT passes along the west edge of high-relief hummocky topography (SB 11) of the Kettle Moraine (fig. 133). West of CTH H the trail is on pitted outwash. Note the small esker (SB 13) 0.2 miles to the west of the trail. For about a

Figure 132. View westward from Bald Bluff across the bed of glacial Lake Scuppernong. Green Bay Lobe drumlins are on the horizon.

mile south of Bluff Rd., the trail follows a broad, shallow valley formed by water flowing from melting buried ice in what is now high-relief hummocky topography just to the east. The stream flowed southward and then westward into glacial Lake Scuppernong. The trail comes onto low, discontinuous parts of the esker mentioned above and then off again before reaching Duffin Rd. Just east of Duffin Rd., spur trails lead to the Oleson homestead cabin and an old lime kiln. Small kilns such as this were common in eastern Wisconsin in the mid-1800s. They were primarily used to roast limestone (calcium carbonate) to produce quick-lime. The quicklime was slaked (water was added) and then used as mortar in construction.

The trail passes around a moulin kame and enters high-relief hummocky topography of the Kettle Moraine just west of Duffin Rd. and remains in this type of topography until the IAT emerges from the forest and goes between farm fields. At this point you are out of the Kettle

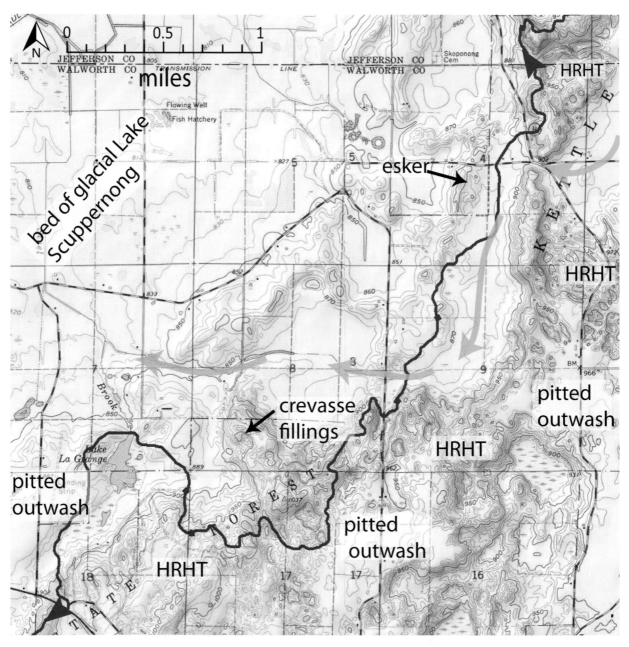

Figure 133. Topography of the Blackhawk Segment. Green arrows show path of meltwater flow. HRHT: high-relief hummocky topography (SB 11). (Map is part of the Little Prairie and Whitewater USGS Quadrangles and was created with TOPO! © 2011 National Geographic Maps.)

Moraine and on outwash. Lake La Grange is in a kettle just west of here. Northeast of the IAT, the edge of the Kettle Moraine is made up of short, closely spaced ridges that extend from the northwest to southeast, mimicking the orientation of the crevasses (SB 2) that were present as the glacier melted away from the Kettle Moraine. These are excellent examples of crevasse fillings (SB 13; fig. 133). The trail skirts the west side of Lake La Grange. Although dammed, this pond is fed by a large spring typical of many along the edge of the Kettle Moraine. The IAT enters high-relief hummocky topography just before the STH 12 trailhead.

## 38. Whitewater Lake Segment
*USH 12 to Clover Valley Rd. (4.7 miles)*

From the STH 12 trailhead, the IAT follows the west, or Green Bay Lobe, edge of the Kettle Moraine (fig. 134). This is high-relief hummocky topography (SB 11). The low area to the west is the bed of glacial Lake Scuppernong. From the utility right-of-way the trail descends onto pitted outwash. This high, pitted-outwash surface (SB 10) makes up the central Kettle Moraine here. Although the plain is quite flat—flat, that is, except for kettles dimpling its surface—it is over 100 feet higher than the flat surface of glacial Lake Scuppernong to the west. The trail remains on this surface across Esterly Rd. Consider leaving the trail to climb to Lone Tree Bluff scenic outlook. The trailhead is about a half mile west on Esterly Rd. (fig. 134). With its top 200 feet above the glacial Lake Scuppernong plain, Lone Tree Bluff provides a good view to the west.

About a third mile west of Esterly Rd., the trail enters high-relief hummocky topography and crosses an esker (SB 13) before crossing CTH P. This esker appears to connect to another one running down the center of Whitewater Lake. West of CTH P the trail rises onto another broad esker. This ridge appears to be the esker that separates Rice Lake from Whitewater Lake. The trail descends into a deep kettle and then onto the sand-and-gravel ridge that makes up the west edge of the Kettle Moraine. This may or may not be a true esker (one that formed in a closed tunnel). The trail remains on this ridge to the point where it turns to the north and west across from the Rice Lake Nature Trail. Two eskers extend into Whitewater Lake and between this lake and Rice Lake (fig. 135). These are two of the best preserved eskers in southern Wisconsin. Water in the glacier tunnels in which the eskers formed flowed toward the southwest. The depressions that hold Whitewater and Rice lakes are kettles, but both lakes are held at their present levels by dams built in the mid-twentieth century. Before the dams, what is now Whitewater Lake was actually two small lakes separated by a wetland, and Rice Lake was a wetland.

West of Kettle Moraine Dr., the trail ascends, then descends, the west side of the Kettle Moraine. It crosses Whitewater Creek, the outlet of Rice and Whitewater lakes just northeast of the trailhead on Clover Valley Rd.

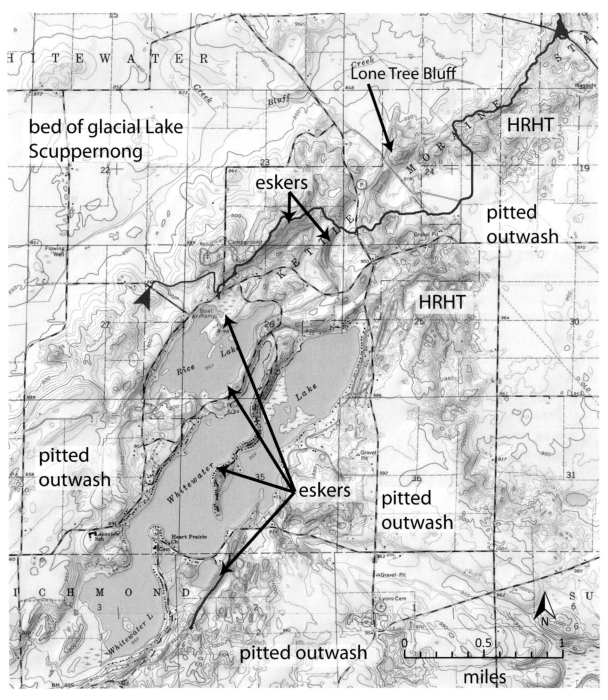

Figure 134. Topography of the Whitewater Lake Segment. HRHT: high-relief hummocky topography (SB 11). (Map is part of the Whitewater USGS Quadrangle and was created with TOPO! © 2011 National Geographic Maps.)

Figure 135. Oblique aerial view looking northwestward across Whitewater Lake. Ridges in the lake are eskers. Ice Age Trail is on far side of lake.

# Southern Green Bay
# Lobe Ice Age Trail Segments

THE GREEN BAY LOBE ADVANCED DOWN the Green Bay–Lake Winnebago lowland into southern and southwestern Wisconsin during the late Wisconsin Glaciation about 30,000 years ago. At its outermost advance position, between 25,000 and 30,000 years ago, the lobe deposited the Johnstown Moraine (SB 6) along much of its ice margin. An exception to this was in northern Dane County, where no moraine formed. In a few places, particularly near Brooklyn (fig. 136), the ice may have extended a little farther than the Johnstown Moraine then retreated to build the moraine. Because of bedrock topography in western Dane County, it can be difficult to identify which landforms are moraines and which are simply bedrock hills. Some of the best places to see the moraine clearly are on the Verona and Valley View segments and at the Cross Plains Unit of the Ice Age National Scientific Reserve (figs. 136, 137). The Johnstown Moraine is also prominent in southern Sauk County, and is responsible for damming Devil's Lake at both ends (fig. 137). The Devil's Lake and Sauk Point segments and the western terminus of the Baraboo Segment include views of and hikes on the moraine, as do other trails in Devil's Lake State Park. The Milton Moraine was deposited during a later advance. This moraine is also more or less continuous across the area that the southern Green Bay Lobe once covered and is just south of the Clover Valley Segment and just north of the Storrs Lake and Milton segments.

In addition to moraines, the glacier left its mark on the landscape here with the help of water. Rivers carried sediment in rushing torrents or meandering streams far from its origin at the ice edge, and other rivers ran through tunnels under the ice itself, depositing sand and gravel in places. At, and just after, the glacial maximum advance, meltwater streams flowed from the ice margin toward the south in the Sugar and Rock rivers and their tributaries

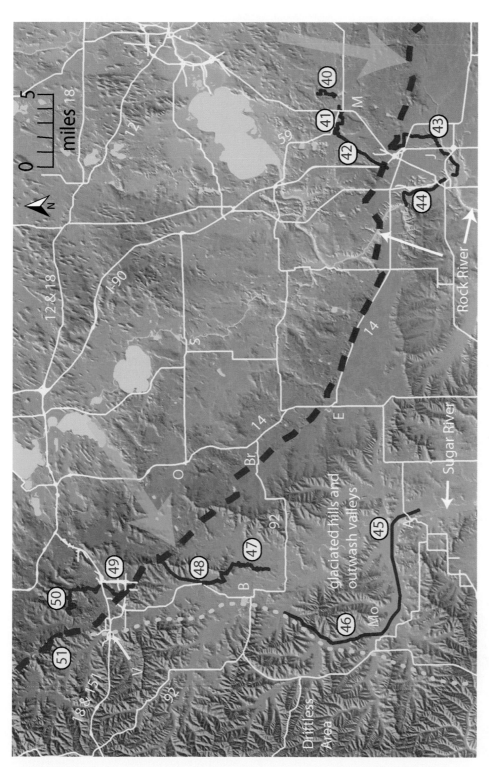

Figure 136. Shaded relief of the southern part of the southern Green Bay Lobe IAT segments (red): (40) Storrs Lake, (41) Milton, (42) Janesville to Milton, (43) Janesville, (44) Devil's Staircase, (45) Albany, (46) Monticello, (47) Brooklyn Wildlife Area, (48) Montrose, (49) Verona, (50) Madison, (51) Valley View. Yellow lines and numbers indicate major highways. Cities shown (yellow): (A) Albany, (B) Belleville, (Br) Brooklyn, (E) Evansville, (J) Janesville, (M) Milton, (Mo) Monticello, (O) Oregon, (S) Stoughton, (V) Verona. Dashed blue line is outer edge of the Johnstown Moraine. Blue arrows show ice-flow direction. Dotted green line is the boundary between glaciated area and Driftless Area. (Base map constructed from USGS National Elevation Dataset and modified by WGNHS.)

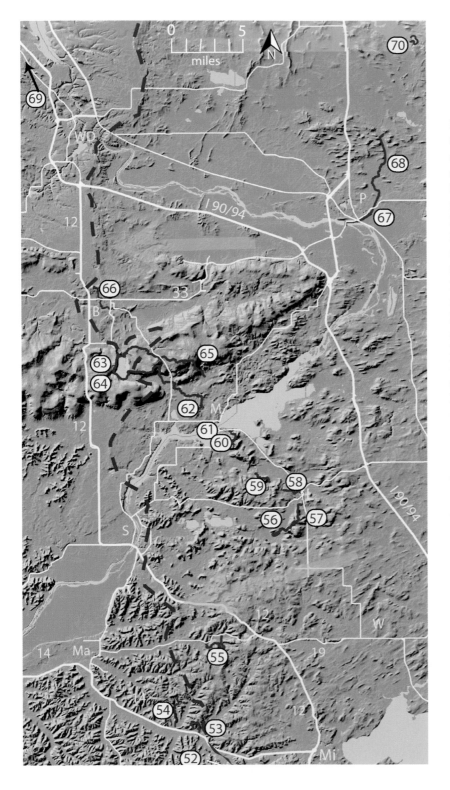

Figure 137. Shaded relief of the northern part of the southern Green Bay Lobe IAT segments (red): (53) Cross Plains, (54) Table Bluff, (55) Indian Lake, (56) Lodi Marsh, (57) Eastern Lodi Marsh, (58) City of Lodi, (59) Groves-Pertzborn, (60) Gibraltar, (61) Merrimac Ferry, (62) Merrimac, (64) Devil's Lake, (65) Sauk Point, (66) Baraboo, (67) Portage Canal, (68) Marquette, (70) John Muir Memorial Park. Yellow lines and numbers indicate major highways. Cities shown (yellow): (B) Baraboo, (L) Lodi, (Ma) Mazomanie, (M) Merrimac, (Mi) Middleton, (P) Portage, (S) Sauk City, (W) Waunakee, (WD) Wisconsin Dells. Ice Age National Scientific Reserve sites: (52) Cross Plains, (63) Devil's Lake, (69) Mill Bluff (northwest of this area—see figure 177 for location). Dashed blue line is the outer edge of the Johnstown advance. Blue arrows show ice-flow direction. The area west of the dashed blue line is part of the Driftless Area. (Base map constructed from USGS National Elevation Dataset and modified by WGNHS.)

(fig. 136). Parts of the trail—the Janesville, Devil's Staircase, and Albany segments—provide excellent examples of this outwash terrain. Farther north, braided streams (SB 8) also carried sand and gravel westward in Black Earth Creek and other Wisconsin River tributaries. The Cross Plains, Table Bluff, Indian Lake, Portage Canal, and John Muir Memorial Park segments all traverse portions of glacial outwash (SB 8; fig. 137). In the largest river valleys, rushing water left huge deposits of outwash sand and gravel, which in some cases piled up to over 300 feet thick. Within the glacier, too, water rushed and constantly transported sand and gravel. Water flowing under pressure in tunnel channels (SB 17) brought huge amounts of sand and gravel to the glacier margin. Several of the IAT segments, including the Milton to Janesville and Verona segments in the south (fig. 136) and the Indian Lake and Lodi segments in the north (fig. 137), cross or parallel these tunnel channels. Behind these moraines thousands of drumlins (SB 14) formed. You can see these drumlins along the Clover Valley Segment (fig. 128) and the Lodi Marsh, Eastern Lodi Marsh, City of Lodi, and Marquette segments (fig. 137). To spot the beds of former lakes dammed by the glacier or a moraine, hike along the Clover Valley and Devil's Lake segments.

There are two other major landscape areas in the vicinity of the IAT in southern Wisconsin. All of the area west of the glaciated area shown in figure 137 and all of the area west of the green dotted line in figure 136 are part of the Driftless Area (SB 21). This means it was never glaciated. It has deeper valleys and steeper valley walls than the glaciated area does. Fragile bedrock features such as tall, thin sandstone monuments or pillars are preserved in the Driftless Area, whereas they have been destroyed by the flow of overriding ice in the glaciated area. In addition, there are no erratics that have been carried long distances by the glacier in the uplands of the Driftless Area. You will have good views of the Driftless Area from the Brooklyn Wildlife Segment as well as from the Verona and Valley View segments (fig. 136). The Table Bluff, Cross Plains, and Devil's Lake segments are at least partly in the Driftless Area, and there are great views from the Cross Plains Unit of the Ice Age National Scientific Reserve (fig. 137). Finally, Mill Bluff State Park, which is a unit of the Ice Age National Scientific Reserve, is a tall column of sandstone that is in the Driftless Area. Its connection with glaciation is that it was an island in glacial Lake Wisconsin (SB 16) when the glacier was at the Johnstown Moraine.

Southern Wisconsin was glaciated at least once before the late Wisconsin Glaciation (see their extent in fig. 136), although no one is sure about the dates of these previous glaciations. These deposits are relatively thin and discontinuous, and in this area you will find most of them on the uplands because meltwater from the late Wisconsin glacier to the north scoured many of the low-lying valleys and refilled them with younger outwash. The Albany, Monticello, Brooklyn Wildlife, and Badger State Trail segments have good examples of this landscape.

All of the southern Green Bay Lobe segments east of Lodi cross Ordovician-age sedimentary rocks (fig. 4). This sequence consists of Prairie du Chien dolomite, which is overlain by St. Peter sandstone, which is in turn overlain by Platteville and Galena dolomite units. Cambrian sandstone is only exposed in outcrops along STH 14 near Cross Plains, in some hills close to the Wisconsin River, around the Precambrian quartzite hills near Baraboo, and at Mill Bluff State Park. The Devil's Lake and Sauk Point segments are mostly on Precambrian Baraboo quartzite.

In summary, the southern Green Bay Lobe area contains twenty-nine IAT segments and a range of landscapes including the Driftless Area and lands that the pre-Wisconsin Glaciation and late Wisconsin Glaciation covered. Glacial landforms in IAT segments that traverse the most recently glaciated area include moraines, tunnel channels, drumlins, outwash plains, meltwater-cut channels, and former lakebeds. In addition to the IAT segments, there are three units of the Ice Age National Scientific Reserve that showcase glacial history: the Cross Plains Ice Age complex just west of Madison, Devil's Lake State Park, and Mill Bluff State Park.

## 39. Clover Valley Segment
*Island Rd. to County Line Rd. (1.6 miles; see fig. 121 in the previous chapter)*

When the glacier retreated from the Milton Moraine, which lies about a half mile south of this segment's western end (fig. 138), small interconnected lakes formed between the ice and the moraine. The flat, poorly drained surface beneath the eastern half of the trail is the site of one of these former lakes. The basin is small and there do not appear to be any definite abandoned beaches. Based on topography, it appears that water level was at about 860 feet, or about as high as the eastern edge of the stand of taller trees along the east-west portion of the trail. From this point eastward to County Line Rd., the trail climbs over the downstream end of a low drumlin (SB 14), drops into a shallow valley, and climbs onto another drumlin whose crest is just west of County Line Rd. Note that the ice-flow direction indicated by the drumlins is not perpendicular to the Milton Moraine as you might expect. This is because the drumlins formed when ice was at its maximum extent at the Johnstown Moraine and the ice-flow direction was slightly different than it was for the Milton advance.

## 40. Storrs Lake Wildlife Area Segment
*Bowers Lake Rd. to Storrs Lake Rd. (2 miles)*

The Milton Moraine (SB 6) is poorly defined in this area, but you can see the outer extent of the Milton advance in spots with somewhat more hummocky topography (SB 11; fig. 139). Here sand and gravel underlie much of the land. During this advance, braided streams (SB 8) flowed away from the ice margin toward the southwest and deposited pitted outwash

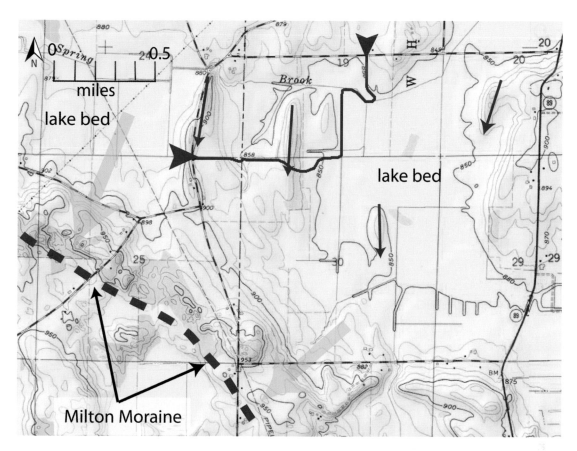

Figure 138. Topography of the Clover Valley Segment. (Location shown in fig. 121.) Dashed blue line shows outer edge of the Milton Moraine and light blue arrows show ice-flow direction to the ice edge at that time. Blue arrows on drumlin crests show ice-flow direction when the glacier was advancing to, or was at, the Johnstown Moraine. When the ice withdrew from the Milton Moraine, shallow lakes were damned between the ice edge and the moraine. (Map is part of the Lima Center USGS Quadrangle and was created with TOPO! © 2011 National Geographic Maps.)

(SB 10) in front of the former position of the ice margin. The outwash covered residual ice masses left behind when the glacier retreated from the Johnstown Moraine. When the buried ice melted, kettles (SB 9) formed to produce the pitted-outwash landscape.

## 41. Milton Segment

*Storrs Lake Rd. to Manogue Rd. (4.7 miles)*

The Milton Segment is entirely on outwash. When the glacier advanced and then retreated from the Johnstown Moraine (SB 6), it deposited outwash and till at elevations between 910 and 950 feet (areas labeled "early till/outwash surface" in fig. 140). Ice then retreated to an

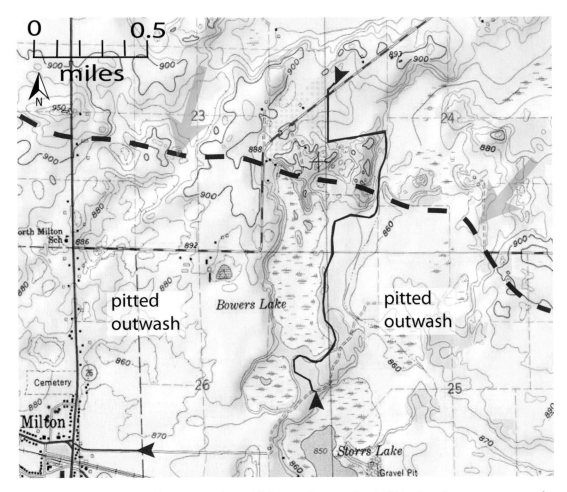

Figure 139. Topography of the Storrs Lake Wildlife Area Segment. Dashed blue line shows outer extent of the Milton ice advance, and blue arrows show ice-flow direction at that time. (Map is part of the Milton USGS Quadrangle and was created with TOPO! © 2011 National Geographic Maps.)

unknown position north of the area shown in figure 140 and readvanced to the Milton Moraine. At that time, the ice margin was just north of the area shown in the figure. Braided streams (SB 8) carried sand and gravel from the Milton ice margin toward the southwest, burying residual ice from the previous advance. Meltout of this ice later produced the pitted outwash (SB 10) in the northeast part of the figure. These streams eventually coalesced into rivers that cut deep outwash channels through the older deposits (the channels are shown with green arrows on fig. 140). The older till and outwash surfaces were left as terraces (SB 8) now standing above the outwash channels. Note that the southernmost outwash channel has no kettles (SB 9) in its bed because there was apparently no buried ice beneath it to melt out after the channel formed.

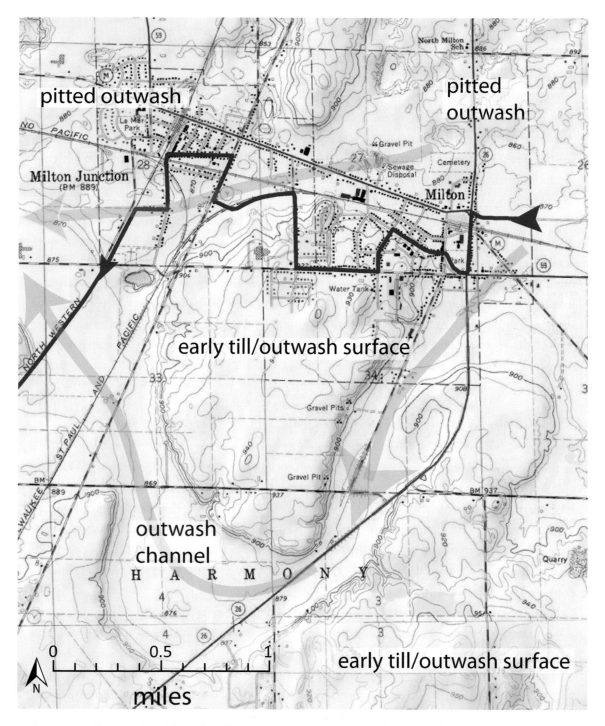

Figure 140. Topography of the Milton Segment. Green arrows show path of water flow from the Milton ice margin. Pitted outwash in the northeast and northwest corners of the map had not collapsed at that time, and a braided stream covered the channel bed. The early, higher till/outwash surface was deposited during the advance to and retreat from the Johnstown Moraine. (Map is part of the Milton USGS Quadrangle and was created with TOPO! © 2011 National Geographic Maps.)

## 42. Janesville to Milton Segment

*Manogue Rd. to Kennedy Rd. (3.7 miles)*

The eastern terminus of this trail segment lies at the edge of an outwash (SB 8) channel that carried water toward the west at the time of the Milton advance (fig. 141). You can see the channel better in figure 140. The IAT rises gently out of this channel onto a higher pitted-outwash surface that the glacier deposited during its retreat from the earlier John-stown advance. West of EMH Townline Rd. the IAT borders a well-developed tunnel channel (SB 17; fig. 141). This tunnel channel carried a large flow of glacial meltwater from beneath the ice and was the source of much of the sand and gravel beneath the huge out-wash plain that surrounds Janesville today. Like most tunnel channels, the bottom of this one is hummocky sand and gravel. The elevation of the bottom of the channel rises from a low of 830 feet just south of Manogue Rd. to over 900 feet at the position of the former ice margin. Thus, water under pressure was flowing uphill in the tunnel channel as the water approached the ice edge. Depressions in the outwash surface outside the maximum position of the Johnstown advance are not kettles, but rather old gravel pits (fig. 141). Figure 142 shows the tunnel channel from the air.

## 43. Janesville Segment

*East Rotamer Rd. to Riverside Park's South Pavilion (9.6 miles)*

Although the trailhead is in a residential area, you can still glimpse the Johnstown Moraine (SB 6) to the north and northeast (fig. 143). This segment is entirely on outwash (SB 8), and its eastern portion mostly follows the path of the modern drainage that flows eventually into the Rock River. The area that this segment covers is on one of the larger outwash plains in Wisconsin, with sand and gravel over 300 feet thick in places. Large gravel pits on the north side of Janesville attest to the quality of this sand and gravel as an aggregate material for con-crete and asphalt (fig. 143).

Once, braided streams (SB 8) covered this whole land surface, allowing the outwash to accumulate. Most of the sand and gravel was deposited as ice advanced to and stabilized at the Johnstown Moraine. Much of the outwash probably came from the tunnel channel (SB 17) to the north (figs. 141, 142). After ice melted away from the Johnstown Moraine, the small valley that the IAT follows began to erode due to flow from springs and from rain-water runoff.

The IAT follows this tributary creek to the Rock River into downtown Janesville, where the trail turns northward and follows the floodplain of the Rock River to the Devil's Stair-case Segment. More discussion of the history of this outwash plain is given in that segment description.

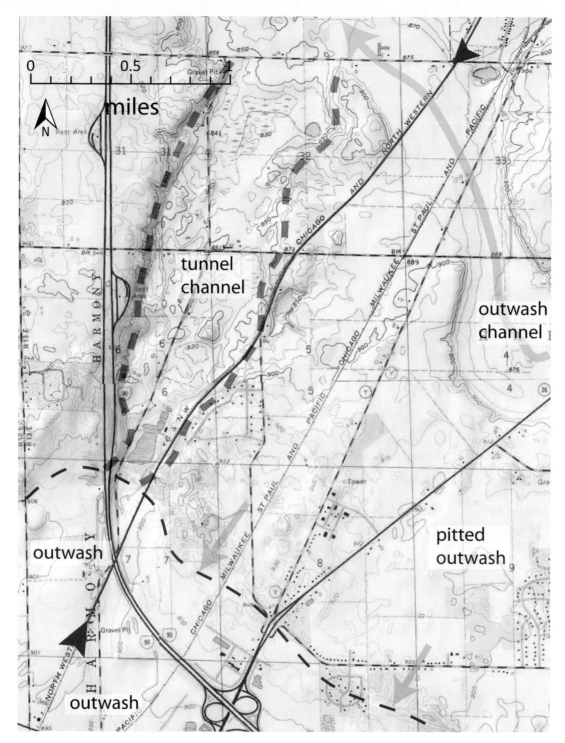

Figure 141. Topography of the Janesville to Milton Segment. Cloverleaf near bottom of map is the intersection of I-90 and STH 26. Green arrow, upper right, is in the bottom of an outwash channel that carried water toward the northwest (see fig. 140). Dashed gray line shows boundaries of a tunnel channel. Water in the tunnel flowed toward the southwest beneath the ice. Dashed blue line shows outer edge of the Johnstown advance. Blue arrows show ice-flow direction. (Map is part of the Milton, Janesville East, and Janesville West USGS Quadrangles and was created with TOPO! © 2011 National Geographic Maps.)

Figure 142. Low oblique aerial view looking north of a tunnel channel (bounded in black) and the Johnstown Moraine, both of which are shown in figure 141. Solid blue line is the approximate maximum extent of the Johnstown advance, and blue arrows show approximate ice-flow direction at that time. Lake Koshkonong is in the distance. The Ice Age Trail more or less follows the eastern (right) side of the tunnel channel for much of its distance.

## 44. Devil's Staircase Segment
*Riverside Park's South Pavilion to Washington St. (1.8 miles)*

From the South Pavilion the trail follows the floodplain of the Rock River northward (fig. 144). Note the flowing wells that are present along the floodplain. These are common wherever there is a distinct difference in the elevation of glacial deposits, particularly in sand and gravel because so much rain can infiltrate to the water table. In this case, water enters the groundwater system higher in the landscape to the west and flows through this sand-and-gravel aquifer toward the river, where it discharges as springs. Putting a shallow well into the sand and gravel at an elevation just above the river taps this underground water and allows it to flow at the surface. Where the trail leaves the floodplain there is a steep climb onto the lower (although it doesn't seem low when you are climbing!) of two terraces. This is the surface on which Riverside Park golf course sits.

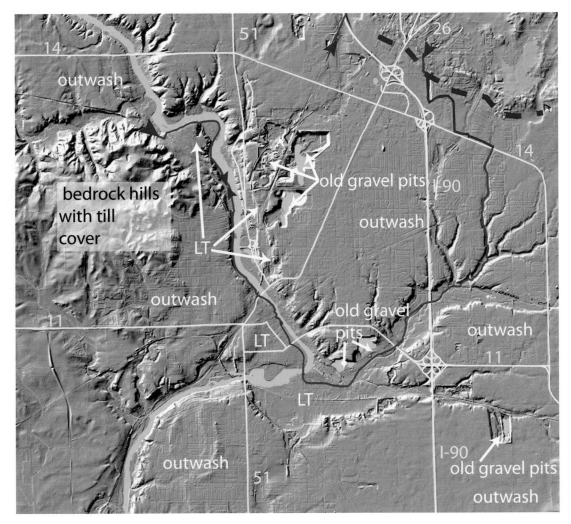

Figure 143. Shaded relief of the Janesville (starting top right) and Devil's Staircase (ending top left) segments. Dashed blue line is outer edge of Johnstown advance. Blue arrow shows ice-flow direction. All labeled outwash is a high terrace. LT: lower terrace (younger). Yellow lines and numbers indicate major highways. (Base map by Brad Blumer, City of Janesville.)

Like many rivers that carried large amounts of outwash away from the glacier margin, the Rock River has terraces (SB 8). When the glacier was at the Johnstown Moraine, the present channel of the Rock River did not exist. A braided-stream system covered much of this area and had filled a deep valley in the bedrock with sand and gravel. Only the hilly area to the west of Riverside Park stood above the extensive outwash plain. When ice began to retreat from the Johnstown Moraine, drainage concentrated into a few channels that joined and began to downcut below the outwash surface. As this downcutting continued, the

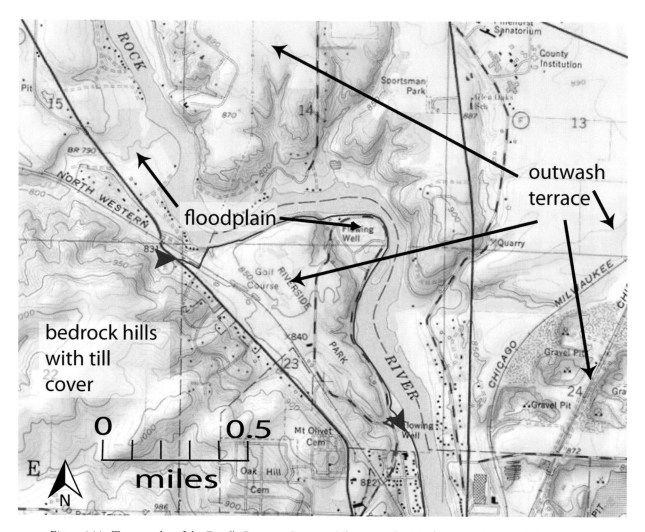

Figure 144. Topography of the Devil's Staircase Segment. The outwash plain (now a high terrace) slopes gently upward to the north where it meets the Johnstown Moraine (see fig. 143). Much of Riverside Park is on a lower terrace. Note the large pits east of the Rock River where sand and gravel were mined. Hills on the west side of the map were glaciated sometime before the late Wisconsin Glaciation. (Map is part of the Janesville West USGS Quadrangle and was created with TOPO! © 2011 National Geographic Maps.)

surface water was directed into a single channel—what is now the Rock River channel—and the more extensive braided-channel system that had once covered the outwash plain dried up. At some point, probably when the glacier edge sat at what is now the Milton Moraine, there was a pause in downcutting and the Rock River channel widened somewhat. At that time it was probably a braided stream again, but it was confined within the walls of the channel and did not cover the whole outwash plain. As ice retreated from the Milton Moraine, the river again became a single channel and downcut to its present level. This left behind two levels of former river bed that are called terraces. The highest terrace is the outwash plain that slopes gently downward away from the Johnstown Moraine. All of the area east of the river in figure 144 is this high terrace, although post-glacial drainage has dissected it along its edges.

Below the high terrace is the less obvious low terrace, probably cut when ice was at the Milton Moraine. There are only small remnants of this terrace remaining, although you can see it beneath the words "Riverside Park" in figure 144. This surface was the streambed when the glacier edge was at the Milton Moraine. Note that the difference in elevation between the high terrace and the intermediate terrace is not great. The elevation of the golf course, which occupies the lower terrace, is 850 to 860 feet above sea level. Directly across the river the high terrace is between 880 and 900 feet.

Glaciers covered the hills on the west side of figure 144 before the late Wisconsin Glaciation, although there are no dates on when this glaciation occurred. The eastern two-thirds of the trail in this segment is on the floodplain. Note that this surface may flood at times, whereas the high and low terraces are well above the modern floodplain.

## 45. Albany Segment
*Bump Rd. at Sugar River State Trail to Monticello (9.3 miles)*

For the eastern 3 miles or so, this segment is on outwash (SB 8) of late Wisconsin age. Albany Lake, just west of the trail, is the dammed Sugar River (fig. 145). The outwash is mostly sand derived from the glacier when it was at the Johnstown Moraine (SB 6) near Verona. Coarse sand and gravel were deposited close to the ice margin, but the size of grains that the stream carried decreased rapidly, and the Sugar River valley in this area was mostly filled with sand. Huge amounts of sand were deposited in the main Sugar River valley at this time (the river flows into fig. 145 from the north). This rapidly accumulating sediment in the main valley partially blocked tributary valleys like the Little Sugar River (flowing into fig. 145 from the west). The Little Sugar River valley, which the trail follows to Monticello (fig. 146), filled with sand, silt, and clay as the flow of the stream slowed. The effect of this filling of the lower section of its valley was to decrease the gradient (steepness) of the Little Sugar River. Streams with abnormally low gradients produce an intricate pattern of tight loops or meanders. Notice how the present stream meanders back and forth within its floodplain (figs. 145, 146). At times of flood, the whole floodplain is covered with water, and in a few places new, straighter channels are eroded, cutting off some meanders. When these old channel meanders are bypassed or cut off, they are called oxbows or oxbow lakes (figs. 145, 146).

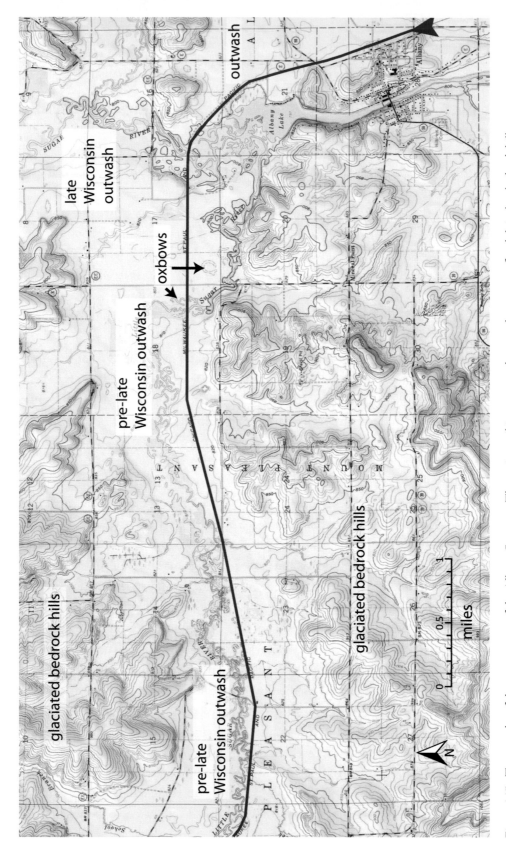

Figure 145. Topography of the eastern part of the Albany Segment. The entire trail is on outwash or modern stream floodplain, but bedrock hills with a thin till cover surround the valley. West of the Sugar River, the Ice Age Trail is on the floodplain of the Little Sugar River. Only two of the many oxbows are labeled. (Map is part of the Albany USGS Quadrangle and was created with TOPO! © 2011 National Geographic Maps.)

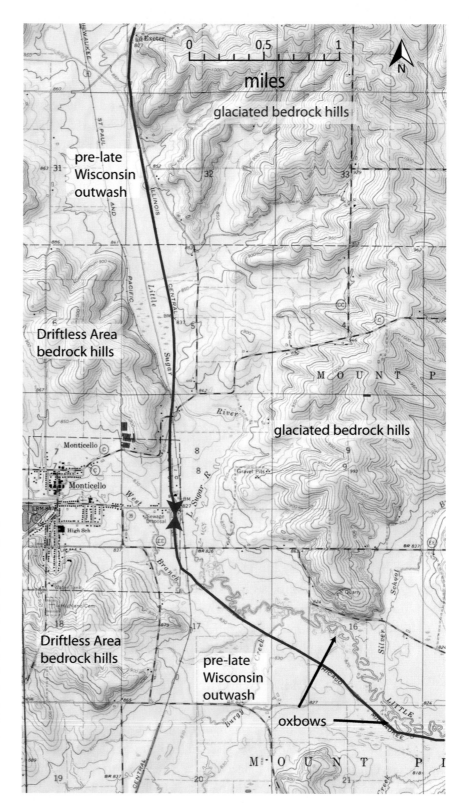

Figure 146. Topography of the western part of the Albany Segment (bottom) and the eastern part of the Monticello Segment (top). Trail follows the valley of the Little Sugar River. Hills to the west are in the Driftless Area. (Map is part of the Monticello and Belleville USGS Quadrangles and was created with TOPO! © 2011 National Geographic Maps.)

The hills north and south of the valley have a thin till (SB 5) cover from pre–late Wisconsin Glaciation. Water erosion erased most signs of these older glacial deposits; the landforms we see today are a result of water acting on the land's surface and the underlying bedrock. When the leaves are off the trees, you can see outcrops of St. Peter sandstone along the sides of the hills, especially to the south of the trail. From this point, the trail follows the Little Sugar River valley to the Monticello Segment.

## 46. Monticello Segment
*Monticello to CTH W (6.5 miles)*

Near its eastern terminus at Monticello, the IAT follows the Badger State Trail and parallels the natural channel of the Little Sugar River (fig. 146). Note meanders that are similar to those along the Albany Segment, although the river here is smaller and the oxbows are more difficult to see (compare with fig. 145). West of CTH C, the river is in a human-made channel. Post-glacial stream sediment and outwash (SB 8) predating the last glaciation underlie the nearly flat valley bottom. Hills rising to the east are bedrock and now have only thin and discontinuous glacial deposits on them that date to a glaciation older than the late Wisconsin. The hills to the west of this trail segment are part of the Driftless Area. They formed in the same way as the hills to the east, but they do not have erratics or other signs that they were ever glaciated. Like the hills to the east, they are underlain by Galena and Platteville dolomites and St. Peter sandstone. Most of the outcrops that you can see in the distance from this segment are St. Peter sandstone.

After crossing Exeter Crossing Rd., the shared IAT and Badger State trails continue northward through the Stewart Tunnel (fig. 147). The tunnel is over 1,200 feet long and was completed in 1887 by what was to become the Illinois Central Railroad. Workers building the tunnel blasted through Platteville dolomite, the unit that overlies the St. Peter sandstone. Inside the tunnel, much of the dolomite is now covered with water, algae, bat droppings, and graffiti, so it is difficult to see the rock very clearly. Also, it is very dark! To see the dolomite, look at the overgrown outcrops of the rock at the tunnel entrances. Find more details about the tunnel construction at the website of the Wisconsin Historical Rail Connection: http://www.whrc-wi.org/History/stewarttunnel.htm.

## 47. Brooklyn Wildlife Segment
*Hughes Rd. to CTH D (3.6 miles)*

Throughout much of this segment the trail follows a north-south-trending highland (fig. 148). This ridge is not a glacial feature but rather a bedrock ridge that also appears to be the westernmost extent of glaciation. There are scattered erratics (SB 5) on this ridge, and it may contain fairly thick till in places, although for the most part any glacial deposits have

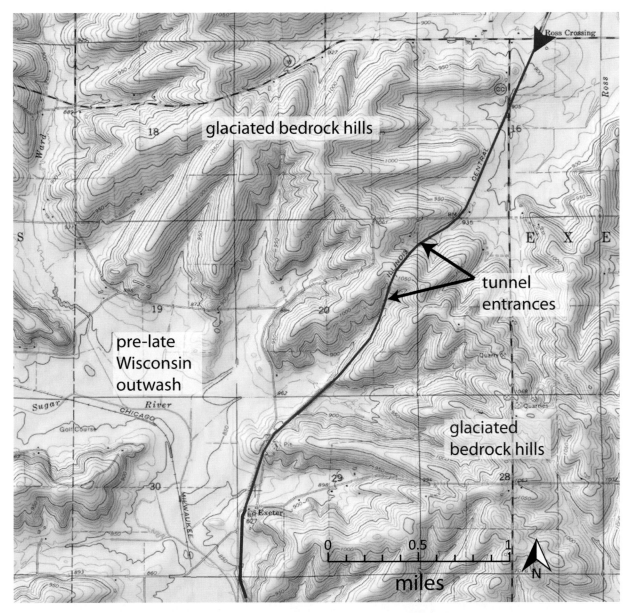

Figure 147. Topography of the western part of the Monticello Segment. Map shows the location of the Stewart Tunnel and surrounding terrain. Location of the entrances is approximate. (Map is part of the Bellville USGS Quadrangle and was created with TOPO! © 2011 National Geographic Maps.)

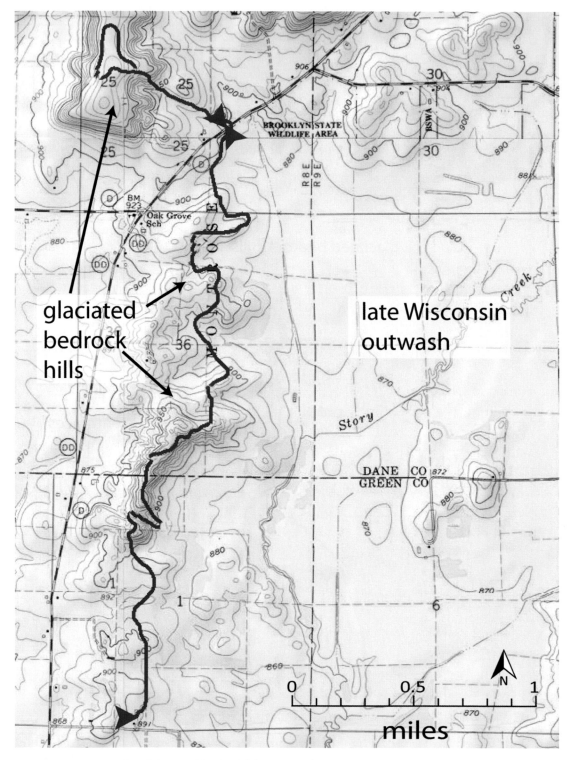

**glaciated bedrock hills**

**late Wisconsin outwash**

Figure 148. Topography of the Brooklyn Wildlife Segment (starting lower left) and the eastern part of the Montrose Segment (top). The Story Creek valley has outwash derived from the glacier when it sat at the Johnstown Moraine, which is to the east, northeast, and north of this location. (Map is part of the Attica and Oregon USGS Quadrangles and was created with TOPO! © 2011 National Geographic Maps.)

been eroded away. From the ridge, you will have excellent views to the west and east. The Sugar River valley, just west of this ridge, contains outwash (SB 8) sand derived from the Johnstown Moraine near Verona. The city of Belleville sits on this outwash (fig. 136).

Beyond the Sugar River valley, the hills and valleys of the Driftless Area extend westward to the Mississippi River. To the east of the trail, the lowland of Story Creek has been partly filled with outwash from the Johnstown Moraine and older glaciers. All of the hills that you can see to the east are bedrock cored, but they have a thin and discontinuous till cover. As you hike along the northern parts of this segment, you will see the Johnstown Moraine on the horizon to the east.

## 48. Montrose Segment
*CTH D to Purcell Rd. (7.4 miles)*

The eastern terminus of this IAT segment is in a low divide in a bedrock ridge (fig. 149). From that point, the trail climbs into and crosses glaciated bedrock hills that offer a great view of the Johnstown Moraine on the horizon to the east and northeast. As in areas to the south, the dates that glaciers covered this region are not known. West of the Sugar River valley are hills of the Driftless Area.

After a short distance following the floodplain of the Sugar River, which is underlain by outwash, the trail climbs again into bedrock hills with a thin till cover. About a half mile south of Purcell Rd., the trail crosses into the area covered by the last glaciation—the late Wisconsin. The Johnstown Moraine (SB 6), a narrow ridge at the very outer edge of this ice advance, has low-relief hummocky topography (SB 11). The IAT shares a route with the Badger State Trail, an abandoned railroad bed, so you may not notice the gradual rise of the moraine as you hike because the grade was engineered to be as gradual as possible. Look off the trail as you hike—you will be able to see the moraine as a higher ridge to either side.

## 49. Verona Segment
*Wesner Rd. to CTH PD (McKee Rd.) (6.9 miles)*

The Verona Segment provides excellent views and opportunities to walk on the Johnstown Moraine (SB 6). From its eastern terminus, the trail climbs onto the moraine and follows its edge for more than 2 miles before turning toward the northeast along an outwash channel (fig. 150). As you hike, notice the cover of mature, open-grown oak trees on parts of the moraine. The hilly topography was not very good for row-crop agriculture, so many farmers left the large oaks to grow. Figure 151 shows the Johnstown Moraine at Prairie Moraine County Park.

The channel that Badger Mill Creek, STH 18, and STH 151 occupy extends northeast to the Milton Moraine. This valley could have been a tunnel channel (SB 17) when the

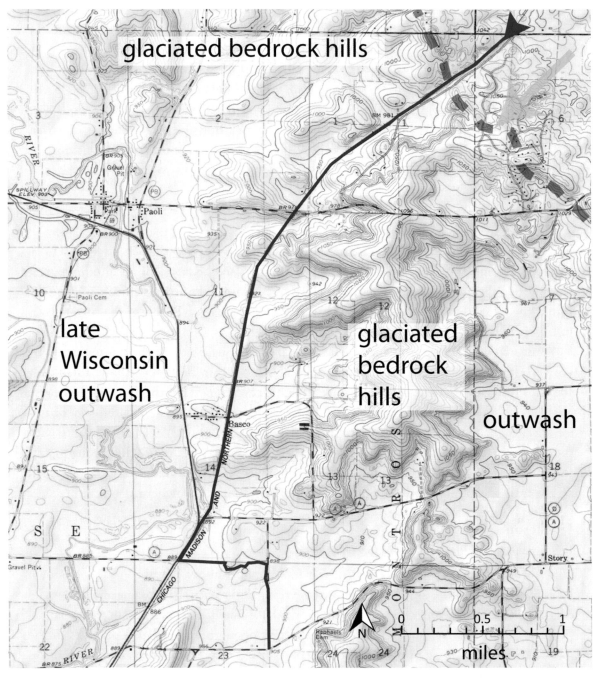

Figure 149. Topography of the western part of the Montrose Segment. Dashed blue line indicates the outermost position of the late Wisconsin glacial advance, which is the outer edge of the Johnstown Moraine. Blue arrow shows ice-flow direction. (Map is part of the Verona and Oregon USGS Quadrangles and was created with TOPO! © 2011 National Geographic Maps.)

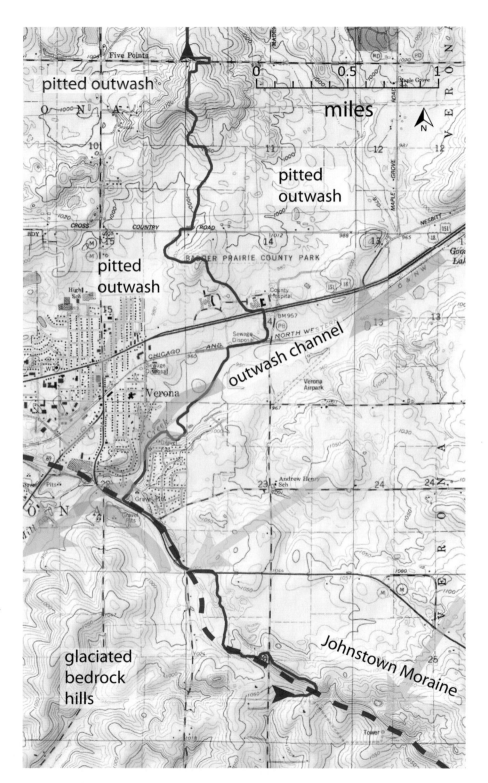

Figure 150. Topography of the Verona Segment. Dashed blue line shows maximum extent of late Wisconsin Glaciation and the outer edge of the Johnstown Moraine. Blue arrows show ice-flow direction. Green arrows show meltwater-flow direction. In much of this area the Johnstown Moraine is about a half mile wide. (Map is part of the Verona and Middleton USGS Quadrangles and was created with TOPO! © 2011 National Geographic Maps.)

Figure 151. Johnstown Moraine near the eastern trailhead of the Verona Segment at Prairie Moraine County Park. Note the large, open grown oaks on low-relief hummocky topography.

glacier sat at the Johnstown Moraine. Unlike most other tunnel channels, this one slopes down toward the Johnstown ice margin (toward the southwest). The river that flowed down the valley away from the ice margin when it sat at the Milton Moraine filled any kettles and eroded any gravel hummocks a tunnel channel would have contained, thus destroying evidence of the tunnel's existence. As the glacier retreated from the Johnstown Moraine, it deposited an outwash plain (SB 10; fig. 150) throughout the area north of Badger Mill Creek valley. Part of Badger Prairie County Park is on this outwash. The hill east of the county hospital buildings is a sanitary landfill. The higher hills you will see south of Cross Country Rd. and as you near the western trailhead of this segment are primarily bedrock (Platteville dolomite) topped with till (SB 5).

## 50. Madison Segment

*CTH PD (McKee Rd.) to Woods Rd. (3 miles)*

Most of the hills in this area have bedrock cores, but within the glaciated area these hills are topped with till (SB 5) or sand and gravel 20 to 50 feet thick (fig. 152). From its eastern trailhead the IAT segment drops from these hills into a valley that was probably a tunnel channel (SB 17). The valley is very narrow where the trail crosses the tunnel channel, but it broadens toward the southwest, the direction in which water flowed to the ice margin. The gravel pits on CTH PD just west of the moraine are in outwash (SB 8) that was deposited by huge flows of water from this tunnel channel. Morse Pond, just south of the IAT, is a kettle (SB 9), as are several other shallow ponds you may see in this former tunnel channel. The Johnstown Moraine (SB 6) is a narrow ridge (generally less than a quarter mile wide), although in places there is a thick till accumulation with kettles (SB 9) behind the actual moraine crest. West of the moraine, north and south of CTH PD, braided streams (SB 8) deposited massive amounts of outwash from the tunnel channel mentioned above, from the outwash channel shown on figure 152, and directly from the ice margin. This outwash has been mined extensively. Richardson Cave formed in Platteville dolomite. Although the entrance is filled with sediment now, it probably carried meltwater from beneath the ice to the ice margin or beyond. It is located just inside the moraine, on private land.

## 51. Valley View Segment

*Shady Oaks Ln. to Ice Age Ln. (2 miles)*

At its eastern trailhead, this segment of the IAT is on a very low part of the Johnstown Moraine (SB 6; fig. 152). There is a little outwash (SB 8) in front of the moraine here, but the hills to the west are driftless. Despite this, they do have a thin cover of windblown silt called loess (SB 8). About a quarter mile west of this segment's eastern end, the IAT rises to the crest of the Johnstown Moraine, which here is a fairly high, narrow ridge standing above the surrounding countryside. East and West Blue Mounds are on the horizon to the west. Note that the moraine is also sitting on a relatively high bedrock hill. The actual thickness of the till making up the moraine here is about 30 to 40 feet, and the rest of the relief is due to the shape of the bedrock surface. The IAT drops off the Johnstown Moraine and into the Driftless Area (SB 21) just west of Mound View Rd. About a quarter mile west of this road, the trail crosses a deep ravine that glacial meltwater probably cut when the ice was at its maximum extent. As soon as the glacier started to retreat from the Johnstown Moraine, this ravine was cut off from carrying meltwater because water could not flow up and over the moraine. The trail crosses through two other, broader valleys that also carried meltwater for a short time when the glacier was at its maximum extent, but the hills you will see between these valleys were not glaciated.

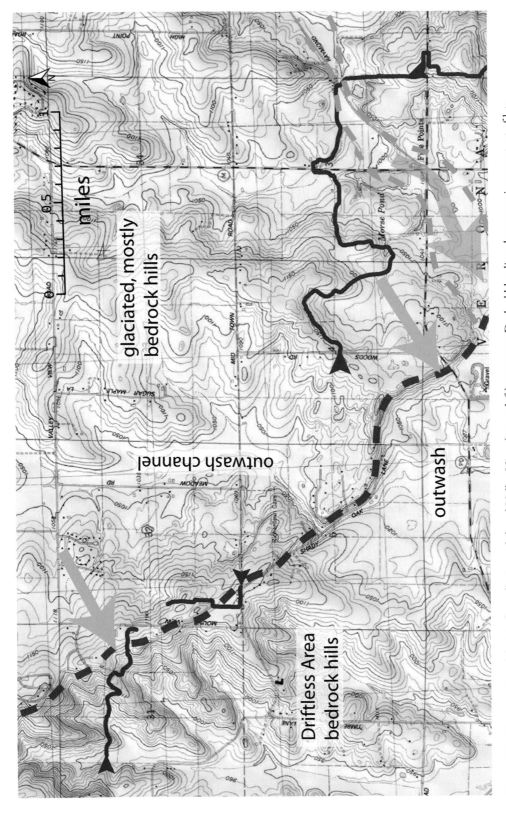

Figure 152. Topography of the Madison (lower right) and Valley View (upper left) segments. Dashed blue line shows maximum extent of late Wisconsin Glaciation and the outer edge of the Johnstown Moraine. Blue arrows show ice-flow direction. In much of this area the Johnstown Moraine is about a quarter mile wide. Dashed green line shows edges of tunnel channel. Green arrow shows meltwater-flow direction in the tunnel. (Map is part of the Middleton USGS Quadrangle and was created with TOPO! © 2011 National Geographic Maps.)

## 52. Cross Plains Unit of the Ice Age National Scientific Reserve

This area straddles two distinctly different landscapes (fig. 153). The northern and eastern edges of the Cross Plains Ice Age complex were covered by the glacier and ice-dammed lakes during the late Wisconsin Glaciation, while the remainder of the site is in the Driftless Area. At the time of this writing there is no IAT segment at the site, but one will be built there as the site develops and more land is acquired. Figure 153 shows the land that is publicly owned at the beginning of 2011.

The bedrock within the Cross Plains complex consists of three main layers of Ordovician rock (table 1): Prairie du Chien dolomite, at the bottom, which is overlain by St. Peter sandstone, and Platteville dolomite that overlies them both (figs. 154, 155, 156). Although you can see each of these three rock units exposed in the reserve, for the most part the exposures are small. Figure 154 shows the distribution of Paleozoic sedimentary rock in the reserve area. Prairie du Chien dolomite is exposed particularly along the walls of Wilkie Gorge (fig. 153) and in a few other places along steep slopes. You can see it better in other places, so please stay off the steep, sensitive walls of the gorge. St. Peter sandstone is exposed in several places in the central complex and near the west side of the complex. Platteville dolomite is well exposed in the roadcut on Mineral Point Rd. west of the Timber Ln. intersection. There are small exposures of it elsewhere. The best spot is at the top of the slope above St. Peter sandstone near the northeastern edge of Shoveler Pond (sinkhole on fig. 153). The vertical distribution of these rock layers is shown diagrammatically in cross section along a north-south line through the complex (fig. 156).

The Platteville dolomite is the youngest bedrock in the Cross Plains complex. Many of the surrounding hills, especially those to the west, are capped with even younger Galena dolomite. West Blue Mound, the highest hill you can see from the complex, wears a cap of even younger Silurian dolomite.

Karst topography—land dotted with underground caves—develops when surface water and groundwater erode and dissolve limestone or dolomite bedrock. Underground caves and caverns are common in these terrains, and you may see signs of these hidden karst features in the form of a sinkhole, or sink, above ground. A sinkhole often just looks like a low spot in the ground, rather than an actual hole, but it is the connection between the surface and these underground caverns. Sinkholes form where surface water finds a path down into the underlying rock. Gradually, the force of the water dripping through this path dissolves the rock, slowly enlarging the opening and allowing even more water to pass through. In some cases a sinkhole can form rapidly when the roof of a cave collapses. There have been several reported cases of a tractor falling into a sinkhole during rainy spring planting seasons when the ground is wet and the groundwater level is high. The Shoveler Pond basin drains into a sinkhole at times of high water (fig. 153). It appears that this sinkhole formed in St. Peter

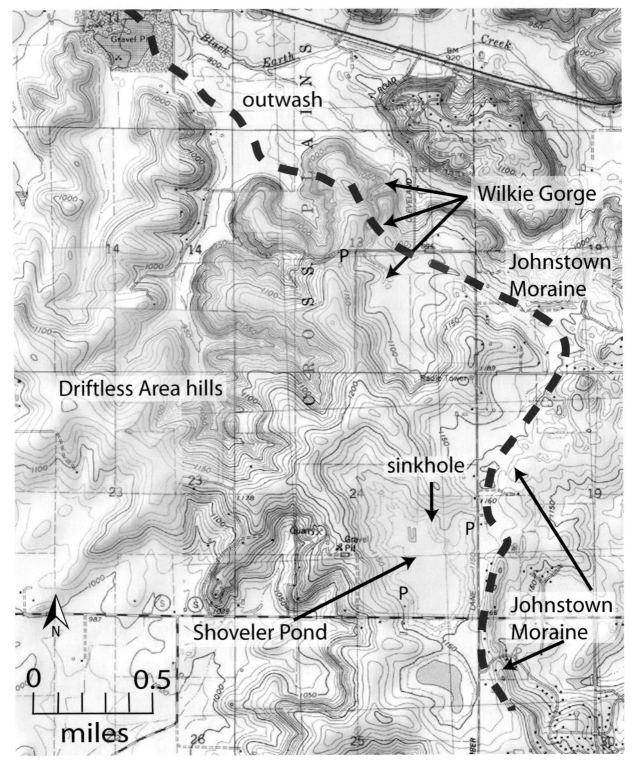

Figure 153. Topography of the Cross Plains Unit of the Ice Age National Scientific Reserve showing the distribution of public land as of January 1, 2011 (blue shading). Note the sinkhole through which Shoveler Pond drains. Dashed blue line is outer edge of Johnstown Moraine. (Base map is part of the Middleton USGS Quadrangle and was created with TOPO! © 2011 National Geographic Maps.)

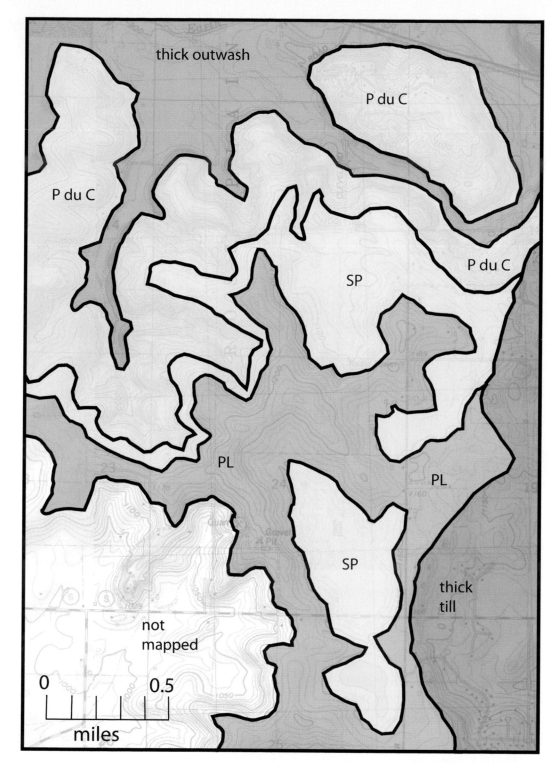

Figure 154. Distribution of near-surface bedrock units at the Cross Plains Ice Age complex. Only three units occur near the surface beneath thin windblown silt (loess): (PL) Platteville dolomite, (SP) St. Peter sandstone, (P du C) Prairie du Chien dolomite. Thick sand and gravel covers the bedrock units in most valley bottoms and under the Johnstown Moraine, so rock is not mapped there. (GIS compilation by Midwest Region Geospatial Support Center, National Park Service.)

Figure 155.  Platteville dolomite above St. Peter sandstone on south wall of sinkhole shown on fig. 153. Sandstone cannot be seen in photo.

sandstone, although it is likely that the collapse itself happened in a layer of dolomite below the sandstone. At the sinkhole, you can see St. Peter sandstone exposed along the north wall and Platteville dolomite above the sandstone high on the south wall.

When the glacier was present just to the east, it is possible that these sinkholes were actually springs where groundwater discharged. Water under pressure would have been forced through the groundwater system and would have been able to come to the surface outside the ice-covered area. There were probably also springs discharging glacial meltwater into the upper part of the Sugar River basin at the southwest edge of the Cross Plains complex.

The Johnstown Moraine (SB 6) borders the site on its eastern side (fig. 157). It is a clearly defined ridge at Mineral Point Rd., but it is lower and harder to see from Old Sauk Rd. to

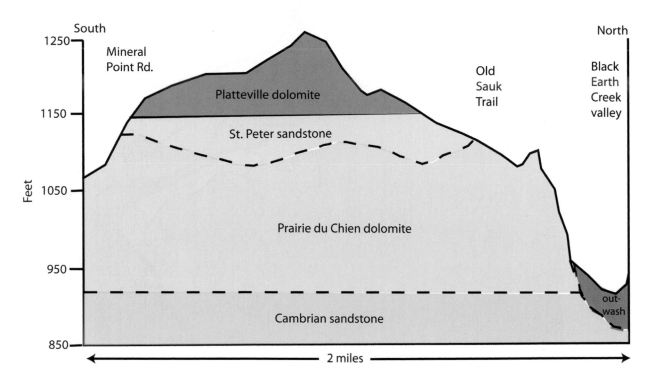

Figure 156. North-south cross section through the Cross Plains Ice Age complex. Sediments above the bedrock are too thin to show except in Black Earth Creek valley. Contacts between bedrock units are mostly projected into the cross section from wells south, east, and west of the reserve. Because most of the contacts are covered, this cross section portrays the likely distribution of rocks. The bottom contact and thickness of St. Peter sandstone is extremely variable over short distances.

where the ice margin crossed Old Sauk Pass. The best place to see the size and nature of the moraine is northwest of the intersection of Cleveland Rd. and Old Sauk Pass. The till is about 50 feet thick above the dolomite bedrock surface. There you can hike the crest westward to Wilkie Gorge (fig. 153).

When the glacier sat at the Johnstown Moraine, the climate was very cold, and permafrost was present in front of and beneath the glacier edge. The glacier probably melted only on a few warm days during a month or two in summer. Meltwater was dammed between the glacier and the Driftless Area landscape to the west. Lake L1 on figure 157 had the highest level. There is still silty lake sediment up to an elevation of at least 1,150 feet and probably slightly higher. The lowest outlet for this lake was at an elevation between 1,150 and 1,160 feet at its northeastern end. The basin that contained lake L1 was apparently a stream valley that drained toward the east, until the glacier and its deposits blocked its flow. A narrow band of outwash (SB 8) separates the finer silty lake sediment from the till in the Johnstown Moraine.

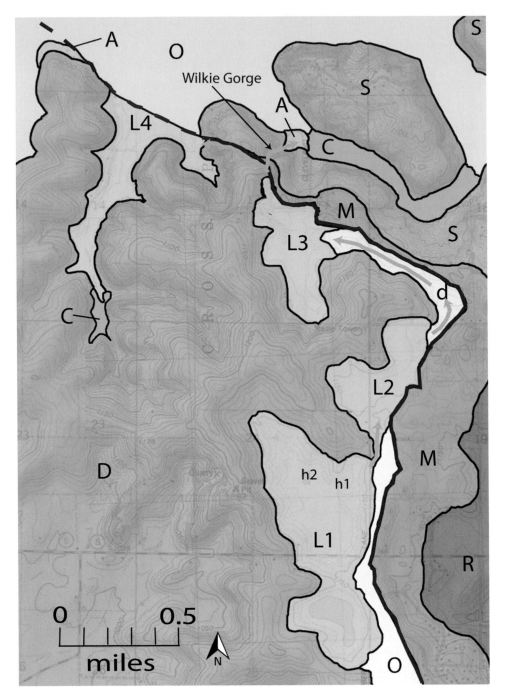

Figure 157. Glacial and related deposits in the Cross Plains Ice Age complex. There is also colluvium at the base of most steep slopes that it is too narrow to map at this scale. C: colluvium accumulated from slope wash; D: Driftless Area with thin silt over bedrock; F: alluvial fan; M: Johnstown Moraine; O: outwash; R: till not in the moraine; S: steep slope controlled by bedrock with patchy till cover. Dark blue line is outer edge of late Wisconsin glacial advance. Blue arrows show meltwater-flow direction. Several shallow lakes were present when the ice was at the Johnstown Moraine. Lake L1 was the highest; L2 was at about the same elevation and was dammed by ice in the present position of Old Sauk Road. Letter *d* indicates a drainage way to the next lake, L3, which was substantially lower than both L1 and L2. Look for a sinkhole at h1, and a possible sinkhole at h2. (GIS compilation by Midwest Region Geospatial Support Center, National Park Service.)

Lake L2 seems to have been at about the same level as lake L1. It seems likely that these lakes were connected for a period of time when the glacier ice advanced against a rock knob and dammed lake L2. This rock knob is just north of the present Old Sauk Rd. With a very small retreat of the ice, lake L2 would have drained or at least dropped in level. This water drained through a drainage way into lake L3, where the elevation of the water surface appears to have been about 1,100 feet. It seems that this lake drained along the ice margin across the ridge north of Old Sauk Pass and into Black Earth Creek valley before Wilkie Gorge was cut. At some point, when the climate warmed enough to allow melting at the bottom and edges of the glacier (SB 3), water found its way under the ice and down the steep slope of Black Earth Creek valley. It was this flow of water that eroded Wilkie Gorge.

As mentioned earlier, throughout much of the time that the Green Bay Lobe was at its maximum extent, temperatures were very cold and the base of the glacier was frozen to its bed. No water could flow out from beneath the ice except in large tunnel channels. When the climate began to warm, ice near the edge of the glacier began to melt, releasing water in and beneath the glacier. Generally, water at the base of a glacier flows from areas of thick ice to areas where the ice is thinner, because the pressure is greater under the weighty, thicker ice. But there are many cases along the edges of modern glaciers where the opposite happens: water flows from the ice margin into or under the ice and then out again at a lower elevation. The channels that form are called submarginal chutes.

Wilkie Gorge is a submarginal chute. Its location was probably determined by a pre-existing weakness or opening in the ice, such as a crevasse. The water in lake L3 was about 200 feet higher than the Black Earth Creek valley, and water would naturally take the most direct path to the bottom of the valley. Once water made its way beneath the ice to the level of the bottom of Black Earth Creek valley, the lake drained suddenly. Water under high pressure and flowing rapidly would have readily cut the deep gorge that you see today.

Lake L4 is in Black Earth Creek valley. It may have been dammed by the ice margin or it may have been dammed by the rising outwash surface in the valley.

## 53. Cross Plains Segment
*Bourbon Rd. to Hickory Hill St. (2.7 miles)*

The Village of Cross Plains lies mostly in the valley of Black Earth Creek (fig. 158). The valley is partly filled with outwash (SB 8) that in places is several hundred feet thick. Before outwash filled the valley, it was substantially deeper than it is today, although exactly how deep it was at its maximum extent no one knows. The river that deposited the outwash you see today removed any moraine that might have been built here.

From Black Earth Creek valley northward to the Wisconsin River, there is no moraine marking the maximum extent of the late Wisconsin Glaciation. The glacier ice that existed

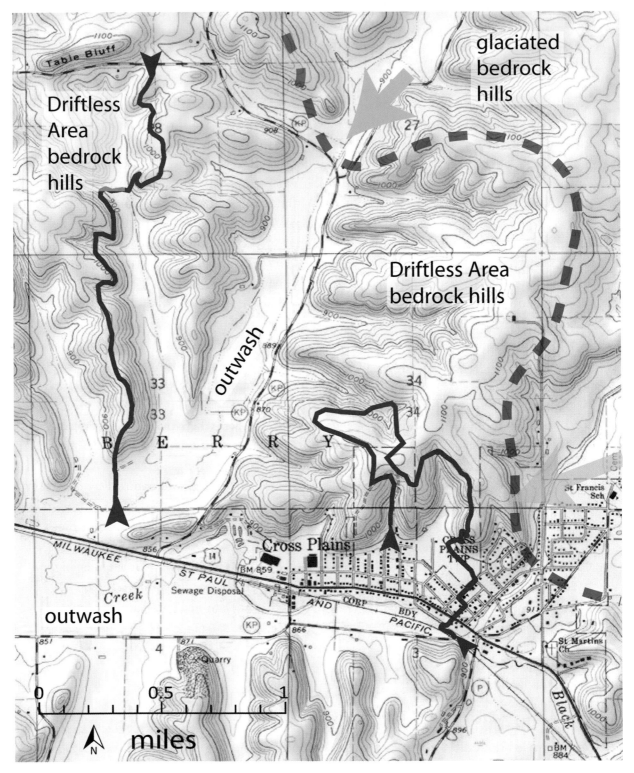

Figure 158. Topography of the Cross Plains (right) and Table Bluff (left) segments. Dashed blue line shows approximate maximum extent of late Wisconsin Glaciation. Blue arrows show ice-flow direction. (Map is part of the Cross Plains and Black Earth USGS Quadrangles and was created with TOPO! © 2011 National Geographic Maps.)

in what is now northwestern Dane County must have been nearly free of debris when it was at its maximum position. The glacier left only a scattering of erratics on the surface, perhaps because subglacial conditions (SB 3) resulted in little erosion in the area up-glacier from northern Dane County. Instead of using the moraine as a marker, we have to trace the maximum extent of the ice through the distribution of erratics (SB 5). West of the maximum extent of the ice is the Driftless Area, where there are no erratics on the uplands.

All of the hills in the area are bedrock. In a few places you can see Cambrian sandstone exposed low on the valley walls and Prairie du Chien dolomite capping the higher hilltops. The headquarters of the Ice Age Trail Alliance is at 2110 Main St. in Cross Plains, just west of where the trail crosses Black Earth Creek. The IAT crosses the valley on outwash then climbs a bedrock ridge on the north side of the valley.

## 54. Table Bluff Segment
*Scheele Rd. to Table Bluff Rd. (2.4 miles)*

Like the Cross Plains Segment, the Table Bluff Segment starts on the nearly flat surface of Black Earth Creek outwash (fig. 158). The valley to the northeast of the trailhead contained an outwash stream (SB 8) when the glacier was at its maximum extent. Because the glacier in this part of Dane County had very little debris in it, this tributary stream, in turn, had very little outwash to deposit. Like all of the tributaries to Black Earth Creek, this one accumulated sediment that came from Black Earth Creek valley into the tributary as well as off the valley walls. At Cross Plains, the accumulated outwash in the main valley measures 150 feet thick. This means that during the last glacier's maximum extent, before the outwash was deposited, the valley was substantially deeper than it is today. As outwash accumulated rapidly (by geologic standards, perhaps several feet a year), its rising surface dammed small tributary streams that came into the main valley.

Cambrian sandstone makes up the lower part of the ridge that the IAT climbs, but there is Prairie du Chien dolomite at the top. This ridge is in the Driftless Area. The valleys on either side of the ridge did not carry meltwater from the glacier but were probably braided streams (SB 8) during the peak of glaciation. Table Bluff, which is just north of this IAT segment, is composed of St. Peter sandstone.

## 55. Indian Lake Segment
*Loop Trail in Indian Lake County Park (2 miles)*

As you hike this trail segment, you will pass Indian Lake, which fills a shallow kettle (SB 9) in outwash (SB 8) in Halfway Prairie Creek valley. It is located between the outermost edge of the last glaciation, which is about 4 miles to the west, and the Milton Moraine (SB 6; fig. 159). The hills on either side of Halfway Prairie Creek valley are primarily made up of

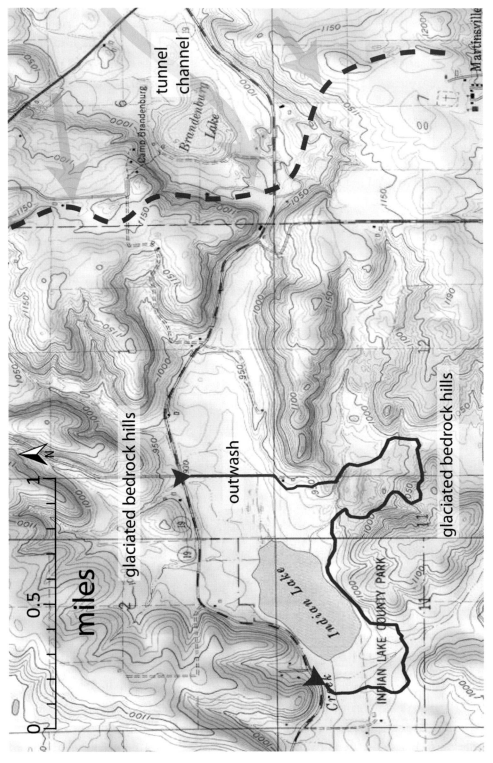

Figure 159. Topography of the Indian Lake Segment. Dashed blue line shows extent of the Milton advance. The Milton Moraine is a low moraine on the ridge tops north and south of Brandenburg Lake, but there is no moraine preserved in the valley bottom. Blue arrows show ice-flow direction. Green arrow shows water-flow direction in tunnel channel. Gorge is located just west of tunnel channel arrow. (Map is part of the Springfield Corners and Black Earth USGS Quadrangles and was created with TOPO! © 2011 National Geographic Maps.)

Prairie du Chien dolomite with a relatively thin cover of till. From the eastern end of this IAT segment, the trail rises onto a gravel terrace of pitted outwash (SB 10), skirts the side of a hill with till over bedrock, and finally drops down onto the outwash surface and circles Indian Lake.

This valley experienced a huge flow of water from a tunnel channel (SB 17) when the ice edge stood at the position of the Milton Moraine, about 1 mile east of the trail (fig. 159). At that time, water under high pressure exited a tunnel from beneath the ice and cut the gorge through which Highway 19 passes between Indian Lake Park and Lake Brandenburg. Rushing water covered the entire valley bottom where Indian Lake Park is now.

How do we know this was a tunnel channel? Prairie du Chien dolomite crops out at road level along Highway 19 in the water-cut gorge. Water wells close to Lake Brandenburg indicate that the top of the rock there is over 200 feet below the surface of the outwash and therefore more than 200 feet below the floor of the gorge. The ridge through which the gorge was cut is the drainage divide separating the water flowing to the Wisconsin River from the water flowing into the Madison lakes and down the Yahara River. Generally, deep valleys don't exist near drainage divides, because large streams were not present to cut them. A valley over 200 feet deep this close to the drainage divide is an anomaly. It must have been cut by subglacial water flowing toward the west under high pressure toward the ice margin. The water in the tunnel probably rose nearly 200 feet in elevation before pouring through the bedrock gorge and down Halfway Creek valley. Lake Brandenburg is a kettle in hummocky sand and gravel that formed as buried glacial ice melted out, long after water in the tunnel channel had stopped flowing. We can trace this tunnel channel across the landscape to the northeast, to Norway Grove in northern Dane County.

## 56. Lodi Marsh Segment
*Loop Trail (1.6 miles)*

In this segment, ice overrode all of the surrounding hills. You can see that many of them have a drumlin (SB 14) shape that indicates the direction the ice flowed over them (fig. 160). This segment consists of a loop trail that crosses a low part of the marsh and a slightly higher area that could be the top of a nearly buried drumlin (not marked on fig. 160). From there, the trail rises onto two higher east-west-trending drumlins, where you will have excellent views of the marsh below. The marsh itself formed after the last glaciation as wetland vegetation slowly filled a shallow pond in part of this valley. The valley itself is a tunnel channel (SB 17). When the valley formed beneath the glacier, water under pressure flowed beneath the ice, cutting a deep valley in the bedrock. Water in the tunnel flowed westward through the valley that now contains Crystal Lake and Fish Lake before erupting from beneath the glacier across the Wisconsin River from what is now Prairie du Sac.

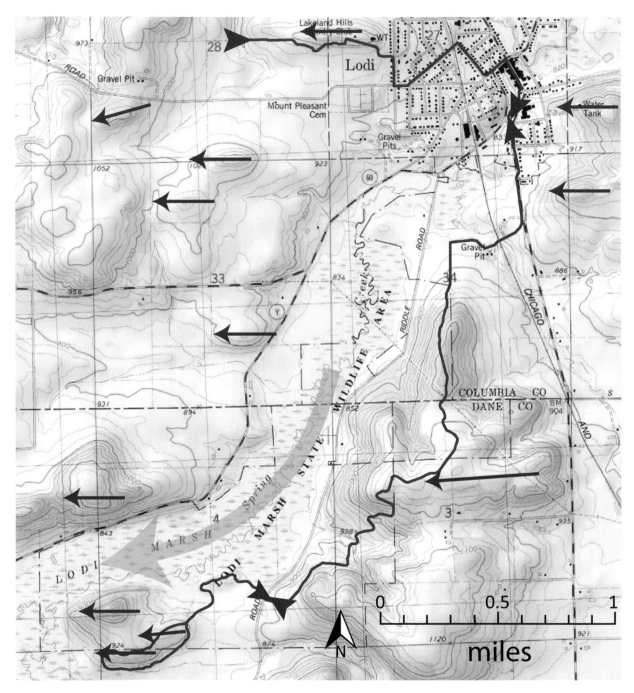

Figure 160.  Topography of the Lodi Marsh (lower left), Eastern Lodi Marsh (center), and City of Lodi (upper right) segments. Blue arrows are on drumlins and show ice-flow direction. Green arrow shows water-flow direction in tunnel channel. (Map is part of the Lodi USGS Quadrangle and was created with TOPO! © 2011 National Geographic Maps.)

## 57. Eastern Lodi Marsh Segment

*Lodi-Springfield Rd. to Pleasant St. (2.4 miles)*

This segment of the IAT traverses till-covered ridges of Prairie du Chien dolomite (fig. 160). While glaciers partially shaped the landscape into drumlin forms, the north-south-trending ridge that the trail follows in its central section is bedrock. From this segment you will have excellent views of Lodi Marsh, which is in a tunnel channel (SB 17; see Lodi Marsh Segment description). You will also see disturbed land from former gravel mining near the railroad grade just west of STH 113.

## 58. City of Lodi Segment

*Pleasant St. to Lodi School complex (1.8 miles)*

This segment traverses outwash (SB 8) in the tunnel-channel floor and drumlin terrain west of the channel (see the Lodi Marsh Segment description). You can see excellent examples of east-west-trending drumlins indicating ice-flow direction (fig. 160). The water tower east of the IAT near the Pleasant St. trailhead is on a drumlin, as is the water tower just north of the trail on the west side of Lodi.

## 59. Groves-Pertzborn Segment

*Bilkey Rd. to CTH J (1.6 miles)*

This fairly short segment skirts around the north side of a drumlin. Platteville dolomite lies under glacial till on the surface. About 1 mile to the northwest of the western end of this segment is Gibraltar Rock County Park. This hill has Cambrian sandstone near the base, Platteville dolomite overlying it, and a thin layer of St. Peter sandstone capping the top. The ice was thick enough at its maximum extent to cover even the top of Gibraltar Rock. Although the overriding glacier shaped all the hills in this area somewhat, the hills are fairly close to the outermost extent of late Wisconsin Glaciation and are little changed from their preglacial distribution and size. Gibraltar Rock is a high point in northern Dane County, but it is almost 500 feet lower than the top of West Blue Mound.

About a half mile west of the western end of this segment there is evidence of a still-stand of the ice margin, as the dashed blue line on figure 161 indicates. This is not a moraine, but an outwash head (SB 8). The glacier margin remained in one place long enough for relatively thick outwash to be deposited in front (west) of it. After the glacier retreated, it left behind a fairly steep ice-contact face of gravel where the ice edge had stood. The land to the east, which was ice covered when the outwash was deposited, is now lower than the outwash surface. North of Chrislaw Rd. the glacier deposited a low moraine at the same time (SB 6; fig. 161). This small moraine may have formed at the same time as the Milton Moraine, although we cannot tell for sure.

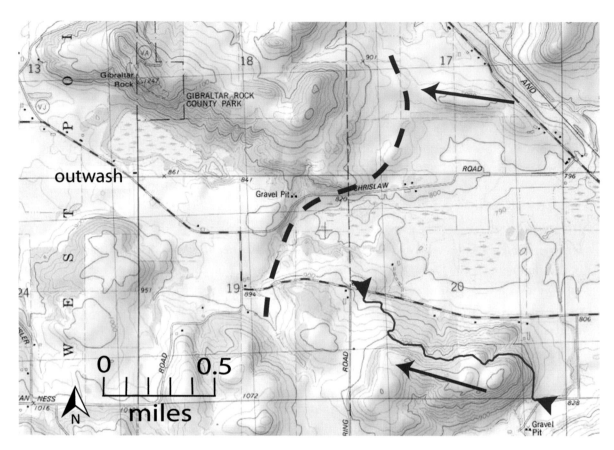

Figure 161. Topography of the Groves-Pertzborn Segment. Dashed blue line indicates an ice-margin position marked by a head of outwash in section 19 and a low moraine in section 17. Blue arrows are on drumlins and show ice-flow direction. (Map is part of the Lodi USGS Quadrangle and was created with TOPO! © 2011 National Geographic Maps.)

## 60 and 61. Gibraltar Segment and Merrimac Ferry Segment

*Slack Rd. to the STH 113 ferry terminal wayside (2.4 miles); STH 113 across the Wisconsin River (0.5 miles, but you can only hike it in winter!)*

The Gibraltar Segment of the IAT climbs a primarily Cambrian sandstone bluff just south of STH 113 (fig. 162). From here, you will have good views of Lake Wisconsin and the outwash beyond it. When the ice began to retreat from the Johnstown Moraine, which lies to the west, glacial Lake Merrimac formed in the lowland now occupied by the river and on the low surface to the north. It lasted until a flood from the drainage of glacial Lake Wisconsin broke through the Johnstown Moraine about where the river cuts through the moraine now.

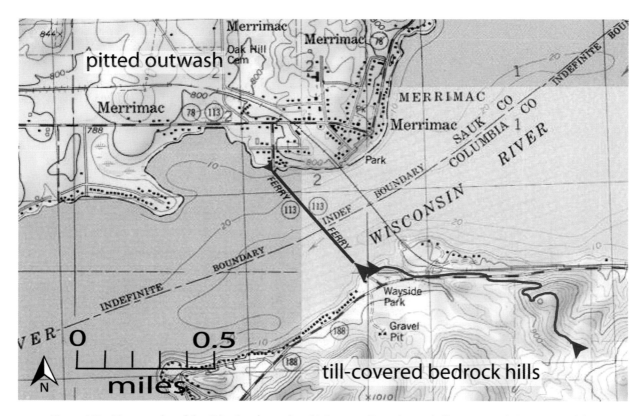

Figure 162. Topography of the Gibraltar (center) and Merrimac Ferry (upper left) segments. (Map is part of the Lodi and Sauk City USGS Quadrangles and was created with TOPO! © 2011 National Geographic Maps.)

Lake Wisconsin is a flowage lake dammed a few miles downstream at Prairie du Sac. Farther north are the Baraboo Hills, which are discussed in detail later in this chapter. The somewhat flat top on the hill at the highest point on this segment has thin Platteville dolomite and thin till on top. Erratics are fairly abundant.

The ferry across the Wisconsin River operates seasonally.

## 62. Merrimac Segment

*Marsh Rd. to South Lake Dr. (4.4 miles)*

From its eastern terminus on Marsh Rd., the trail climbs out of a large kettle (SB 9) onto a high pitted-outwash (SB 10) surface (fig. 163). On the north side of the trail, east of Marsh Rd, you will see several more deep kettles and relatively high sand-and-gravel hummocks. The highest of these hummocks may be a moulin kame (SB 12). West of STH 113 the trail crosses a nearly flat area that is a former lakebed. At its maximum extent, glacier ice flowed directly westward into this valley, as shown in figure 163. When ice withdrew, water was

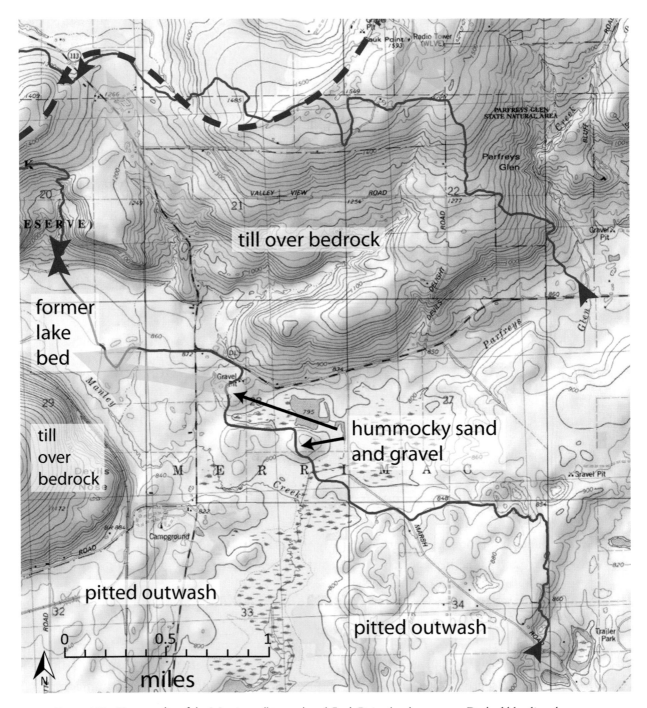

Figure 163. Topography of the Merrimac (bottom) and Sauk Point (top) segments. Dashed blue line shows extent of the last glaciation. Blue arrows show ice-flow direction at the time of the glacial maximum. (Map is part of the Baraboo USGS Quadrangle and was created with TOPO! © 2011 National Geographic Maps.)

dammed between the ice on the eastern side, high valley walls on the northern and the southern sides, and the moraine at the westernmost extent of the advance (this is shown on the map for the Devil's Lake Segment). Once the ice edge had retreated far enough to allow water to escape southward around Devil's Nose, the lake drained.

## 63. Devil's Lake Unit of the Ice Age National Scientific Reserve

The bedrock in southern Wisconsin is mostly horizontal sandstone, shale, and limestone (SB 21) that accumulated in shallow seas covering the area during the Paleozoic Era. But in the region around Baraboo (fig. 164), you'll see much older rocks that tell us something about conditions that existed here far back in Precambrian time, some 1,700 million years ago. At about that time layers of quartz sand gradually accumulated in an ocean basin underlain by older igneous and metamorphic rocks. As sand accumulated on the shallow bottom, currents swept in more sand to pile layer upon layer. Some of this sand was cross-bedded (as in fig. 61, but this photo is of St. Peter sandstone) and some had ripple marks (fig. 62), both of which are preserved. The end result was several thousand feet of sand. The quartz grains must have been washed from the continent after long weathering and erosion of even older Precambrian silicate rocks. The sand was covered by other layers of sediment, some of which contained iron oxide, a red substance that stained the underlying sand grains various shades of red and purple.

The thick sand covered a large part of what is now Wisconsin and neighboring states. The sand was deeply buried by other Precambrian sediments, hardened into sandstone, then metamorphosed under great heat and pressure to become the metamorphic rock quartzite. Quartzite the same age as and with a similar appearance to the Baraboo quartzite is found in several areas of Wisconsin (Waterloo and Barron quartzites), and Minnesota and South Dakota (Sioux Quartzite). Several Chippewa Lobe segments of the IAT cross the Barron Quartzite in the Blue Hills.

During metamorphism, the rocks were bent into a series of folds called anticlines and synclines. At Baraboo, rocks were folded down into a syncline (fig. 165). The Precambrian rock layers above the quartzite at Baraboo have been mostly eroded away, leaving the down-folded part that looks on a map like a partially submerged rowboat (fig. 164). The quartzite dips nearly vertically on the north side of the syncline, and you can cross the exposed edge of the quartzite in less than a mile. The quartzite of the south limb dips less steeply (20 to 35 degrees), and its outcrop is much wider.

Some remnants of younger Precambrian rocks are preserved in the trough of the syncline. Iron ore from these sediments was mined through underground shafts in the late nineteenth and early twentieth centuries. Erosion continued until the late Cambrian (table 1), producing a relief similar to that of today's, but of course, there were no land plants or animals at the

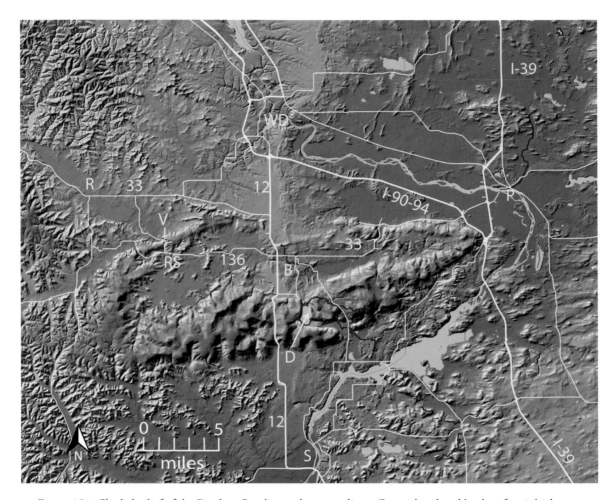

Figure 164. Shaded relief of the Baraboo Syncline and surroundings. Greenish-colored land surface is higher elevation than reddish-colored land surface. Yellow lines and numbers indicate highways. Cities and other features shown (yellow): (B) Baraboo, (D) Devil's Lake, (P) Portage, (R) Reedsburg, (RS) Rock Springs, (S) Sauk Prairie, (V) Van Hise Rock. Ice Age Trail segments (red) are labeled on figure 137. (Base map constructed from USGS National Elevation Dataset and modified by WGNHS.)

time. We know that the valley that Devil's Lake is in today was formed during late Precambrian, because a well in the valley bottom penetrated Cambrian sandstone that is beneath sand and gravel. The valley had to have been there when the sandstone was deposited.

The sea level then rose, eventually covering much of the continent. As mentioned in the introduction to this book, Wisconsin was 10 degrees south of the equator in Cambrian time and was rotated about 90 degrees to its orientation today (fig. 6), so that the syncline's present north margin was exposed to the strong trade winds that blew from the east most of the year. The resulting waves beating against the quartzite cliffs rounded the talus blocks

## Baraboo Syncline

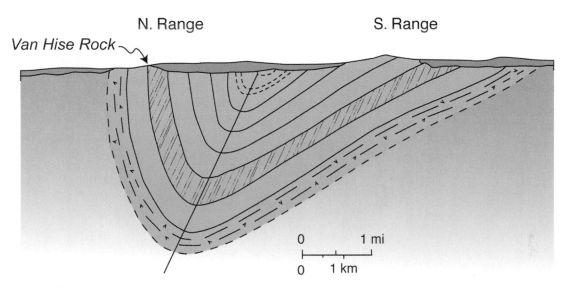

Figure 165. Cross section of the Baraboo Syncline looking toward the east. Baraboo quartzite in lavender. Brown areas represent Cambrian sandstone and glacial deposits. (Modified from Dott 1999; drafted by Mary Diman.)

that had accumulated in large quantities along the shores. Grand exposures of these rounded quartzite boulders can be seen at Parfrey's Glen, which is accessible from the east end of the Sauk Point Segment of the IAT. Eventually the remnants of the Baraboo quartzite's syncline were completely buried by Cambrian sand and younger Paleozoic sediments.

The Baraboo region experienced erosion during much of the Mesozoic Era, Tertiary Period, and Quaternary Period (table 1). The old quartzite was finally exposed again when rivers stripped away the overlying Paleozoic sediments from the Baraboo area. The potholes you see (Devil's Lake Segment) on the east bluff (fig. 166) may have formed at this time. When a stream encountered the edge of the hard quartzite, eddies swirled pebbles around and cut a series of potholes. There is more information on pothole formation in the Interstate State Park discussion.

During the late Wisconsin Glaciation, the Green Bay Lobe covered the eastern portion of the Baraboo Syncline (fig. 167), blocking off both ends of the ancient Precambrian river valley. Meltwater was dammed between the two tongues of ice where Devil's Lake is today. Figure 168 shows a location in Greenland that looks very much like it looked here 25,000 cal. years ago. The glacier lobe is building a narrow, sharp-crested moraine with a lake in front. Although the bluffs at Devil's Lake were never covered by glacial ice, they were subjected to an extremely harsh climate. Freezing and thawing for much of the year loosened

Figure 166. Water-worn quartzite with potholes. These are located close to the Potholes Trail. The pothole in the left foreground is about 1 foot across.

blocks of bedrock on the cliff that tumbled down to form an accumulation of boulders called talus (fig. 169).

Devil's Lake is 1.3 miles long and 0.6 miles wide, and it is about 40 feet deep. Its level is higher than the land outside the blocking moraines; it contains little or no limestone debris, so it is a soft-water lake, and it has preserved pollen blown in since the glacier melted back. During the spring of 1970, students and staff from the University of Wisconsin–Madison set up equipment on the ice to sample the lake sediment for pollen analysis. The work required a portable metal framework and strap hoists to push a two-inch-diameter piston corer into the lake sediment to retrieve the core segments. The coring penetrated 20 feet into the sediment before reaching underlying glacial sand. Measured volumes of sediment were taken every few centimeters in the core. The sediment was then dissolved in

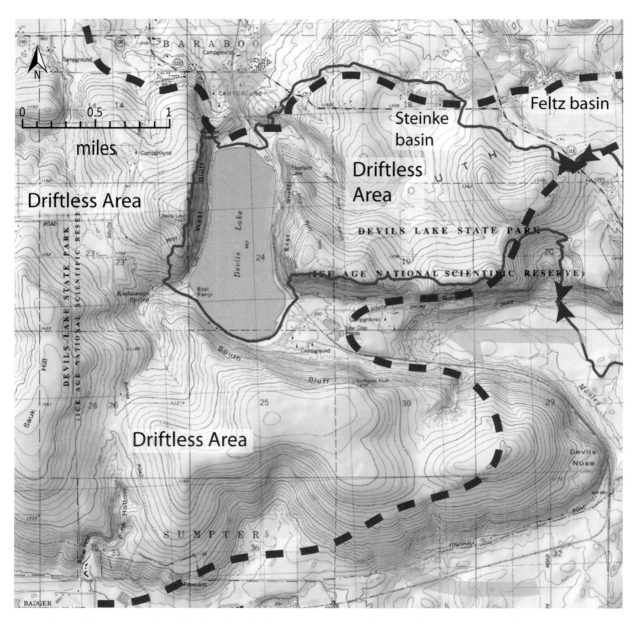

Figure 167. Topography of the Devil's Lake Segment and the Devil's Lake Unit of the Ice Age National Scientific Reserve. Dashed blue line shows maximum extent of glaciation. Blue arrows show ice-flow direction at the time of the glacial maximum. Note that the Devil's Lake valley was dammed at both ends by the ice sheet and is now dammed by moraines. (Map is part of the Baraboo and North Freedom USGS Quadrangles and was created with TOPO! © 2011 National Geographic Maps.)

Figure 168.  View of Greenland showing a setting similar to that at Devil's Lake about 25,000 years ago. Note that the bluffs are not glaciated. The white area in front of the glacier is a frozen lake. (Photo by R. P. Goldthwait.)

hydrofluoric acid, which does not harm the tough pollen-grain walls. The results of this study are shown in figure 170. Note how the pollen types changed through time as new plants migrated into the area with fluctuations in the climate. Zone D-2, when there was relatively little pollen being produced, represents the Younger Dryas cooling recorded in the Greenland Ice Sheet core as shown in figure 79. This confirms that when the glacier advanced into northeastern Wisconsin and killed the Two Creeks forest, the climate also cooled here at Devil's Lake, well away from the glacier's edge. Also note the huge increase in ragweed pollen in zone D-6 that accompanied European forestry and agricultural practices.

Figure 171 shows the valley of the Baraboo River as it flows into the Baraboo Syncline just north of Rock Springs (fig. 164). Of course, this is outside the glacial border and off the IAT, but it is worth the time to see it when you are in the area. The valley shows vertical

Figure 169.  Low oblique aerial view showing the south side of the East Bluff at Devil's Lake State Park. The Ice Age Trail runs along the top of the bluff. Devil's Lake is in the background. Note the large talus accumulation at the base of the slope.

quartzite with wonderful ripple-mark patterns (fig. 62) in the abandoned quarries on the west side of the road. The valley also contains Van Hise Rock (V in fig. 164), a remnant of vertical quartzite that is famous for what it shows about the structure of the north limb of the syncline. Not far south of Van Hise Rock, but on a path to the west of the road, there is an outcrop of quartzite that looks highly polished. It was probably sandblasted by strong winds blowing through the valley during the Ice Age when there was little protective vegetation. You can see other evidence of wind polishing of quartzite on the Sauk Point Segment.

## 64. Devil's Lake Segment
*South Lake Dr. to STH 113 (10 miles)*

For fuller details of the geology and geologic history of this area, turn to the description of the Devil's Lake Unit of the Ice Age National Scientific Reserve. From the eastern terminus

# Devil's Lake, Sauk County, Wisconsin

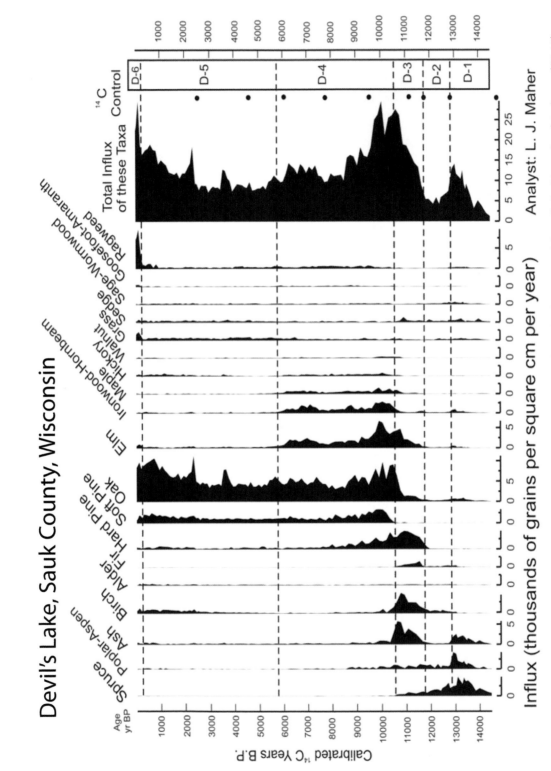

Figure 170. Abundance of pollen in a core collected from the bottom of Devil's Lake. Calendar-year age (also called calibrated age) is shown (SB 18). Black areas show abundance of pollen for each species. Note that spruce (far left column) was abundant right after deglaciation, but by 10,500 cal. years ago there was little evidence of it in the pollen record. Oak became common at that time.

Figure 171. Low oblique aerial view of the Baraboo River valley where it and STH 136 cut through the northern limb of the Baraboo Syncline. View is toward the west. A large, active quarry in Baraboo quartzite is in the foreground (private property!). Van Hise Rock is labeled but not visible in the photo. Rock Springs is a short distance to the south.

of the Devil's Lake Segment of the IAT, the trail rises steeply into East Bluff basin (fig. 167). To the right (east), you will see a Cambrian sandstone outcrop, but to the left (west) the slope is made up of Baraboo quartzite. This small basin held an ice-marginal lake at the time the glacier was retreating from its maximum position higher on the slope. A small, nearly flat area is the bed of that former lake.

From here, the trail climbs the wall of East Bluff basin to the top of the East Bluff. The point where the trail turns directly to the west marks the approximate maximum extent of glacier ice, although no moraine marks that position. From that point westward, as it follows the southern edge of the East Bluff, the trail is in the Driftless Area. From the bluff you will have good views into the valley below. In particular, note the Johnstown Moraine (SB 6) that dams Devil's Lake where it crosses the valley bottom. In addition to distant views, there are several sites of geologic interest not far from the IAT. On the Potholes Trail, you can see several examples of potholes on a steep slope (fig. 166), and on the Devil's Doorway Trail you can see even more.

Potholes form at the bottom of fast-moving rivers in places where gravel or larger rocks spin in an eddy or whirlpool, wearing a more or less cylindrical hole in solid rock on a river bottom. At the Interstate State Park Unit of the Ice Age National Scientific Reserve you can see excellent examples of large potholes that formed during the last glaciation. Their formation is discussed in the Interstate State Park description in this book. Geologists believe that the potholes at Devil's Lake formed at the bottom of a rushing river long before the last glaciation.

Several features along the IAT where it follows the south edge of the East Bluff are evidence that the bluff tops here were not glaciated: Balanced Rock, Devil's Doorway, and other rather fragile features of the quartzite would presumably have been pushed over had glacier ice overridden the East Bluff. Also, there are no erratics on the summits of the bluffs.

Before descending on the Balanced Rock Trail, consider going north a little over a mile on the East Bluff Trail to Elephant Cave. This is not a true limestone cave, but it is a place where differential weathering has opened a cavity at the unconformity (defined in the introduction) between the Baraboo quartzite below and the Cambrian sandstone above. This gap in the rock record is longer than 400 million years! Come back south on the trail a few hundred feet from Elephant Cave and you can see the unconformity on the rock wall east of the trail.

Where the trail meets the flat ground near the south end of Devil's Lake, it goes onto sandy lake sediment. This sediment was deposited when the glacier dammed both ends of the valley and the lake was more extensive than it is now.

On the hike around the south end of Devil's Lake, you will see large blocks of quartzite forming talus at the base of the South Bluff (fig. 169). Most of the blocks here apparently accumulated as talus before the glacier retreated from its maximum position. Ten- to twenty-foot-wide shallow depressions in the talus are called grottos. There are good views of talus from the IAT where it follows the crest of the West Bluff. There are also excellent views of the Johnstown Moraine, which dams the north and south ends of the valley. From the bluff, the IAT descends onto the Johnstown Moraine and then down its front to a flat lake sediment surface.

Where the trail climbs back into the upland, it follows the former maximum position of the ice margin. The ridge just to the southeast appears to be the moraine, although the landform is smaller and narrower than the moraine just to the west. The deep gully just to the east of the trail as it skirts Northern Lights Campground was an ice-marginal channel that carried glacial meltwater into the higher-level Devil's Lake. After passing under CTH DL, the trail follows the Johnstown Moraine, which here is not very tall, although it does have kettles (SB 9) and hummocks (SB 11) typical of an end moraine. The low area south of CTH DL, called Steinke basin, was occupied by a lake when the ice margin was at the moraine. After the IAT recrosses CTH DL, it leaves the glaciated area and borders the southern edge of Feltz basin, which held another ice-dammed lake.

## 65. Sauk Point Segment
*STH 113 to CTH DL (4.2 miles)*

The STH 113 trailhead is at the Johnstown Moraine (SB 6; figs. 163, 164). To the northwest of the moraine you will see a flat area. This area, called the Feltz basin, is the former bed of a lake that filled the area in front of the ice dam. East of STH 113, the trail crosses the moraine and climbs a quartzite ridge in the Driftless Area before going back onto the moraine crest. Note that some of the quartzite boulders are faceted (shaped) and polished by the wind (fig. 172). These are called ventifacts and are fairly common on surfaces just outside the margin of the late Wisconsin glacier. While the ice was there, there was little or no vegetation and lots of blowing sand.

From the moraine crest you will have excellent views to the south. On a clear day, you'll be able to spot Blue Mounds, and under just the right conditions, the state capitol building in Madison in the far distance.

From this high point on the moraine, the trail descends to the parking lot at Parfrey's Glen. This is a State Natural Area, and if it is open, you may enjoy a hike up this lush, beautiful glen. Check the DNR website for current status. Recent rainstorms flooded the area and destroyed walkways, which will not be replaced. If you do hike up, notice the rocks on the walls of the glen, which are Cambrian sandstone and conglomerate, and also note how the abundance and size of quartzite rocks in the sandstone increase as you walk toward the head of the glen. The sandstone and conglomerate were deposited in a shallow ocean as waves pounded against the quartzite bluff not far to the north. The gorge itself was probably a creek valley in preglacial time but was deepened when glacial meltwater poured through it from the upland above.

## 66. Baraboo Segment
*Effinger Rd. to UW–Baraboo campus (5.2 miles)*

The eastern trailhead of the Baraboo Segment is on the floodplain of the Baraboo River. The trail remains on the floodplain as it follows the northern side of the river through the city (fig. 173). Note two large river meanders, one on each side of the city. Lake sediment deposited in glacial Lake Baraboo underlies slightly higher areas south of the river and under

Figure 172. Polished, grooved, and faceted ventifact formed by windblown sand during very cold conditions near the edge of the ice sheet. There are many ventifacts on Baraboo quartzite. This one was located along the Sauk Point Segment. Pen shows scale.

the main part of the city. When the glacier sat at the Johnstown Moraine (SB 6), the glacier dammed the Baraboo River valley to the west, and rivers deposited extensive lake sediment in the basin in glacial Lake Baraboo. As the glacier retreated from the Johnstown Moraine, the lake expanded, depositing even more sediment under what is now the city of Baraboo and areas to the east between the Baraboo Hills. It drained at the same time as glacial Lake Wisconsin, which is discussed in the chapter on the western Green Bay Lobe IAT segments. Just south of Eighth Ave. the trail rises off the floodplain onto a lake-sediment surface, and then a short distance north of Eighth Ave. it crosses onto a till surface. As it enters the University of Wisconsin–Baraboo campus, the IAT rises onto the Johnstown Moraine.

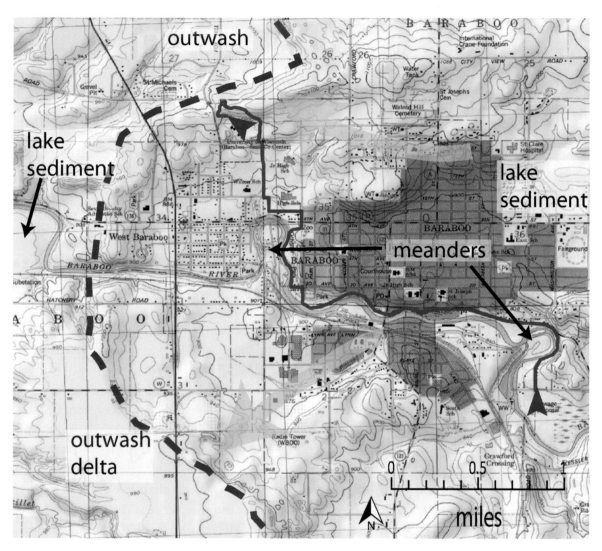

Figure 173. Topography of the Baraboo Segment. Dashed blue line shows outer edge of the Johnstown Moraine. Hills west of the dashed blue line are in the Driftless Area. Blue arrows show ice-flow direction. Outwash outside the moraine is partly a delta built into glacial Lake Baraboo. Lake sediment under the city of Baraboo was deposited in glacial Lake Baraboo after ice retreat from the Johnstown Moraine. Note meanders of the Baraboo River on the modern floodplain. (Map is part of the Baraboo and North Freedom USGS Quadrangles and was created with TOPO! © 2011 National Geographic Maps.)

## 67. Portage Canal Segment

*STH 33 at Pauquette Park to Agency House Rd. (2.8 miles)*

This area has a rich geologic and cultural history. The French explorers Father Jacques Marquette and Louis Joliet arrived here in 1673, and the area of course has a Native American history that stretches back thousands of years before that. The end of this IAT segment (fig. 174) at Pauquette Park is on a terrace of the Wisconsin River. The depressions you see in the park are not natural; they are probably old excavations for silt and clay for bricks. The old Agency House at the other terminus of this segment has brick that was made here, according to signage. The hills that you can see just north of the park are drumlins (SB 14; fig. 174). The large boulder in the northeast corner of the park is quartzite that was carried by the glacier. By looking closely at the rock, you can see faint striations (SB 4) on its surface. You can also tell that the rock once tumbled along in fast-moving water, because its surface has percussion marks (small curving and concentric cracks) that formed when the rock collided with other rocks. It seems likely that people moved the rock to this park, although it could have rolled off the drumlin to the north.

The low area around the city of Portage is one of a few continental divides that exist in the United States. This divide separates water in the Wisconsin River basin that flows to the Mississippi, and ultimately the Gulf of Mexico, from water in the Fox River basin that drains to the St. Lawrence River and eventually the Atlantic Ocean. The low, marshy land east and north of the city has flooded at times of high water, allowing Wisconsin River water into the Fox River watershed. The U.S. Army Corps of Engineers constructed artificial levees along the banks of the Wisconsin River in this area to contain floodwater and minimize flooding in the low parts of the city. This also prevents Wisconsin River water from flowing into the Fox River drainage at times of flood. The IAT follows a levee along one of the locks that used to allow boats to go from the Fox River into the slightly higher Wisconsin River (fig. 174).

The low area east and north of the city (fig. 174) was a less-than-two-mile portage between the two rivers for Native Americans and early European settlers. The portage was only a short distance south from, and roughly parallel to, the part of the IAT that runs along the old Portage Canal. Fort Winnebago, built in 1828, is on STH 33 east of the trail (fig. 174). The Indian Agency House, which is close to the north end of this segment, was built in 1832 to house the agent to the Ho-Chunk (at that time called Winnebago) Tribe. Historic markers in the area indicate features of interest. The *Ice Age Trail Companion Guide* also provides in-depth historical information.

From the point where the IAT crosses the canal at Wisconsin Street to the segment's eastern end, you will cross sediment deposited in slow-moving water. The sediment is mostly sand and silt with a thin cover of marsh deposits in many places. The surface has a

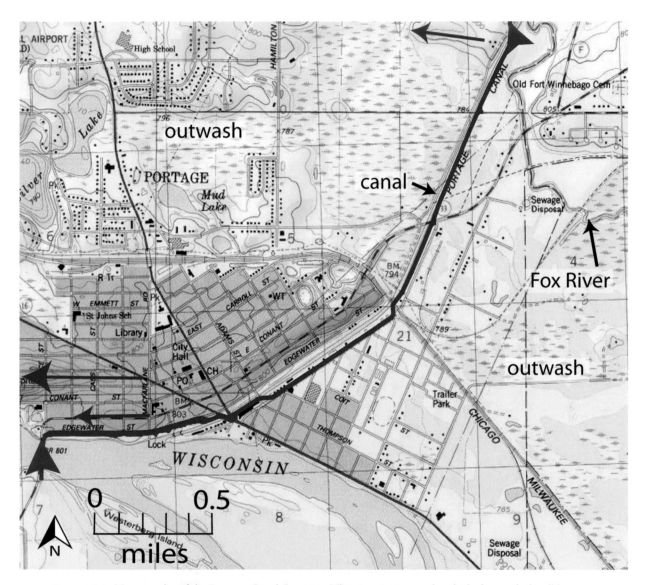

Figure 174. Topography of the Portage Canal Segment. Thin peat covers sand and silt that underlie all low areas. Blue arrows are on drumlins and show ice-flow direction. Much of the surface close to the canal has been modified by human activity. (Map is part of the Portage USGS Quadrangle and was created with TOPO! © 2011 National Geographic Maps.)

very low gradient. The trail itself is on an artificial levee created during construction of the canal (fig. 175). Almost all the hills you can see from here are drumlins, except for the Baraboo Hills, which you may glimpse in the distance to the southwest.

## 68. Marquette Segment
*Agency House Rd. to Lock Rd. (4.3 miles)*

From the park at the end of Agency House Rd., the IAT follows the old Portage Canal before merging with the Fox River just a short distance along the path (fig. 176). The ridge the path follows is not an esker, although it may look like it. Instead, dredgings of silt and sediment from the river have created this feature. In places the segment also follows an old railroad grade. All the land surface here before human modification was peat over sand and

Figure 175.　View of Portage Canal looking toward the southwest. The Ice Age Trail is on the levee across the canal. (Photo by Vin Mickelson.)

silt. The hills surrounding the trail here are drumlins, created as ice flowed from east to west through the area. Just south of where the trail crosses Clark Rd., there is a drumlin (on private land!) west of the IAT with a small exposure of till where the road cuts through it. Most of the time, vegetation probably covers this exposed till. Across the valley you can spot a low, broad ridge, which is a small, discontinuous end moraine (labeled on fig. 176).

North of Clark Road, the Fox River is confined to a valley about a quarter mile wide (fig. 176). This valley was the outlet of glacial Lake Oshkosh whenever the Green Bay Lobe blocked northward drainage into the Fox River valley. When it was functioning as an outlet, this valley was filled side to side with water flowing toward the south and into the Wisconsin River. For further explanation of glacial Lake Oshkosh, see SB 16 and figure 54.

## 69. Mill Bluff Unit of the Ice Age National Scientific Reserve and Glacial Lake Wisconsin

When the Green Bay Lobe advanced to its maximum position 25,000 to 30,000 cal. years ago, it dammed east-flowing drainage along its west edge in numerous places, forming many lakes along the ice margin. The largest of these lakes was glacial Lake Wisconsin. Clayton and Attig (1989) have summarized much of what is known about the lake. As soon as the glacier advanced to the east end of the Baraboo Hills near the present city of Portage, the present-day Wisconsin River dammed drainage from the north and west (fig. 177). The lake grew until it drained out the East Fork of the Black River on its northwest edge (fig. 177). When ice sat at the Johnstown Moraine, smaller lakes were present in the Baraboo River basin and other basins in the Baraboo Hills and north of the Baraboo Hills. When the lake was at its maximum size, it was 140 feet deep at the present Wisconsin Dells (WD).

At its maximum extent, the lake flooded between pillars of sandstone bedrock that had been produced through millions of years of stream erosion. Sand and finer sediment slowly settled on the lake bottom, filling valleys and producing a nearly flat lake bottom. The sandstone pillars stood as islands in the lake. Mill Bluff, one of these pillars, is located within the bed of glacial Lake Wisconsin very close to Interstates 90 and 94 (fig. 177). From the top of Mill Bluff you will have an excellent view of the bed of glacial Lake Wisconsin to the east, the Driftless Area hills to the south and west, and other nearby sandstone pillars (fig. 178).

As you look at Mill Bluff, consider that its pillar shape comes from its composition: it and the other rock around is Cambrian sandstone, a rock that is not particularly resistant to erosion. It appears that these pillars formed in a few places where, over millions of years, water slowly percolating from the ground surface carried calcium carbonate that cemented the grains of sand together into much tougher rock. This formed a cap on top of the softer sandstone beneath, and over time it protected the underlying soft sandstone from eroding away as it did in the areas between the pillars. Look for examples of cross-bedding (fig. 61)

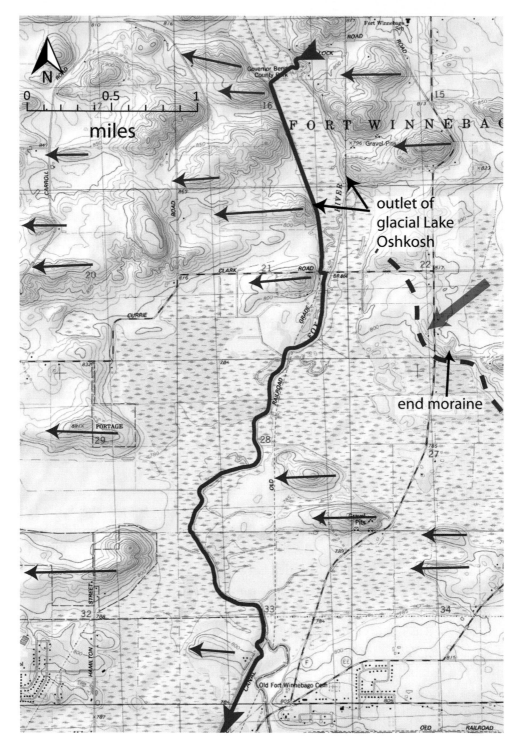

Figure 176. Topography of the Marquette Segment. Thin peat covers sand and silt that underlie all low areas. Thin blue arrows are on drumlins and show ice-flow direction. Dashed blue line shows front of small moraine. Thick blue arrow shows ice-flow direction at the time the moraine was formed. Trail follows the Portage Canal in the southern mile and the Fox River north to Governor Bend County Park. (Map is part of the Portage USGS Quadrangle and was created with TOPO! © 2011 National Geographic Maps.)

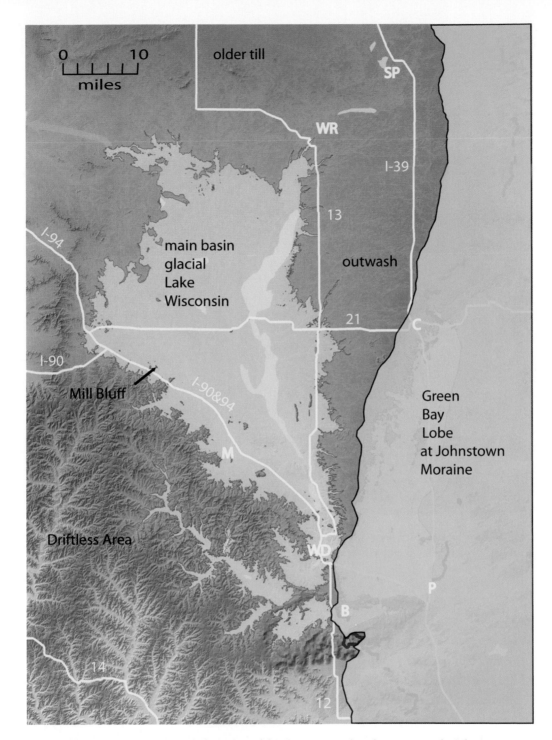

Figure 177. Shaded relief showing the extent of the Green Bay Lobe when it was at the Johnstown Moraine (greenish blue). Glacial Lake Wisconsin is shown in very light gray and the light blue is present-day water. The blue arrow (upper left) shows the outlet of the glacial lake down the East Fork of the Black River. Yellow lines and numbers indicate highways. Cities shown (yellow): (B) Baraboo, (C) Coloma, (M) Mauston, (P) Portage, (SP) Stevens Point, (WD) Wisconsin Dells, (WR) Wisconsin Rapids. The IAT (red) on the bluffs at Devil's Lake, just south of Baraboo, is in the Driftless Area. (Base map constructed from USGS National Elevation Dataset and modified by WGNHS.)

Figure 178. View looking north from the northern end of Mill Bluff at sandstone pillars standing above the bed of glacial Lake Wisconsin. It appears that Camel Bluff is actually two pinnacles separated by a low area.

in some of these sandstone layers. There are also trace fossils—for instance the casts of the burrows of animals that lived in the Cambrian sea (fig. 179).

When the ice retreated from the Johnstown Moraine, glacial Lake Wisconsin expanded into the area behind the moraine. This is called the Lewiston basin of glacial Lake Wisconsin (fig. 52). Water continued to drain out the East Fork of the Black River, as you see in figure 177, until the ice margin retreated to the eastern end of the Baraboo Hills. At this point water could escape through a much lower outlet around the east end of the Baraboo Hills and down the Wisconsin River, as you can see on figures 52 and 180. Because the ice dam was fragile and had a huge wall of water pushing against it, when the dam broke there was a massive flood of water into a much smaller glacial lake called glacial Lake Merrimac (see Merrimac Segment description) and then down the Wisconsin River valley. The short-lived flood carried boulders far down the Wisconsin River, and water filled the valley from side to side. Geographers Jordan and Knox (2008) estimated that the peak flood 60 miles

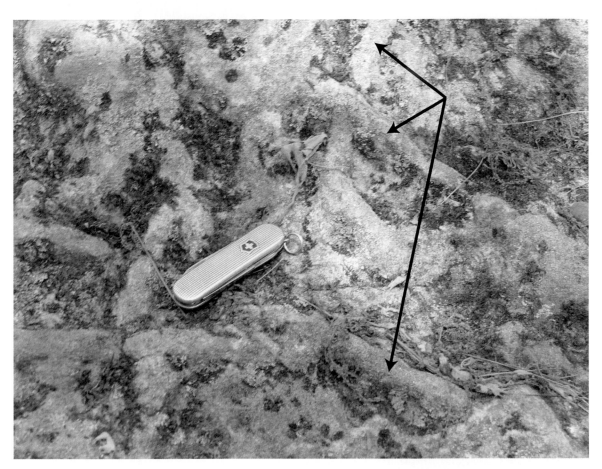

Figure 179. Surface of boulder along trail north of the southern parking lot showing casts (fillings) of animal (likely trilobite, see fig. 7) burrows preserved in the sandstone. These formed in the intertidal zone or somewhat farther offshore in the shallow Cambrian sea. Knife is about 2 inches long. This boulder, which fell from above, is along the trail from the eastern parking lot.

downstream (between Muscoda and Boscobel) was 45,000 cubic meters per second (almost 12,000,000 gallons per second), more than 10 times larger than the historic peak flood recorded in 1938 at that location. They calculated that the flood at the outlet near Portage (fig. 180) was as much as 150,000 cubic meters per second (that's about 40,000,000 gallons per second), enough to fill a typical NBA basketball arena in about 10 seconds (Jordan, personal communication, 2007)!

The torrent of water broke through the Johnstown Moraine, and as a result the overall water level dropped rapidly. You probably know about the landforms that result: the rush of water from the main basin of glacial Lake Wisconsin was instrumental in cutting the narrow channels in Cambrian sandstone that we now call the Wisconsin Dells. Probably within a

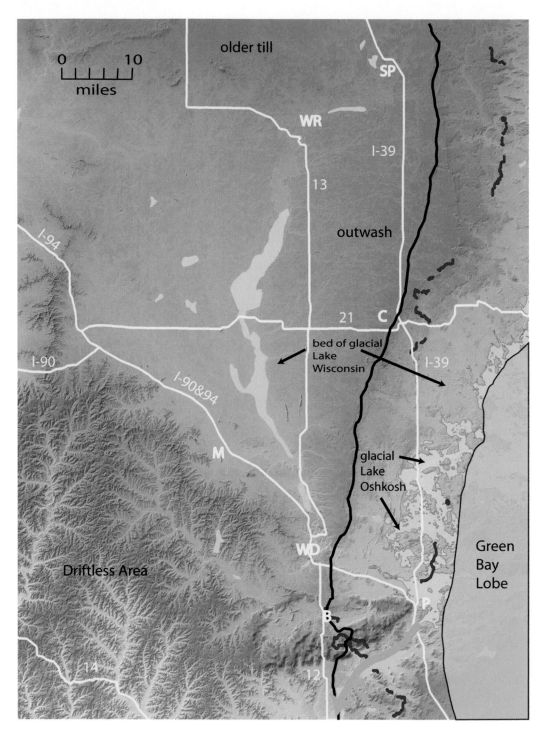

Figure 180. Shaded relief showing the Green Bay Lobe (greenish blue) after drainage of glacial Lake Wisconsin and the beginning of formation of glacial Lake Oshkosh (light gray). Glacial Lake Oshkosh drained down the Wisconsin River as indicated by the blue arrow. Light blue color indicates present-day lakes. Red lines are trail segments labeled in figures 137 and 183. Yellow lines and numbers indicate highways. Cities shown (yellow): (B) Baraboo, (C) Coloma, (M) Mauston, (P) Portage, (SP) Stevens Point, (WD) Wisconsin Dells, (WR) Wisconsin Rapids. (Base map constructed from USGS National Elevation Dataset and modified by WGNHS.)

few days Lake Wisconsin was gone, and glacial Lake Oshkosh began to form along the edge of the Green Bay Lobe (fig. 180). This new lake continued to expand and contract as the Green Bay Lobe retreated and readvanced several times during its overall retreat (SB 16).

## 70. John Muir Memorial Park Segment

*Loop Trail (2 miles)*

The trail loops around Ennis Lake (figs. 181, 182), which you may recognize as a kettle in pitted outwash (SB 10). At this trail segment there are excellent educational displays relating

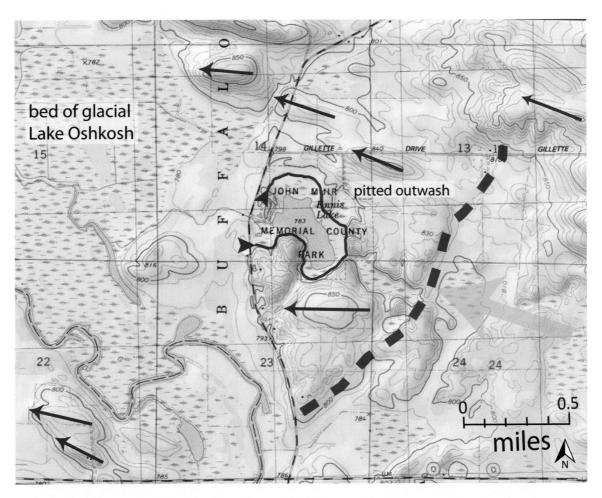

Figure 181. Topography of the John Muir Memorial Park Segment. Dark blue arrows show location of drumlins and ice-flow direction when they formed. Dashed blue line is an ice-margin position marked not by an end moraine but by an outwash head, and light blue arrow shows ice-flow direction when ice was at that position. (Map is part of the Endeavor USGS Quadrangle and was created with TOPO! © 2011 National Geographic Maps.)

to the life of John Muir and his early days exploring the region. Muir was not only an early proponent of a system of national parks, but he was also an early champion of the glacial theory that was just developing in the mid-1800s while he was living here. Muir later went on to explore present-day glaciers in the Sierras, the Rockies, and Alaska. As a result of his explorations and advocacy, several landmarks and trails bear his name: there is a Mount Muir in California and another in Alaska; the Muir Glacier and Muir Inlet in Glacier Bay National Park; a Muir Trail in the Sierras; and Muir Woods National Monument within Golden Gate National Recreation Area near San Francisco. Muir wrote eloquent accounts of his adventures hiking the glaciers of Alaska and other western landscapes, as well as impassioned proposals for land preservation.

Figure 182. Ennis Lake (see fig. 181) looking toward the west at hummocky sand and gravel and pitted outwash beyond. (Photo by Vin Mickelson.)

Surrounding Ennis Lake you will see low drumlins (SB 14). The Green Bay Lobe here flowed toward the west or slightly toward the northwest (fig. 181). This site is east of the major Johnstown and Milton moraines, but the ice margin paused during its retreat just east of the lake. Outwash was deposited by a braided stream (SB 8) on a surface that slopes gently toward the west. Isolated masses of glacier ice that were buried by the sediment later melted out, producing the kettles (SB 9) in the pitted outwash that you see today. Note that the east edge of this pitted outwash plain drops steeply off to the east. This is where the glacier was at the time the outwash plain was deposited. The steep slope is known as an outwash head (SB 8). These are common features along the west edge of the Green Bay Lobe, and you can use these landforms to recognize brief pauses of the retreating ice even where there is no moraine. As the glacier retreated, meltwater was dammed in front of the glacier in what is now the Fox River basin. The lake that formed is called glacial Lake Oshkosh (SB 16).

About 2 air miles northeast of Ennis Lake you can see glacial striations etched into the rock near the summit of Observatory Hill, which is a State Natural Area. The striated summit is Precambrian rhyolite (SB 21) that juts above Paleozoic rock surrounding it. From the scratches on the rock and its position we can tell that the glacier once covered the top of this hill and flowed toward the southwest.

# Western Green Bay
# Lobe Ice Age Trail Segments

The nature of glacial deposits changes dramatically as you walk northward along the western edge of the former Green Bay Lobe. You'll be able to spot many obvious tunnel channels (SB 17) here because there are fewer bedrock hills to confuse with glacial landforms than there are in the southern Green Bay Lobe. You'll see the exception to this in a few places where residual bedrock hills stick up through the glacial sediment. As you hike these segments you'll also see end moraines (SB 6) and outwash heads (SB 10), which are easier to recognize here than they are in much of southern Wisconsin.

In the late Wisconsin glacial advance, the Green Bay Lobe scraped and scoured the landscape, leaving drumlins (SB 14) as reminders of the glacier's path and delivering sediment to the ice edge. In this part of Wisconsin, you can see signs of the outermost extent of this lobe: the Johnstown Moraine (SB 6) stands above the surrounding landscape and contains the accumulated debris that the massive ice sheet carried to its edge (fig. 183). This moraine continues northward almost to the village of Coloma (C on fig. 183), where it appears to split into two moraines: the Hancock and the Almond moraines. Most interpretations, and the one shown in figure 184, indicate that the Hancock Moraine is older than the Johnstown Moraine, whereas the Almond and Johnstown moraines are about the same age. We can't prove this, however, because the till (SB 5) in all three moraines is very similar and there is no way to distinguish one from the other except spatially. Presumably all are older than the Milton Moraine in southern Wisconsin, but there is little continuity of the Milton Moraine north of Portage, so their relative ages cannot be demonstrated either. The Greenwood, Bohn Lake, and Deerfield segments are all in the Almond Moraine and as a result have lots of ups and downs (fig. 183).

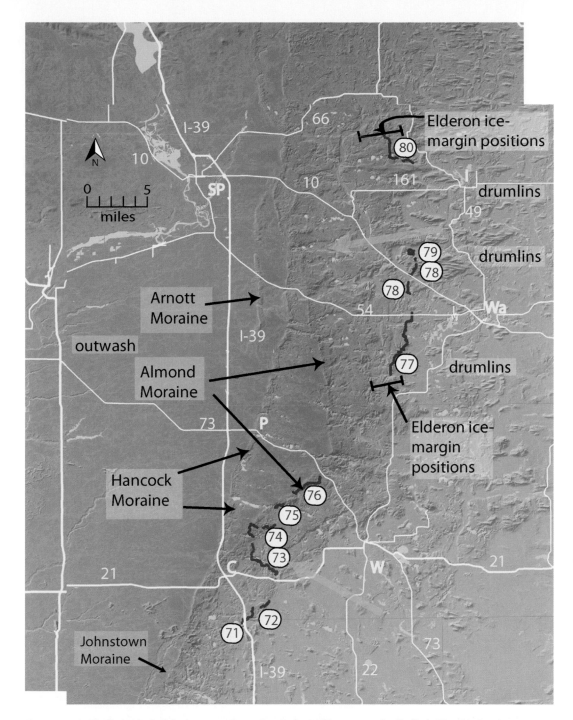

Figure 183. Shaded relief of the western Green Bay Lobe IAT segments (red): (71) Chaffee Creek, (72) Wedde Creek, (73) Mecan River, (74) Greenwood, (75) Bohn Lake, (76) Deerfield, (77) Belmont–Emmons–Hartman Creek, (78) Waupaca River (south and north), (79) Skunk and Foster Lakes, (80) New Hope–Iola Ski Hill. Yellow lines and numbers indicate major highways. Cities and villages shown (yellow): (C) Coloma, (I) Iola, (P) Plainfield, (SP) Stevens Point, (W) Wautoma, (Wa) Waupaca. Prominent landforms are labeled. Blue arrows show ice-flow direction. Discontinuous Elderon ice-margin positions occur from the northern to southern edges of the map in the zone indicated. (Base map constructed from USGS National Elevation Dataset and modified by WGNHS.)

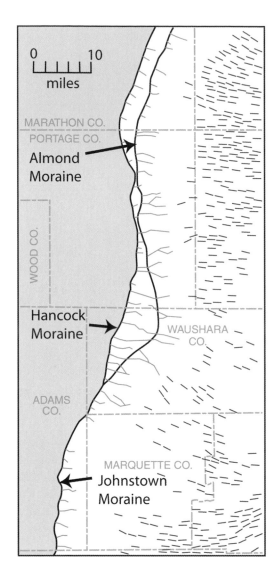

Figure 184. Moraines (black), tunnel channels (green), and drumlins (blue) along the west edge of the area covered by the Green Bay Lobe in central Wisconsin. Arnott, Elderon, and younger moraines not shown. Area is somewhat larger than that shown in figure 183. (Modified from Clayton et al. 1999; drafted by Mary Diman.)

Behind the distinct ridges of the Johnstown, Almond, and Hancock moraines are a series of irregular, discontinuous ridges. These form steps in the landscape that indicate pauses in the retreat of the ice margin or even small readvances. These irregular ridges are referred to as the Elderon Moraine system, although they are typically outwash heads (SB 8) rather than actual moraines. The ridges presumably are about the same age as the Lake Mills Moraine and perhaps the Milton Moraine, both of which occur farther south.

Figure 185 shows a profile of the land surface from just west of Coloma (C on fig. 183) to the city of Wautoma (W on fig. 183). Note that when the ice edge sat at the Hancock and Almond moraines, the land sloped away from the glacier. Water flowed down this slope

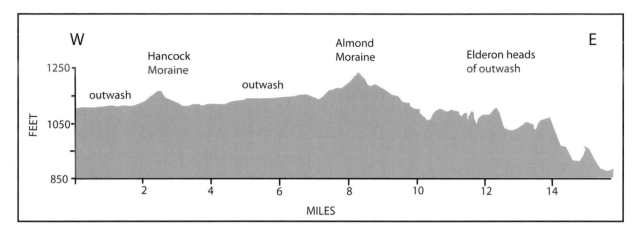

Figure 185. Profile of the land surface from a point just west of Coloma to Wautoma (see fig. 183). Note that the Hancock and Almond moraines stand high in the landscape and lower outwash surfaces are between and west of them. From the Almond Moraine eastward, the land slopes eastward. This difference in slope produced a distinctly different glacial record east and west of the Almond Moraine. (Profile created with TOPO! © 2011 National Geographic Maps.)

toward the west across an outwash plain and into glacial Lake Wisconsin. Once the glacier's edge began to retreat from the Almond Moraine, the land sloped back toward the glacier, to the east. Meltwater could not flow directly away from the ice anymore but instead ponded in valleys or flowed on outwash plains and in channels toward the south, roughly parallel with the ice margin. In many places braided streams (SB 8) deposited substantial thicknesses of sand and gravel on the ice itself. Later, as the buried ice melted, it produced pitted outwash (SB 10) or high-relief hummocky topography (SB 11).

In spots where the ice margin sat in one place for a period of time, thick outwash was deposited along the glacier front. When the glacier retreated, it left behind an outwash head with a steep face where the sand and gravel sat against the ice (the ice-contact slope, fig. 36 in SB 10) instead of a moraine. The Belmont–Emmons–Hartman Creek, Waupaca River, and New Hope–Iola Ski Hill segments are all in the area of Elderon ice-marginal features. All the segments have some kettles, but there are particularly deep ones near the IAT along the Greenwood, Bohn Lake, and Deerfield segments. In particular, look for the deep kettles in high-relief hummocky topography in the Skunk and Foster Lakes Segment. This is a great example of high-relief hummocky topography.

You can see many of the tunnel channels (SB 17) that carried huge flood waters at the bed of the ice up-slope to the glacier edge when the ice was at the outermost moraine (figs. 184, 185). For especially good views of tunnel channels, go for a hike along the Mecan River, Greenwood, Bohn Lake, and Deerfield segments.

The eastward slope of the land not only shaped the topography as the glacier retreated but also continues to determine water movement, and therefore landforms, today. Rain that falls on the sandy till (SB 5) and outwash in and to the west of the Almond and Hancock moraines enters the groundwater system and flows toward the east, down the regional slope. Groundwater-fed springs bubble into the lakes and headwaters of nearly every creek in the area.

Big boulders occur here and there all along the IAT, but from the Waupaca River Segment northward there are a bevy of boulders. These are different from most other erratics you'll see along the trail. While ice excavated, churned, and dropped assorted rock types and forms indiscriminately as it ground across the landscape, the boulders that ice deposited in this segment are almost all one rock type: granite (SB 21).

Paleozoic sandstone and dolomite underlies nearly all of southern and eastern Wisconsin (fig. 4). From Door County all the way around the Green Bay Lobe deposits, the trail has been on sedimentary rocks, except in the roller-coaster segment of the Baraboo Hills. This is about to change: near the northern end of the Waupaca River Segment, the trail crosses Precambrian granitic rocks (figs. 4, 186). From here to Interstate State Park at the western terminus of the IAT, the trail goes over mostly Precambrian rocks that in places poke up through much younger glacial deposits. You'll also notice many boulders as you hike in Marathon County and eastern Langlade County over and west of what is called the Wolf River batholith, which was briefly described in the overview of Wisconsin geology in the introduction. The granite is fractured and coarse grained, both characteristics that allow water to penetrate it. With exposure to water, minerals, especially mica, are soon broken down by a process called weathering. This involves expansion of mineral grains that forces the grains apart, producing a thick layer of loose grains of feldspar, quartz, and other minerals on the land surface. Solid granite core stones remain because water was unable to penetrate in areas between widely spaced fractures. Thus the granite here was already weathered to a considerable depth before the glacier advanced across the landscape. The ice easily picked up the loose mineral grains and the core stone boulders as it slid overtop. You can see evidence of the weathered granite as you pass "rotten granite" quarries in central and western Marathon County. This material is commonly used to surface gravel roads and road shoulders. It has an orange or pink cast because of the iron released by weathering and the abundant pink feldspar (SB 21).

You'll also see boulders of Wolf River granite along the Skunk and Foster Lakes Segment and in segments to the north. Try to identify feldspar and quartz in the coarser-grained boulders (SB 21; fig. 187). Notice also how grain-by-grain weathering of the granite is taking place today as water between the grains freezes and expands. Moss and lichen colonize as they take advantage of the moisture and nutrients in the cracks (fig. 187). More water then gets in between the grains, and it freezes and expands, popping the grains apart even more.

# Bedrock Geology

**Legend**

*Ordovician Rocks*

 Opc: Prairie du Chien dolomite with some sandstone and shale

*Cambrian Rocks*

 Є: sandstone with some dolomite and shale

*Middle Proterozoic Rocks*

 Wolf River Rocks—
   g: mostly granite
   a: mostly gabbro

*Lower Proterozoic Rocks*

 q: quartzite

gr: granite, diorite and gneiss

s: meta-sedimentary rocks including iron formation
vo: basaltic to rhyolitic metavolcanic rocks; some metasedimentary rocks
ga: meta-gabbro, diorite

*Lower Proterozoic Rocks or Upper Archean Rocks*

gn: granite, gneiss

50 mi

*Ice Age Trail*

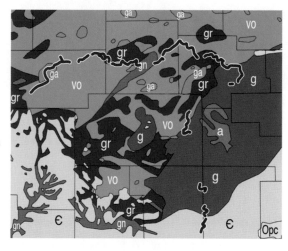

Figure 186. Bedrock geology of central Wisconsin showing the path of the Ice Age Trail. Wolf River granite is shown in dark brown. The Waupaca River Segment crosses the contact between that and Cambrian sandstone. (Modified from WGNHS.)

Figure 187. Coarse-grained Wolf River granite showing pink-to-white orthoclase feldspar grains and gray quartz grains. Dark areas are mostly weathered biotite mica. Note moss and lichens growing on the rock. These contribute to weathering and breaking apart of the mineral grains. Quarter shows scale. Photo was taken along the Skunk and Foster Lakes Segment.

## 71. Chaffee Creek Segment

*I-39 southbound wayside to Czech Ave. (2.2 miles)*

This segment of the IAT mostly crosses pitted outwash (SB 10) deposited in front of the retreating Green Bay Lobe (fig. 188). The high hill just south of the rest area is a drumlin (SB 14) that may have a rock core. When glacier ice sat with its ice margin at the position shown in figure 188, small ponds were dammed in front of the ice. The IAT crosses the bed of one of these east and west of the I-39 crossing. Note that there is sand and gravel on the creek bottom, not clay, indicating that the small pond had a sandy bottom. The trail then climbs onto pitted outwash, where it remains to Czech Ave. The ice margin shown does

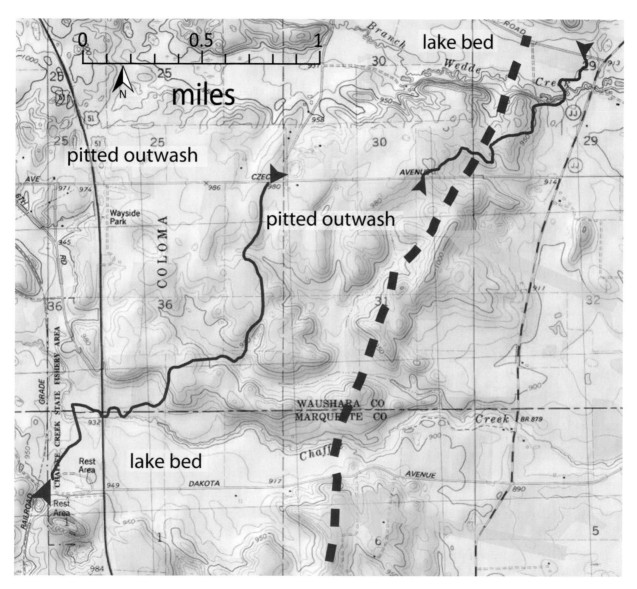

Figure 188. Topography of the Chaffee Creek (left) and Wedde Creek (upper right) segments. Dashed blue line shows an end moraine built by retreating Green Bay Lobe. Blue arrows show ice-flow direction. (Map is part of the Westfield East, Westfield West, and Richford USGS Quadrangles and was created with TOPO! © 2011 National Geographic Maps.)

feature a small end moraine (SB 6) that is discontinuous north and south of here. It is probably part of the Elderon Moraine system shown in figure 183, but there is no single, distinct moraine.

## 72. Wedde Creek Segment
*Czech Ave. to Cypress Rd. (1.2 miles)*

As you hike a short distance from the trailhead on Czech Ave. (fig. 188), you will climb off the pitted-outwash (SB 10) surface and onto a small end moraine (SB 6) that is one of the Elderon-phase ice margins that formed as ice retreated (fig. 188). The trail descends onto more pitted outwash before dropping into the Wedde Creek floodplain. The valley here is the bed of a small lake that ice dammed when it sat at, then retreated from, the moraine just described.

## 73. Mecan River Segment
*STH 21 to Buttercup Dr. (6.8 miles)*

The Mecan River, like nearly all the streams that flow downslope (east) toward the former glacier's retreating edge, is spring fed. Rainwater trickles through the sandy outwash west of the Johnstown Moraine and till in the moraine into an underground aquifer. The streams flow on sand and gravel and therefore are clear, fast-flowing streams.

From STH 21 (fig. 183) the trail follows the river on outwash (SB 8) then climbs onto pitted outwash (SB 10) for about 2 miles before crossing an Elderon-phase ice margin. Not far west from there, the trail enters a tunnel channel (SB 17). This is a great example of the tunnel channels that formed all along the western edge of the Green Bay Lobe during its maximum extent at the Hancock Moraine. Note in figure 189 that the trace of the tunnel channel goes through the Almond Moraine and out to the Hancock advance ice margin. This trace shows that the tunnel channel was active during the Hancock phase and that debris-covered ice remained buried along its path during the Almond advance. Once the Almond-phase ice retreated, this buried ice melted out, leaving a series of kettles along the path of the tunnel channel (fig. 190). As you enjoy your hike along the south side of the lake, you'll have excellent views of this deep tunnel channel along the trail. After the trail leaves the tunnel channel, it climbs up and onto pitted outwash. Look on the map for the large outwash fan at the mouth of the tunnel channel that formed outside the Hancock Moraine.

## 74. Greenwood Segment
*Seventh Ave. to Ninth Ave. (4.9 miles)*

This segment of the IAT parallels the edge of a very deep, long kettle (SB 9) not far from the trail's Seventh Ave. terminus. This kettle appears to be a narrow part of a tunnel channel

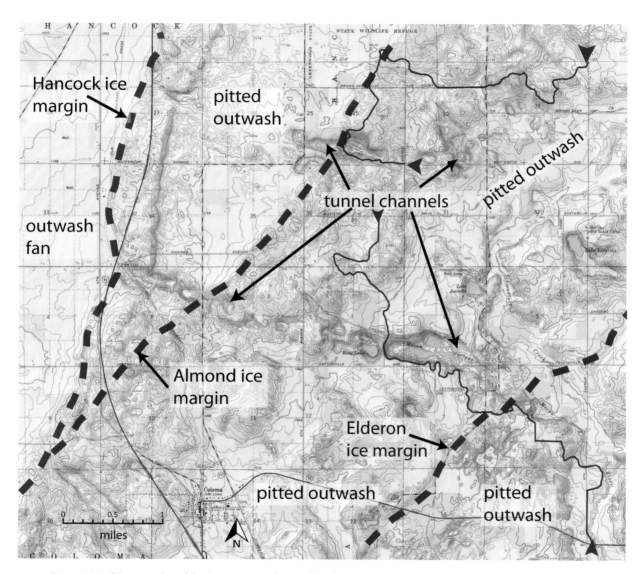

Figure 189. Topography of the Mecan River (bottom) and Greenwood (top) segments. Dashed blue lines indicate ice-margin positions of the Green Bay Lobe. In some cases a distinct moraine is present, and in other cases there is an outwash head (SB 10). Two tunnel channels are indicated. The southernmost tunnel contains the Mecan Springs, and the northernmost one is characterized by a deep kettle hole where it crosses the Almond Moraine. (Map is part of the Coloma and Richford USGS Quadrangles and was created with TOPO! © 2011 National Geographic Maps.)

Figure 190. Low oblique aerial view of part of the Mecan Springs. The Ice Age Trail is along the south (left) side of the lakes. Chain of lakes is in tunnel channel. View is looking west.

(SB 17) that ran to the Hancock Moraine (fig. 189). To see evidence of this, look at the pattern of contour lines on figure 189 that interrupts the otherwise regular pattern of contours on the Hancock Moraine. This tunnel channel is not as well developed as the Mecan River tunnel channel (fig. 189), which was carved about 2 miles south of here. It could be that this tunnel channel was simply not as deep as the Mecan River tunnel channel when it formed. The alluvial fan to the west of the Hancock Moraine at the mouth of this tunnel channel is almost nonexistent, indicating that this tunnel never carried very much sediment. It is clear, however, that the large kettle that the trail passes through west of Seventh Ave. is in a tunnel channel that formed when ice was at the Hancock Moraine (fig. 191).

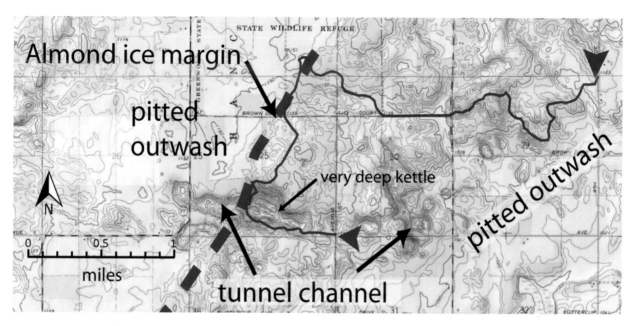

Figure 191. Topography of the Greenwood Segment. Dashed blue line indicates the front of the Almond Moraine. Note that the deep kettle in the path of the tunnel channel extends west of the moraine. This indicates that the tunnel channel was active during building of the Hancock Moraine, but that the buried ice that filled the tunnel channel after water flow stopped did not melt until after the glacier had withdrawn from the Almond Moraine. See figure 189 for a broader view. (Map is part of the Richford USGS Quadrangle and was created with TOPO! © 2011 National Geographic Maps.)

After emerging from the large kettle, the trail runs parallel to the front of the Almond Moraine for about a mile before turning eastward across the moraine and crossing pitted outwash close to its terminus at Ninth Ave.

## 75. Bohn Lake Segment

*Ninth Ave. to Ninth Ln. (1.2 miles)*

On this relatively short segment of the IAT, the trail goes through a well-developed tunnel channel (SB 17) that runs more or less east-west across pitted outwash (SB 10), the Almond Moraine (SB 6), and more pitted outwash to the west of the Almond Moraine (fig. 192). The trail descends the east side of the deep kettle that contains Bohn Lake, passes around the north side of the lake, then climbs out of the tunnel channel to its Ninth Ln. terminus.

Many kettle lakes vary considerably in extent and water level over the years, and even decades, depending on rainfall. This is because many have no stream inlet or outlet and have a water level the same as the groundwater level. Therefore, you can tell if the groundwater level is up or down, because the lake levels directly reflect this. A combination of dry years

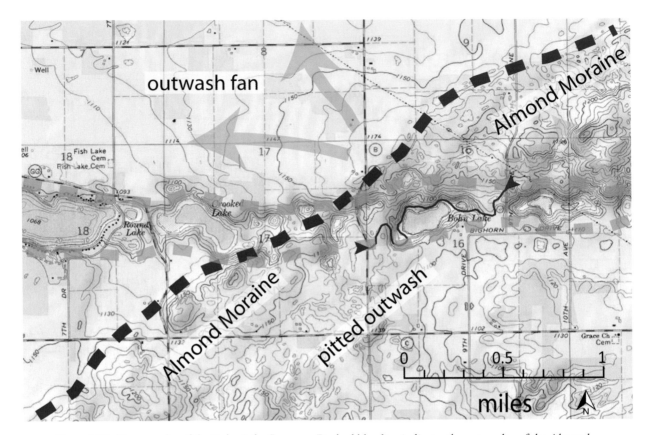

Figure 192. Topography of the Bohn Lake Segment. Dashed blue line indicates the outer edge of the Almond Moraine. Dashed green lines indicate the sides of a tunnel channel. Water flowed to the west in the channel, up the regional slope. Green arrows indicate meltwater-flow direction on a large outwash fan built in front of the Almond Moraine. Blue arrow shows ice-flow direction. (Map is part of the Plainfield and Richford USGS Quadrangles and was created with TOPO! © 2011 National Geographic Maps.)

and pumping of groundwater for large-scale irrigation to the west has lowered the water table in recent years, resulting in low lake levels and exposing an extensive flat lake bottom. The impacts of combined climate- and human-caused water loss are visible on the lake bottoms, where native wetland and forest vegetation is reclaiming the lake (fig. 193). We tend to think of water shortages and loss of water habitat as occurring in deserts, but it can happen anywhere and lead to landscape-level changes such as this if conditions persist.

## 76. Deerfield Segment

*Beechnut Dr. to CTH O (3.7 miles)*

From Beechnut Dr., the Deerfield Segment drops rapidly downward into a deep kettle (SB 9) at the bottom of a tunnel channel (SB 17; fig. 194). This is right at the outer edge of the

water in April 2009          former shoreline

Figure 193.  Part of the bottom of Bohn Lake kettle showing substantially lower water level than has been the norm through the late 1990s. The IAT borders the lake just uphill from the former shoreline.

Almond Moraine. You might notice a small, very low outwash fan in front of the tunnel channel. There is no trace of the tunnel channel continuing westward to the Hancock Moraine (SB 6). Outwash may have covered the tunnel channel while the glacier was sitting at the Almond Moraine. After crossing the tunnel channel, the trail runs along the crest of the Almond Moraine to 12th Ave., where it descends onto pitted outwash. The hummocks (SB 11) that the trail crosses about a quarter mile before its terminus on CTH O are likely underlain by till (SB 5) released from the glacier as underlying ice melted out.

## 77. Belmont–Emmons–Hartman Creek Segment
*Second Ave. to STH 54 (7.6 miles)*

Unlike trail segments just to the south, this segment is on terrain formed as the massive glacier retreated down the regional slope (fig. 195). The entire hiking area is characterized

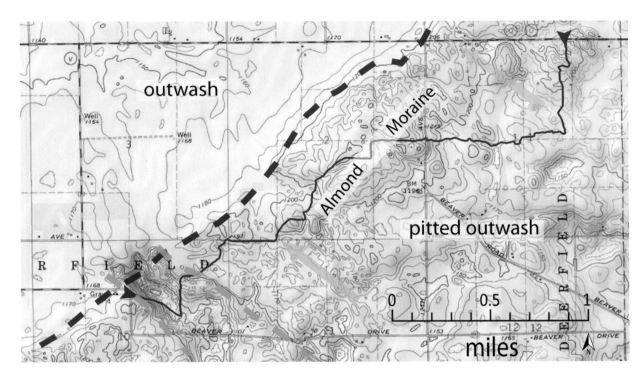

Figure 194. Topography of the Deerfield Segment. Dashed blue line indicates the outer edge of the Almond Moraine. Dashed green lines indicate the sides of a tunnel channel. Water flowed west in the channel. Blue arrows show ice-flow direction. (Map is part of the Plainfield and Wautoma Northeast USGS Quadrangles and was created with TOPO! © 2011 National Geographic Maps.)

by a series of steps called outwash heads (SB 8, 10), which were created during pauses in that retreat. The steps differ from moraines (SB 6) because they are not composed of till (SB 5). Instead, they contain sand and gravel deposited in braided rivers flowing mostly southward along and on top of the ice margin. You can see a diagram of the stepped topography in figure 196. Most of the kettles in the pitted outwash formed as buried ice melted out from beneath the sand-and-gravel layer.

From the Second Ave. terminus, the trail rises out of a narrow creek valley onto a pitted-outwash surface (SB 10; fig. 195). Note that north of Second Ave. the trail is on land that is substantially higher than the areas to the east. Just to the east of the trail is a steep ice-contact slope (SB 10) of one of these outwash heads. The glacier edge sat approximately where the trail is now, and outwash flowed westward and southward across buried glacier ice. When the glacier melted back from this position, it left a substantially lower land surface east of here. About a half mile north of Hartman Lake, the trail will descend the ice-contact face of the easternmost outwash head. After crossing Allen Creek, it will travel over several hills

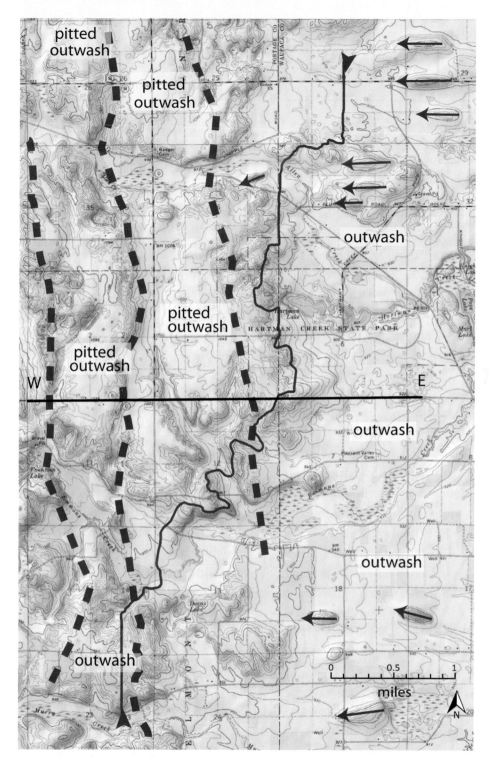

Figure 195. Topography of the Belmont–Emmons–Hartman Creek Segment. Dashed blue lines show former ice-margin positions. Ice-contact slopes lie just to the east of the dashed lines. Blue arrows are on drumlins and show ice-flow direction. W–E black line is the location of the profile shown in figure 196. (Map is part of the King USGS Quadrangle and was created with TOPO! © 2011 National Geographic Maps.)

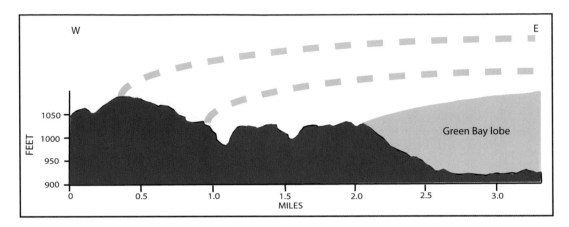

Figure 196.  Profile of the land surface along the W–E line shown in figure 195. Solid blue shows where the glacier was when the easternmost outwash head shown in figure 195 formed. This IAT segment more or less follows this ice-margin position throughout most of its length. Dashed blue lines indicate earlier glacier surfaces. (Profile created with TOPO! © 2011 National Geographic Maps.)

that appear to be composed of sand and gravel. The trail levels out and crosses about a half mile of nearly flat outwash before its STH 54 terminus.

Because of the lack of true end moraines, it is difficult to trace individual ice-margin positions over any significant distance. The ice margins shown are part of the Elderon phase of deglaciation, but no one has named the individual ice margins. Notice on figure 195 the distinctly different landscape west of the trail, where pitted outwash piled up, and east of the trail, where there was relatively little sediment deposited from the ice surface as the ice retreated (SB 11). Instead, drumlins (SB 14), which formed at the glacier bed, are scattered across the surface. Numerous springs trickle to the surface along the east side of the pitted outwash. The springs feed Hartman Lake and the east-flowing creeks in this area.

## 78. Waupaca River Segment
*STH 54 to Grenlie Rd. (6.2 miles)*

This segment begins on STH 54 but is then discontinuous or on roads to the Waupaca River at Cobb Town (fig. 197). Where the trail starts, it immediately crosses pitted outwash (SB 10). Drumlins (SB 14) rise just to the east; look for a spectacularly shaped drumlin that STH 54 crosses about a half mile east of the trail head. North of Cobb Town the trail follows the floodplain of the Waupaca River before rising onto a pitted-outwash surface. The steep slope to the east of the trail down to Foley Rd. is an ice-contact face of an outwash head (SB 10). When the glacier sat here, braided streams flowing away from the ice and then southward along the ice margin deposited sand and gravel outwash (SB 8), which then partly collapsed as buried ice melted out.

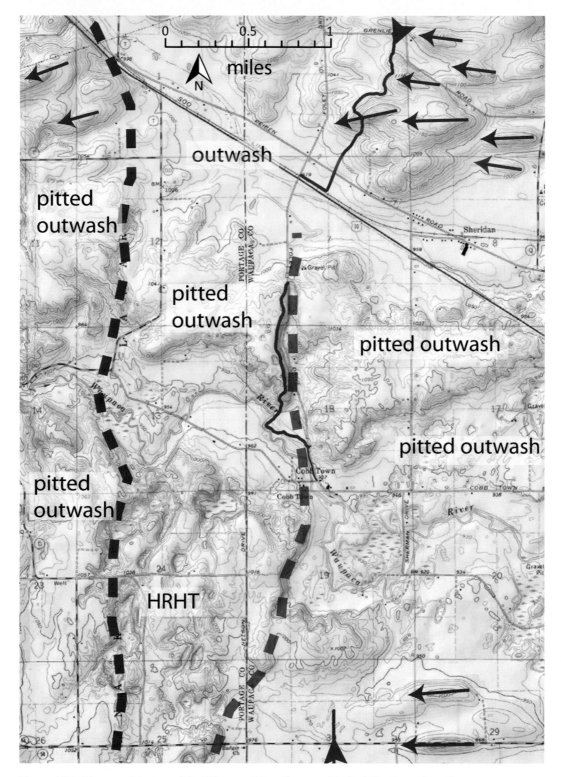

Figure 197. Topographic map of the Waupaca River Segment. Dashed blue lines show former ice-margin positions. Blue arrows are on drumlins and show ice-flow direction. HRHT: high-relief hummocky topography SB 11). (Map is part of the King and Scandinavia USGS Quadrangles and was created with TOPO! © 2011 National Geographic Maps.)

After the trail descends to Foley Rd. (which runs north to STH 10), it crosses outwash. Unlike other outwash traversed, this is not kettled and was deposited later than the pitted outwash encountered earlier in the trail. Look for the very small creek running through this valley: this is an underfit stream (SB 8), meaning that the present stream is too small to have cut the valley. When you see this, you know that the flow of water that cut this valley was substantially larger. This wide valley is the lower end of the large outwash channel shown on figure 198. By the time meltwater ran through this outwash channel, the ice edge had retreated about 2 miles eastward from figure 197, and a lake had formed between the ice and the higher land to the west. Water flowed toward the east in this channel and into that unnamed ice-dammed lake (SB 16). The Village of Sheridan, on STH 10 about a mile east of the IAT, sits on a delta built into this lake (fig. 197).

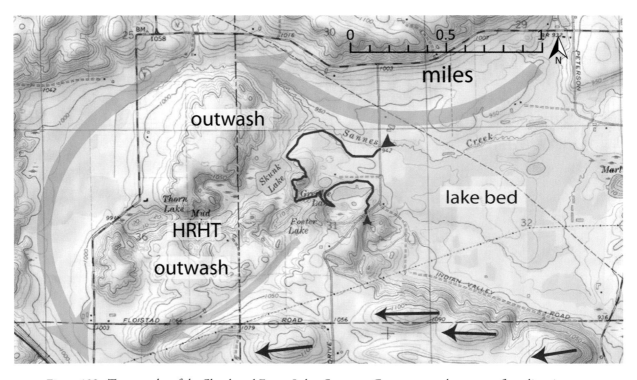

Figure 198. Topography of the Skunk and Foster Lakes Segment. Green arrows show water-flow direction in outwash channels. Light blue overprint shows location of glacier before it retreated to allow the lake to form in its place. Water flowing from the north was diverted to the west by that ice. Ice then retreated, leaving an ice-contact face north of Sannes Creek. Blue arrows are on drumlins and show ice-flow direction. HRHT: high-relief hummocky topography (SB 11). (Map is part of the Scandinavia USGS Quadrangle and was created with TOPO! © 2011 National Geographic Maps.)

## 79. Skunk and Foster Lakes Segment

*N. Foley Rd., south trailhead to N. Foley Rd., north trailhead (2 miles)*

On this trail segment (fig. 198), which starts and ends on Foley Rd., you will hike through spectacular high-relief hummocky topography (SB 11) and pitted outwash (SB 10). All of the lakes sit in deep kettles (SB 9), and a large meltwater channel wraps around the area on the west and north. When this hummocky area was still gravel over buried ice and higher than it is today, ice-marginal lakes to the north fed meltwater into the roaring channel that skirted this ice mass. Glacier ice also sat in the area of figure 198 where you see a light blue color. The ice diverted the meltwater and prevented the low area from being filled with the sand and gravel carried in the meltwater channel. Just north of Sannes Creek you can see a steep ice-contact slope (SB 10) from the floor of the channel down to the lakebed, where the sand and gravel in the channel piled up against the frozen wall of the glacier that occupied the future lakebed. As the ice continued to retreat, the channel was abandoned and a lake formed. Note also the abundance of Wolf River granite boulders.

## 80. New Hope–Iola Ski Hill Segment

*CTH MM to Sunset Lake Rd. (5.5 miles)*

On this segment (fig. 199) are excellent examples of high-relief hummocky topography (SB 11) throughout much of the trail. The stretch between the ski jump and Stolenberg Rd. is all on this extreme topography, with scattered boulders of Wolf River granite dotting the landscape. Technically, these are not erratics (SB 15) because Wolf River granite is the local bedrock. Many of these boulders remained on the ice surface near the glacier edge, and they were dropped as it melted back.

In addition to seeing thousands of boulders, you'll also be able to deduce how extremely thick the deposits of sand and gravel that once coated the surface of the ice were. To do this, consider the depth of kettle lakes, which approximates the minimum thickness of the debris cover on the ice. In this case, the layer of sand, gravel, and till (SB 5) was over 100 feet thick. The two ice-margin positions shown do not represent major readvances of the ice and, like most ice margins in the Elderon series, are discontinuous and hard to trace. North of Budsburg Lake, the trail climbs onto a pitted-outwash surface. This sand and gravel was deposited when the ice edge was at the westernmost position shown in figure 199. About a half mile south of CTH Z, the IAT again enters high-relief hummocky topography and remains in this dramatic landscape. Look here, too, for those abundant granite boulders.

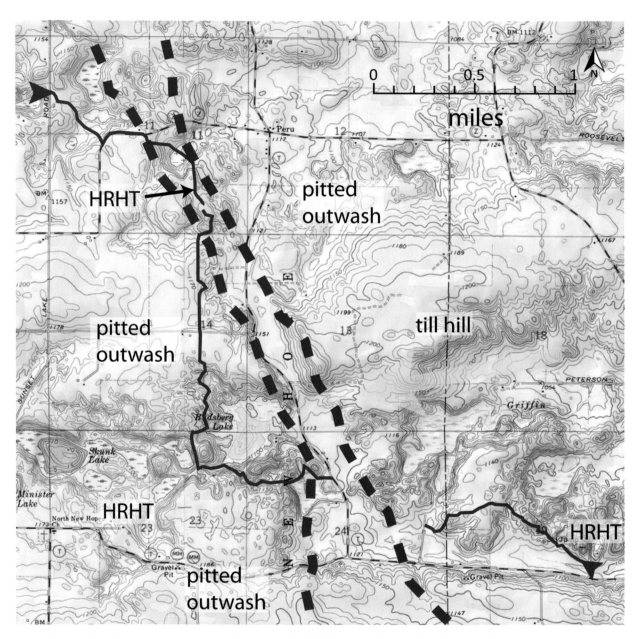

Figure 199. Topography of the New Hope–Iola Ski Hill Segment. Dashed blue lines show former ice-margin positions. Blue arrow shows ice-flow direction. HRHT: high-relief hummocky topography (SB 11). (Map is part of the Iola and New Hope USGS Quadrangles and was created with TOPO! © 2011 National Geographic Maps.)

# Northern Green Bay Lobe and
# Langlade Lobe Ice Age Trail Segments

THIS GROUP OF EIGHT TRAIL SEGMENTS follows end moraines (SB 6) of the Green Bay Lobe (figs. 1, 3), which flowed toward the northwest from the Green Bay lowland, and of the Langlade Lobe, which flowed southwestward out of the Lake Superior basin (figs. 200, 201). The western end of these segments is at the boundary between the Langlade Lobe and the Wisconsin Valley Lobe, which flowed southeastward into this area. From the eastern edge of the Langlade Lobe westward (fig. 201), the moraines are characterized by high-relief hummocky topography (SB 10). There are some tunnel channels (SB 17), but they are not as abundant as they are along the western side of the Green Bay Lobe. Conversely, ice-walled-lake plains (SB 15) are a more common landform in the moraines across northern Wisconsin than they are to the south. Extensive peat bogs are also more frequent in the moraines across northern Wisconsin than they are to the south, partly because of climate and partly because of the more acidic nature of the soils formed on the glacial deposits of the northern lobes.

The flow direction of the Green Bay and Langlade lobes is indicated by the orientation of drumlins (SB 8) that formed behind the ice edge (fig. 200). Although there are no precise dates on when these lobes advanced to their maximum positions, relative ages are distinguishable at the junctions of the lobes. Figure 201 shows the crosscutting relationships between various ice margins in this area where the three lobes come together. The Hancock Moraine, the outermost end moraine of the Green Bay Lobe in the area south of Antigo, appears to be overridden by the Almond Moraine. The present interpretation is that the Almond Moraine is the outermost moraine from there northeastward to where Green Bay Lobe deposits are covered by Langlade Lobe deposits. The maximum advance of the Langlade

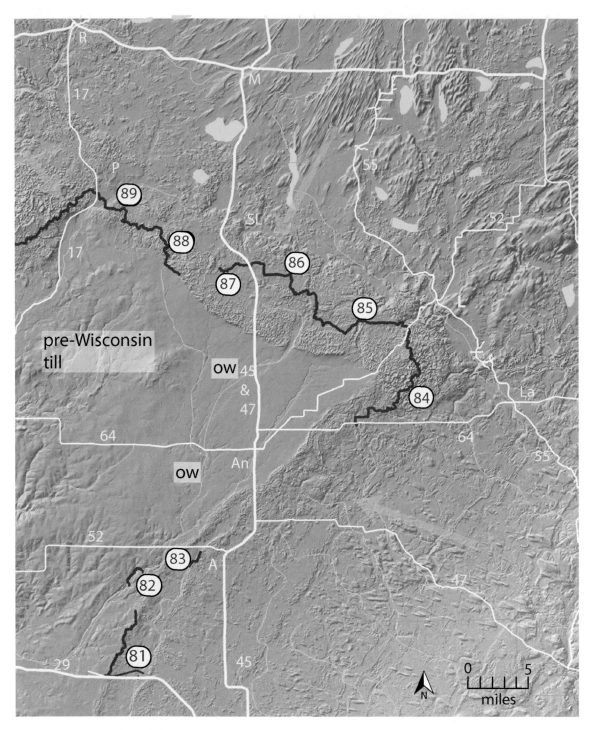

Figure 200. Shaded relief of the northern Green Bay Lobe and Langlade Lobe IAT segments (red): (81) Ringle, (82) Eau Claire Dells, (83) Plover River, (84) Kettlebowl, (85) Lumbercamp, (86) Old Railroad, (87) Highland Lakes Eastern, (88) Highland Lakes Western, (89) Parrish Hills. Blue arrows show ice-flow direction. Yellow lines and numbers indicate major highways. Cities and villages shown (yellow): (An) Antigo, (A) Aniwa, (C) Crandon, (L) Lily, (La) Langlade, (M) Monico, (P) Parrish, (R) Rhinelander, (SL) Summit Lake. Names of geologic features are shown on figure 201. (Base map constructed from USGS National Elevation Dataset and modified by WGNHS.)

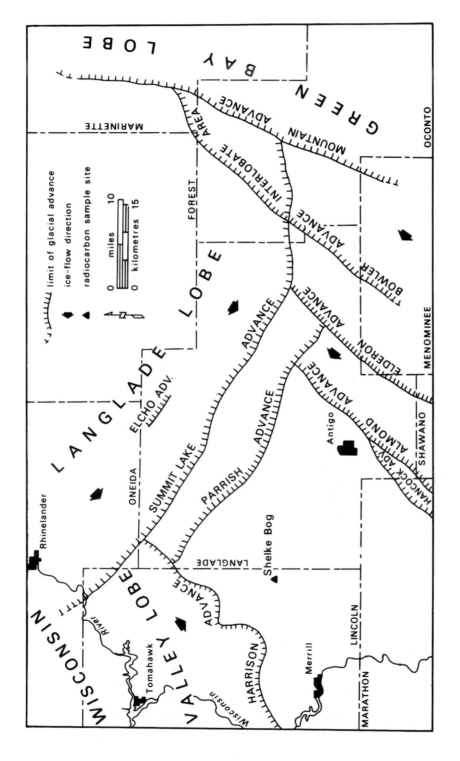

Figure 201. Extent of late Wisconsin glacial advance in north-central Wisconsin. The Green Bay Lobe ice flowed into the area from the southeast, the Langlade Lobe ice flowed from the northeast, and the Wisconsin Valley Lobe ice flowed from the north and northwest. Compare the ice-margin positions with the shaded-relief of figure 200. (From Mickelson 1986.)

Lobe built the Parrish Moraine, which overrides the Almond Moraine (fig. 201), so the Parrish Moraine is somewhat younger than the Almond Moraine.

The Green Bay Lobe advanced into this area across Paleozoic dolomite and sandstone before flowing over Precambrian rock. As much as 50 percent of the pebbles in the Green Bay Lobe till are dolomite, whereas dolomite pebbles are almost absent in the till and outwash of the Langlade Lobe. Both lobes picked up boulders of Wolf River granite (discussed in the introduction to the chapter on the western Green Bay Lobe segments), so both areas are characterized by many granite boulders at their surfaces.

When the glaciers sat at the Almond and Parrish moraines, drainage was restricted by uplands to the west, and outwash streams flowed southward into what is now the Eau Claire River and through the Eau Claire Dells. Outwash more than 100 feet thick filled the area between the two moraines. This large outwash plain is called the Antigo "flats" (figs. 201, 202). At that time, the land was still mostly free of vegetation, and silt-sized (SB 21) particles were picked up and blown across the flats. Eventually several feet of this loess (SB 8) accumulated here. This fine-grained sediment holds moisture and, when draped over well-drained sand and gravel, makes excellent farmland. Cultivation for potatoes and other crops is widespread. In fact, Antigo silt loam, the soil covering much of the Antigo flats, is the state soil of Wisconsin. The thick sand and gravel also contains a large amount of groundwater that is readily available for irrigation.

The area to the west of the Antigo flats is bedrock topography covered by older, pre–late Wisconsin, glacial deposits (fig. 200), and this land is not nearly as productive for row crops as is the loess-covered sand and gravel.

Much of the Ringle Segment parallels the Hancock Moraine and shows excellent end-moraine topography (fig. 200). Immediately to the west of the Hancock Moraine, the Eau Claire Dells Segment shows the effects of glacial meltwater as it drained from the ice margins in the Antigo area and cut through Precambrian granite and rhyolite to create the Dells of the Eau Claire River. All of the remaining segments are on deposits of the Langlade Lobe. The Kettlebowl, Highland Lakes Western, and Parrish Hills segments are all in the Parrish Moraine and show high-relief hummocky topography. The Lumbercamp, Old Railroad, and Highland Lakes Eastern segments are mostly on pitted outwash behind the Parrish Moraine and in front of a recessional moraine called the Summit Lake Moraine (fig. 200).

## 81. Ringle Segment

*CTH Y (Curtis Ave. in Hatley) to Helf Rd. (9.3 miles)*

This segment of the IAT is in a geologically complex setting (figs. 200, 203). The Hancock Moraine (SB 6) lies a little more than 2 miles west of the trailhead in Hatley. The Almond Moraine lies a half mile to the east. Where STH 29 and the Mountain-Bay State Trail cross

Figure 202. Low oblique aerial view looking toward the southeast across the Antigo flats. The Parrish Moraine, where it is cut by the East Branch of the Eau Claire River, is in the foreground. The Almond Moraine of the Green Bay Lobe is in the distance. CTH A is the road just beyond the river.

through the Almond Moraine just east of Hatley, they are in a tunnel channel (SB 17). The IAT crosses through the Hancock Moraine in the same tunnel channel. At the trailhead in Hatley, the IAT is on a pitted outwash (SB 10) plain dissected by the Plover River. Although there are kettles interrupting the outwash surface, it is otherwise nearly smooth and slopes gently toward the southwest. This outwash was deposited by an ice-marginal braided river flowing away from the ice edge and toward the southwest when the glacier was at the Almond Moraine. This outwash, and later river, channel erased the evidence of the tunnel channel between the two moraines.

From the trailhead, the trail remains on pitted outwash for about 2 miles before climbing into the Hancock Moraine. At the east edge of the Hancock Moraine, the IAT leaves the Mountain-Bay State Trail and turns north, climbing back into the Hancock Moraine. Outwash (SB 8) lies to the west. The moraine has hummocky topography (SB 11), but it does

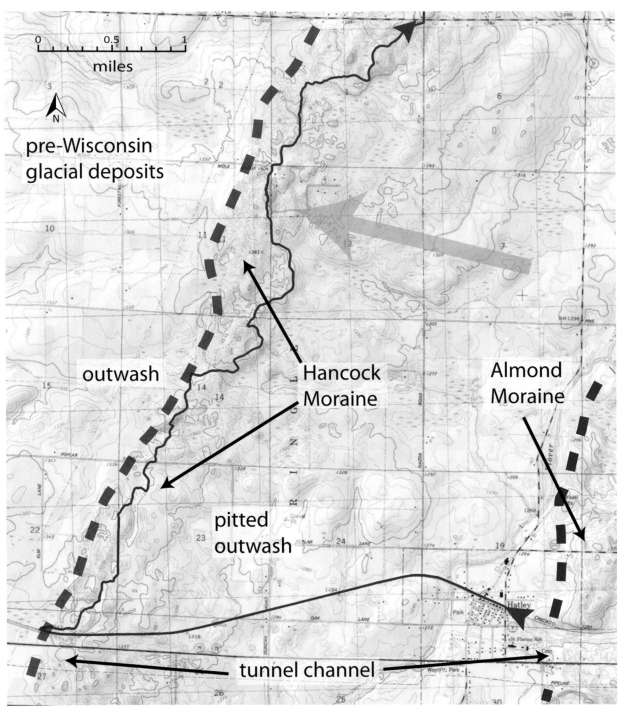

Figure 203. Topography of the Ringle Segment. Dashed blue lines indicate the outer edges of the Hancock and Almond moraines. Blue arrow shows ice-flow direction. Except where it crosses the Almond and Hancock moraines, the tunnel channel is not visible because of later outwash deposition. (Map is part of the Ringle and Hogarty USGS Quadrangles and was created with TOPO! © 2011 National Geographic Maps.)

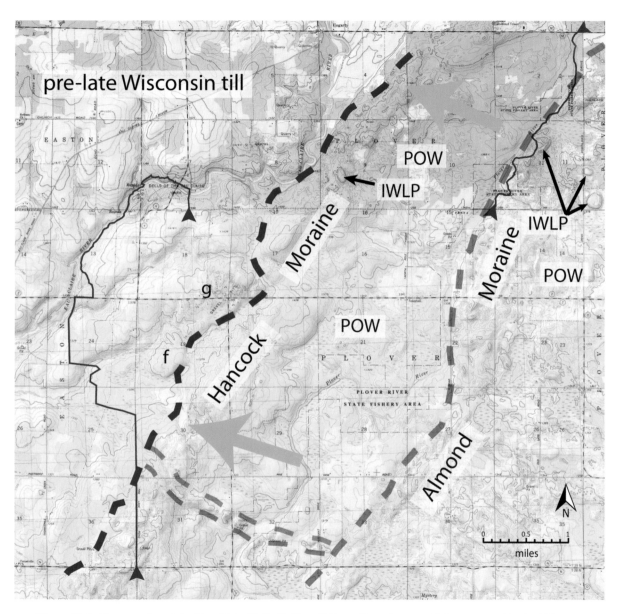

Figure 204. Topography of the Eau Claire Dells (left) and Plover River (upper right) segments. Dashed blue line shows outermost edge of the Hancock Moraine. Blue arrow shows ice-flow direction. The dells are near the northern section of the trail, at the Dells of the Eau Claire County Park. (Map is part of the Ringle and Hogarty USGS Quadrangles and was created with TOPO! © 2011 National Geographic Maps.)

not have as much high-relief topography as places in the moraine to the south or north. Some of the flatter-topped hummocks are ice-walled-lake plains (SB 15). The Marathon County landfill is in the moraine, and sediment here is dominated by till and till-like sediment (SB 5). Boulders of Wolf River granite are abundant (see the introduction to the chapter on the western Green Bay Lobe segments). It appears that most hummocks here are composed of deposits accumulated from the ice surface as the ice beneath melted out. The trail remains in this high-relief hummocky topography all the way to the Helf Rd. trailhead.

## 82. Eau Claire Dells Segment
*Helf Rd. to Sportsman Dr. (7.5 miles)*

The Helf Rd. trailhead is right in the Hancock Moraine (SB 6; fig. 204). Most of the sediment is collapsed supraglacial sediment, and it forms hummocky topography (SB 11). Many of the boulders are Wolf River granite. Where the trail leaves the Hancock Moraine on Fire Lane Rd., there is a small area of pitted outwash, and from there the trail is on thin pre–Wisconsin-age glacial deposits. These sediments have not been mapped in detail, but they are composed of till, sand, and gravel deposited well before the last glaciation. It seems likely, based on the amount of weathering on these deposits, that they formed before the last interglacial period, about 130,000 years ago. Most of the depressions on topographic maps that are labeled "gravel pit" in this area are actually in weathered granite (called "rotten" granite locally). The weathering of this granite took place well before the last glaciation and is still taking place today. See the discussion in the chapter on the western Green Bay Lobe segments.

The Dells of the Eau Claire River is the narrow channel that water cut through granite and rhyolite (fig. 205). The channel was not covered by ice during the last glaciation, but it was eroded by southwest-flowing glacial meltwater when the glacier sat east of here at the Hancock Moraine. All of the water draining onto the outwash plain of the Antigo flats collected and drained through this valley, cutting the gorge. There are some potholes similar to those found at Interstate State Park, at the western terminus of the IAT, but the Eau Claire River potholes are not nearly as large, because they were not formed by a giant torrent of water. In the modern river (fig. 205), smaller potholes are still forming.

## 83. Plover River Segment
*Sportsman Dr. to STH 52 (2.7 miles)*

The Plover River Segment (fig. 204) follows on and off the front of the Almond Moraine for its whole length. At the Sportsman Dr. trailhead the Almond Moraine is to the east and the Plover River flows on pitted outwash (SB 10) just to the west. The relief in the moraine (SB 10) is quite low here, but it increases farther to the north. Most boulders on the surface are Wolf River granite. An almost symmetrical ice-walled-lake plain (SB 15) lies just north

Figure 205. Dells of the Eau Claire River. Channel is cut into granitic rocks. (Photo by Vin Mickelson.)

of Sportsman Dr., about 1 mile to the east of the trailhead (fig. 204). About 2.5 miles from the trailhead, the IAT crosses a small tributary to the Plover River and climbs into higher relief end moraine topography. Part of this topography is produced by collapsed ice-walled-lake plains. An arrow in figure 204 points to the rim of one of these low-relief ice-walled-lake plains. There is a break in the moraine that was produced by meltwater as ice receded from the Almond Moraine. From here to the STH 52 trailhead the trail follows the Plover River on outwash.

## 84. Kettlebowl Segment
*Sherry Rd. to STH 52 (9.5 miles)*

The Sherry Rd. trailhead is on pitted outwash (SB 10) that was deposited by braided streams in front of the Parrish Moraine (SB 6). The trailhead is right at the outer edge of Langlade Lobe deposits. To the south and west are deposits of the Green Bay Lobe (fig. 200).

Just east of Sherry Rd. and a short distance south of the trailhead is a very large boulder. This is one of the locally derived Wolf River granite boulders that are so abundant in this part of Wisconsin. There are many more along this segment of the IAT. A short distance north of Sherry Rd., the trail climbs into the Parrish Moraine (figs. 200, 206). From here northward for several miles it winds around and over hummocks and kettles. This is classic high-relief hummocky topography (SB 11), which typifies the outermost late-Wisconsin moraine across much of northern Wisconsin (fig. 207). All of the depressions are kettles (SB 9).

Local relief along the IAT in the moraine is more than 80 feet, indicating that the debris layer on top of glacier ice was at least that thick. For several thousand years after the Langlade Lobe retreated from this moraine, ice beneath the thick accumulation of rock, sand, and gravel slowly melted out in an irregular pattern. Figure 37 shows what this area probably looked like during the meltout. The sediment on the glacier surface slipped and flowed to low points, protecting the underlying ice from incoming solar radiation, in contrast to what was happening in the surrounding areas. Because the debris-covered ice melted more slowly, it eventually stood above the surrounding cleaner ice that melted faster. More slipping and sliding of the wet sediment off the high spots concentrated debris in the low area surrounding it, allowing the high spots of ice to melt. This process is called a reversal of topography. What was a high spot on the debris-covered ice becomes a kettle, and what was a low area on the debris-covered ice becomes a hummock (fig. 38).

Kent fire tower is on the highest hummock in the moraine (fig. 206) and at 1,903 feet is the highest point in Langlade County. It is about a half mile east of the IAT, and is approached by a trail that splits off the IAT. Approximately a half mile northwest of that trail junction, the IAT leaves the moraine and moves onto a pitted-outwash plain. On the plain is Big Stone Hole, an elongate kettle in which a large concentration of boulders accumulated as the buried ice melted out (fig. 206). Many of these and the large boulders along the trail are Wolf River granite.

The IAT climbs into the Summit Lake Moraine, where it crosses Kent Fire Tower Rd. This moraine also has high-relief hummocky topography, but the kettles are not as deep. Their shallowness suggests that there was less debris on the ice here than at the Parrish Moraine. The ski slope is on a high hummock in the Summit Lake Moraine.

## 85. Lumbercamp Segment
*STH 52 to CTH A (12 miles)*

For almost a mile west of STH 52, the IAT traverses hummocky topography (SB 11) near the front of the Summit Lake Moraine (SB 6; figs. 200, 208). Baker Lake is one of a series of kettles (SB 9) that may represent the path of a subglacial channel when ice sat in this position. About a quarter mile west of Baker Lake, the trail drops out of the moraine onto a

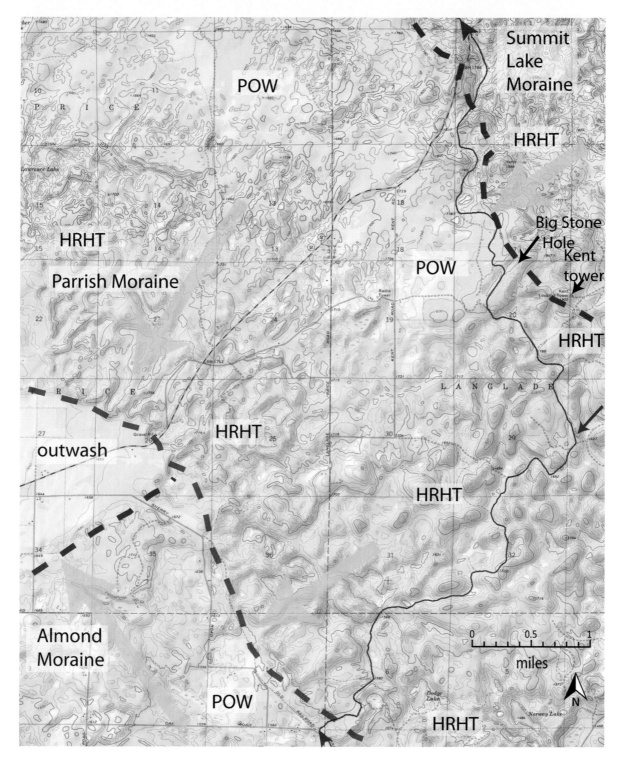

Figure 206. Topography of the Kettlebowl Segment. Dashed blue lines indicate the outer edge of the Parrish and Summit Lake moraines of the Langlade Lobe and the Almond Moraine of the Green Bay Lobe. Blue arrows show ice-flow direction. Red arrow indicates location and orientation of figure 207. HRHT: high-relief hummocky topography (SB 11); POW: pitted outwash (SB 10). (Map is part of the Polar and Pickerel USGS Quadrangles and was created with TOPO! © 2011 National Geographic Maps.)

Figure 207. Low oblique aerial view of high-relief hummocky topography in the Parrish Moraine. Red arrows point to the Kettlebowl Segment. Location is shown in figure 206.

pitted-outwash plain (SB 10). This nearly level surface, which is interrupted by relatively few kettles, continues for several miles to the west. It was deposited by braided streams (SB 8) flowing away from the glacier when the edge of the ice was at the Summit Lake Moraine. Where the trail crosses CTH S, it skirts the northern edge of a till (SB 5) surface with ridges aligned a little east of north to a little west of south (fig. 209). These ridges were shaped by sliding ice at the bottom of the glacier when the ice edge sat at the Parrish Moraine. The ridges are probably too small to be considered drumlins (SB 14), but they do indicate the direction of the ice flow. There are also many kettles in this complex topography.

In about a quarter mile, the trail drops out of this topography back onto pitted outwash and remains on this surface through the Peters Marsh State Wildlife Area. This outwash was deposited by streams flowing from the Summit Lake Moraine toward the southwest.

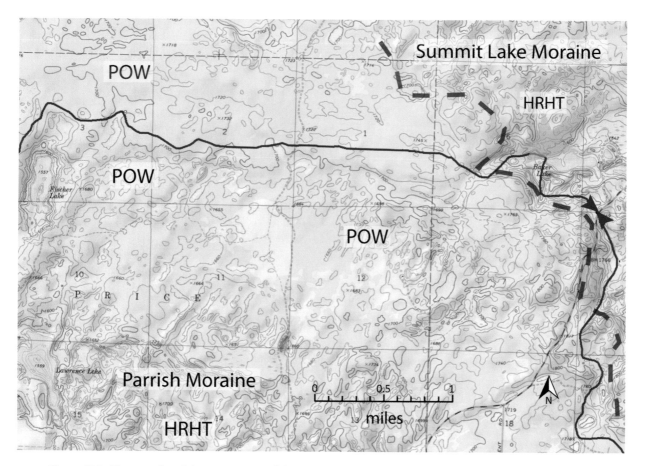

Figure 208.  Topography of the eastern part of the Lumbercamp Segment (lower right) and the northern part of the Kettlebowl Segment (top). Dashed blue line is outer edge of the Summit Lake Moraine. HRHT: high-relief hummocky topography (SB 11); POW: pitted outwash (SB 10). (Map is part of the Pickerel and Pearson USGS Quadrangles and was created with TOPO! © 2011 National Geographic Maps.)

Water collected into what is now the East Branch of the Eau Claire River and flowed through the Parrish Moraine, more or less following the present path of CTH A (fig. 202). Springs are abundant in this area because the groundwater system is fed by rainwater percolating into the sand and gravel east and north of here. The water in Peters Marsh is an expression of the water table, which is at or slightly above the land surface.

## 86. Old Railroad Segment

*CTH A to USH 45 (9.5 miles)*

From the CTH A trailhead the IAT crosses pitted outwash (SB 10) just west of Ventor Lake and Upper Ventor Lake (fig. 210). Peat bogs are extensive throughout the area between the

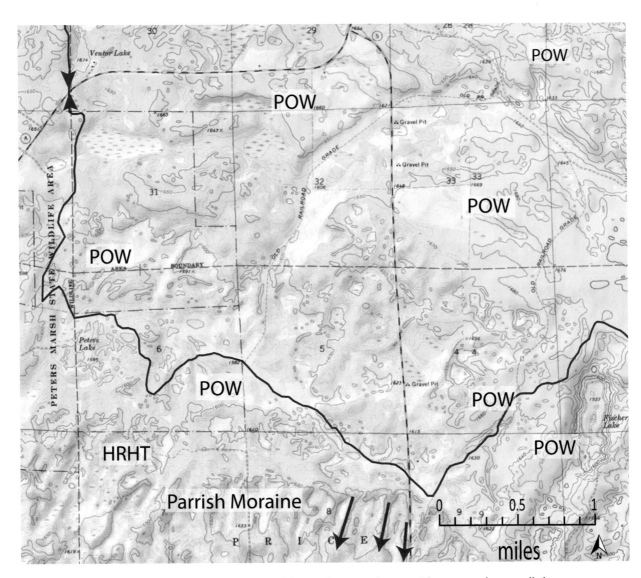

Figure 209. Topography of the western part of the Lumbercamp Segment. Blue arrows show small elongate ridges parallel to ice flow. HRHT: high-relief hummocky topography (SB 11); POW: pitted outwash (SB 10). (Map is part of the Pickerel and Pearson USGS Quadrangles and was created with TOPO! © 2011 National Geographic Maps.)

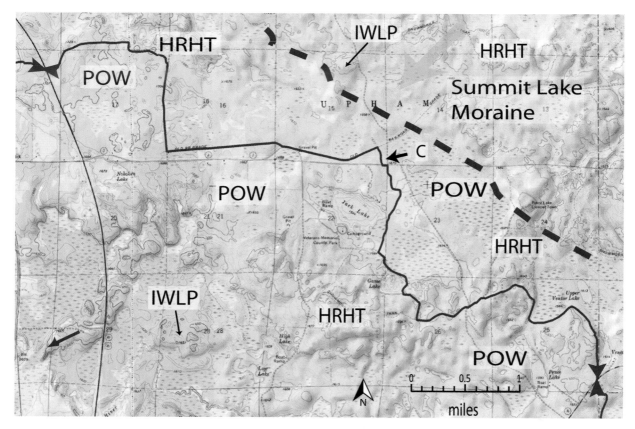

Figure 210.   Topography of the Old Railroad Segment. Dashed blue line indicates the outer edge of the Summit Lake Moraine. Light blue arrow shows ice-flow direction. Dark blue arrow near lower left-hand corner of map marks what appears to be an elongate ridge (a drumlin? SB 14) shaped by flowing ice. C: Clearwater Stone Hole; HRHT: high-relief hummocky topography (SB 11); IWLP: ice-walled-lake plain (SB 15); POW: pitted outwash (SB 10). (Map is part of the Pearson and Kempster USGS Quadrangles and was created with TOPO! © 2011 National Geographic Maps.)

Summit Lake and Parrish moraines. Although deep kettles (SB 9) are abundant in the pitted outwash, many broad areas of peat, such as the one skirted by the trail west of Upper Ventor Lake, are probably shallow blanket bogs. These form where the water table is very close to or even at the surface. Once the bog is formed, it helps retain a high water table. The IAT then crosses high-relief hummocky till surface (SB 11) and deep kettles (SB 9) before passing Game Lake and entering Veterans Memorial County Park. High, Low, Jack, and Game lakes are all kettles. Clearwater Stone Hole (C on fig. 210) is an interesting group of rocks. From what we can see, it looks like it was produced by human activity. Perhaps these boulders were piled here when the pine trees were planted. Note the shallow furrows that indicate the area had been plowed before the pines were planted. The trail remains on pitted outwash to the trailhead on USH 45. The Summit Lake Moraine is to the north and the Parrish Moraine is to the south.

## 87. Highland Lakes Eastern Segment

*USH 45 to Forest Rd. (3.1 miles)*

This segment of the IAT crosses pitted outwash (SB 10) and skirts around the north side of a peat bog just west of an old railroad grade and CTH B (fig. 211). From there, it enters high-relief hummocky topography (SB 11). Notice that there are quite a few boulders on this surface, which is underlain by till and supraglacial debris (SB 5) that was deposited slowly as debris-covered ice melted away. Bogus Swamp, which lies about a quarter mile south of the IAT, is a large kettle filled with peat (fig. 211). Around the kettle you can see ridges of debris that sloughed off the mass of ice as it melted to produce the depression now occupied by Bogus Swamp. Smaller blocks of ice that melted out, creating Susan Lake and other kettles nearby, also shed sediment that piled up as ridges in the low areas between the

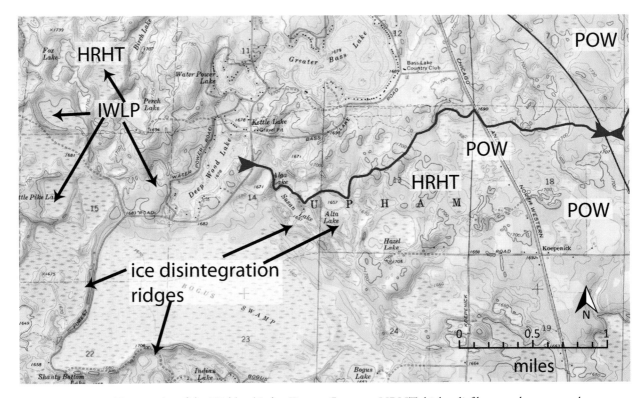

Figure 211. Topography of the Highland Lakes Eastern Segment. HRHT: high-relief hummocky topography (SB 11); IWLP: ice-walled-lake plain (SB 15); POW: pitted outwash (SB 10). (Map is part of the Kempster USGS Quadrangle and was created with TOPO! © 2011 National Geographic Maps.)

ice blocks. After the ice was gone, these ridges were left behind. The northwest-southeast-trending ridges, including the one that the IAT follows along the north side of Susan Lake, formed in this way (fig. 211). They are composed mostly of till-like sediment, not sand and gravel. After leaving this trail segment, if you follow Forest Road to the south around Bogus Swamp, you will be on another of these disintegration ridges. The depression occupied by Bogus Swamp, as well as some of the other basins, may well be the remains of short-lived ice-walled lakes (SB 15) that did not fill completely with sediment.

## 88. Highland Lakes Western Segment

*Kleever Rd. at Lowells Rd. to CTH T (5.8 miles)*

This segment features the highest outer moraine ridge anywhere along the IAT (fig. 212). In addition to debris accumulating as it slid off the ice surface (SB 6), the glacier may also have been bulldozing sediment into this ridge. Figure 213 shows a cross section through and behind the moraine here. Starting from the left (west), the profile crosses the outwash plain (SB 8), an outwash apron, the outer Parrish Moraine ridge (SB 6), high-relief hummocks (SB 11), and an ice-walled-lake plain (SB 15). Note the difference between this profile and that of the Green Bay Lobe shown in figure 185. Unlike the Green Bay Lobe, the Langlade Lobe was not flowing up a slope, so the outwash streams could flow away from the ice margin instead of along it. Also, this ice margin had thicker sediment on top of the ice as deglaciation took place. Thus, the assemblage of landforms is quite different in the two moraines.

The Kleever Rd. trailhead is at the edge of an outwash fan built by water flowing toward the southwest through a break in the outermost ridge of the Parrish Moraine (fig. 212). The outer edge of the moraine here is a high, continuous ridge except where it is broken by melt-water drainage channels. The IAT parallels this ridge on a steep outwash surface called an outwash apron, which was deposited by braided streams flowing from the glacier edge. Where the trail crosses Second Ave. (also Kokomo Lane), there is another break in the moraine front. The West Branch of the Eau Claire River flows through the outermost moraine ridge here. This break in the moraine front was cut by flows of glacial meltwater much larger than the flow of the river today.

Although much of the flat area south and west of this point is now covered by thin peat, it is an outwash plain underlain by sand and gravel. The trail itself climbs into the moraine just west of the West Branch of the Eau Claire River. Several of the nearby flat-topped hills, including the two about a half mile east of the trail, are ice-walled-lake plains (SB 15; not all of them are labeled on fig. 212). From here to CTH T the trail crosses several low ridges that may be the edges of very low ice-walled-lake plains and other disintegration ridges in this high-relief hummocky topography. All of the hummocky topography in figure 212 is in the Parrish Moraine.

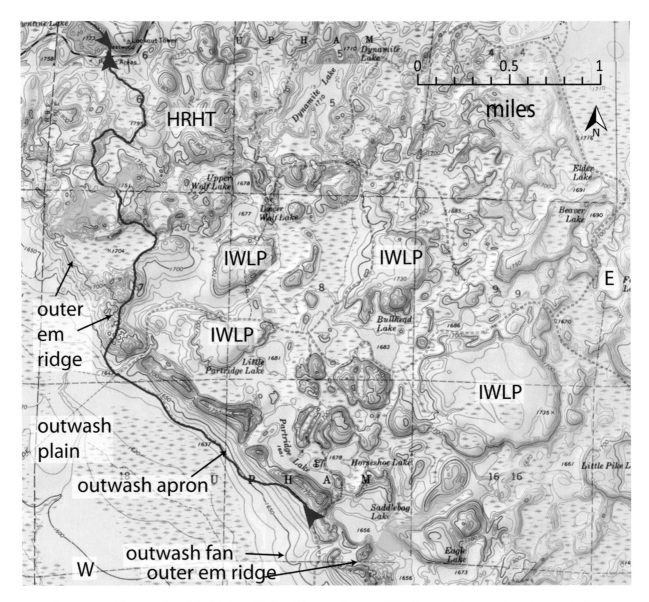

Figure 212. Topography of the Highland Lakes Western Segment. The distinct ridge near the eastern end of this segment is the outer edge of the Parrish Moraine. Blue arrows show ice-flow direction. HRHT: high-relief hummocky topography (SB 11); IWLP: ice-walled-lake plain (SB 15). The profile shown in figure 213 extends from points W to E on this map. (Map is part of the Bavaria and Enterprise USGS Quadrangles and was created with TOPO! © 2011 National Geographic Maps.)

## 89. Parrish Hills Segment

*CTH T to First Lake Rd. (12 miles)*

From the CTH T trailhead the IAT stays in high-relief hummocky topography (SB 11) all the way to the Prairie River (figs. 214, 215). This is in the Parrish Moraine (SB 6; fig. 200), which gets lower in relief as it approaches the Prairie River. This is likely because there was less sediment on the ice surface to be let down as the hummocky topography developed. Some of the ridges that separate depressions may be the rims of ice-walled-lake plains (SB 15), but the lake plains are not as high as they are near the east end of the Highland Lakes Western Segment. There is also a difference in the appearance of the moraine front here; it is so low that it is difficult to discern. Compare this moraine front with the one a few miles to the southeast, where a high-relief ridge forms the outer edge of the moraine (fig. 212). For some reason, there was less debris in and on the ice here.

The Prairie River carried outwash (SB 8) toward the southwest from the glacier when it was at the Summit Lake Moraine (fig. 200). The moraine to the west and southwest is the

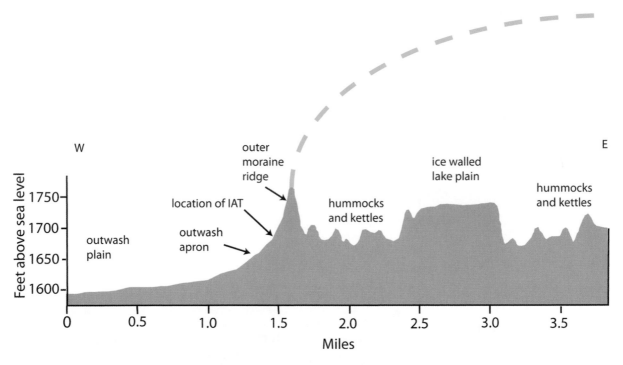

Figure 213. Profile from the outwash plain across the outwash apron, marginal moraine ridge, hummocks and kettles, and ice-walled-lake plain. Dashed blue line is approximate ice-surface profile at glacial maximum. The profile extends approximately between points W and E on figure 212. Contrast this with figure 185, the profile across the outermost Green Bay Lobe Moraine. (Profile created with TOPO! © 2011 National Geographic Maps.)

Harrison Moraine, which was built by the Wisconsin Valley Lobe. West of the river the trail climbs gently up an outwash fan (SB 8). Water that deposited the fan flowed southeast, away from the Wisconsin Valley Lobe. Baldy Hill, just above and west of the fan, is the outer edge of the Harrison Moraine. Note the large erratic boulders (SB 5) on the outer edge of the moraine, and the lack of them on the outwash fan. We know that the glacier sat at the Harrison Moraine after debris-covered ice had melted out of the Parrish Moraine at this location. Note that the fan does not have kettles in it. If the outwash fan had been built out over buried ice of the Parrish Moraine, later meltout of the ice would have caused the outwash fan to become pitted. Based on this reasoning, it seems likely that the Harrison Moraine was built at about the same time as the Summit Lake Moraine, but there are no radiocarbon dates that can determine just when the glacier was at its maximum extent in either lobe.

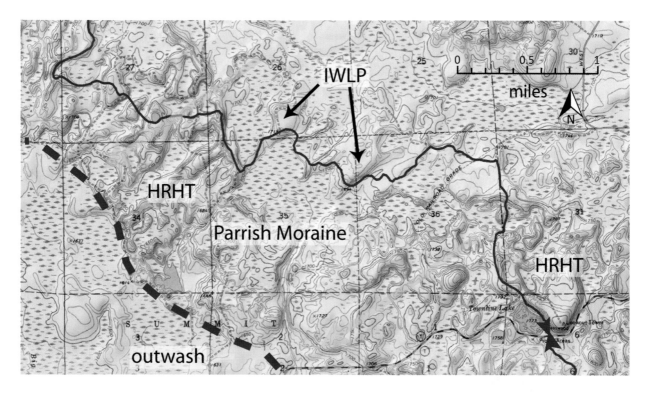

Figure 214. Topography of the eastern part of the Parrish Hills Segment. Blue arrows show ice-flow direction. Dashed blue lines show outer edge of the Parrish Moraine of the Langlade Lobe and the Harrison Moraine of the Wisconsin Valley Lobe. HRHT: high-relief hummocky topography (SB 11); IWLP: ice-walled-lake plain (SB 15). (Map is part of the Enterprise and Parrish USGS Quadrangles and was created with TOPO! © 2011 National Geographic Maps.)

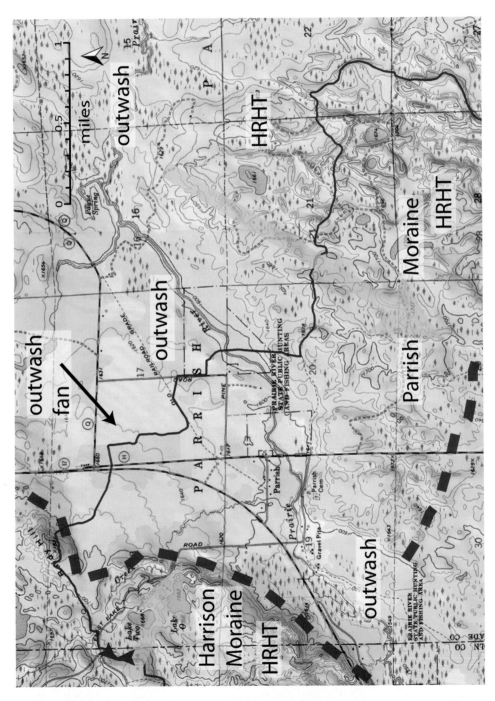

Figure 215. Topography of the western part of the Parrish Hills Segment. Blue arrows show ice-flow direction. Dashed blue lines show outer edge of the Langlade Lobe on the east (Parrish Moraine) and Wisconsin Valley Lobe on the west (Harrison Moraine). Water on the outwash fan (SB 10) flowed toward the southeast, away from the Harrison Moraine. HRHT: high-relief hummocky topography (SB 11). (Map is part of the Enterprise and Parrish USGS Quadrangles and was created with TOPO! © 2011 National Geographic Maps.)

# Wisconsin Valley Lobe
# Ice Age Trail Segments

The WISCONSIN VALLEY LOBE advanced into northern Wisconsin out of the Lake Superior basin between 30,000 and 25,000 cal. years ago (fig. 1). It deposited reddish-brown, sandy, gravelly till (SB 5) and supraglacial sediment (SB 11). Look for the boulders strewn on the ground surface as well as the moraines (SB 6) that this advance left behind: the Harrison Moraine on the east and the Wood Lake Moraine on the west side of the lobe (figs. 216, 217).

Outside the Harrison and Wood Lake moraines are glacial deposits that predate the late Wisconsin Glaciation. Radiocarbon dating shows that the Shelke bog, which is organic sediment under sand but on top of older till in eastern Lincoln County, formed between 36,000 and 40,800 radiocarbon years BP, so the till beneath it is even older than that (Mickelson 1986). The landscape covered by this earlier glaciation has been eroded, and most glacial landforms are gone. However, there is what appears to be older hummocky topography along the west side of the Wisconsin Valley Lobe. In places it is hard to distinguish exactly where the late Wisconsin advance reached its maximum extent in that landscape (Ham and Attig 1997; Attig 1993).

There are no radiocarbon dates that can give us a precise time frame for the late Wisconsin advance of the Wisconsin Valley Lobe, but we can use the crosscutting relationships of moraines to figure out the relative ages of the advances (fig. 217). Drumlins (SB 14) in northern Lincoln County show there was ice flow from the northwest, but a narrow zone in back of the Wood Lake Moraine indicates a flow of ice to that moraine from the northeast (fig. 217). Therefore, the earliest advance was from the northeast to the Wood Lake Moraine, and a later flow from the northwest removed the evidence except in the zone close to the Wood Lake Moraine. It is possible that debris-covered ice existed in the Wood

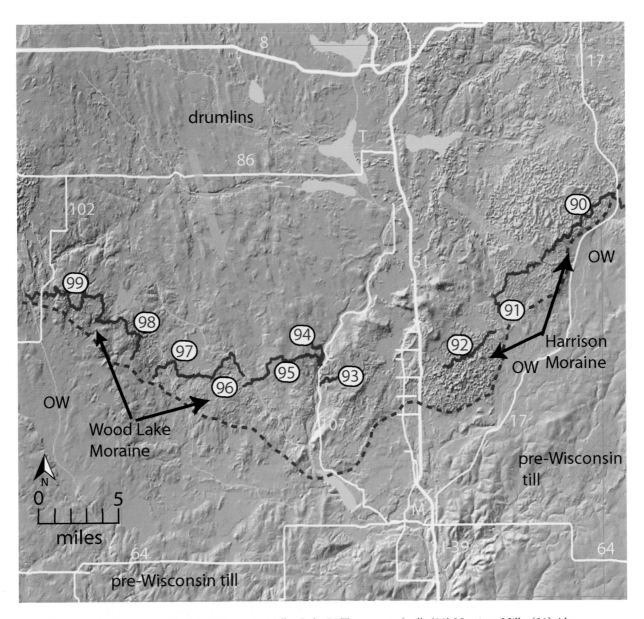

Figure 216. Shaded relief of the Wisconsin Valley Lobe IAT segments (red): (90) Harrison Hills, (91) Alta Junction, (92) Underdown, (93) Grandfather Falls, (94) Turtle Rock, (95) Averill–Kelly Creek Wilderness, (96) Newwood, (97) Camp 27, (98) Timberland Wilderness, (99) Wood Lake. Yellow lines and numbers indicate major highways. Cities and villages shown (yellow): (M) Merrill, (T) Tomahawk. OW: outwash (SB 8). Blue arrows show ice-flow direction. Dashed blue line indicates outer extent of late Wisconsin glacial advance. (Base map constructed from USGS National Elevation Dataset and modified by WGNHS.)

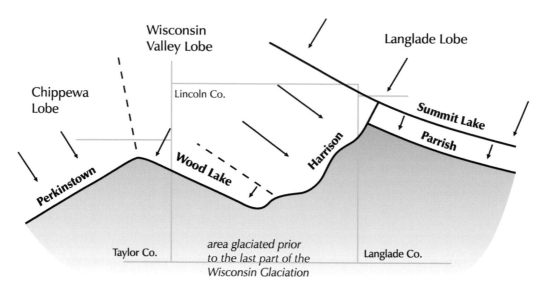

Figure 217. Crosscutting relationships of moraines in north central Wisconsin. (Modified from Ham and Attig 1997.)

Lake Moraine when ice readvanced from the northwest to the Harrison Moraine, producing the northwest-to-southeast-oriented drumlins. We can also tell that the Harrison Moraine formed after the ice retreated from the Parrish Moraine, but before the Langlade Lobe advanced to the Summit Lake Moraine (fig. 201) because of the crosscutting relationships.

It has been suggested that the advance to the Harrison Moraine was a glacial surge, a rapid advance of part of a glacier that is often caused by changes in water pressure at the glacier bed (Ham and Attig 1997). Modern surging glaciers often accumulate a thick layer of sediment on their ice surface, particularly near their edges, resulting in high hummocks (SB 11) and deep kettles (SB9) when the ice melts away. This might account for the higher-relief hummocky topography along the Harrison Hills and Underdown segments of the IAT, which are on the east side of the lobe.

The hummocky topography in these two IAT segments is a world-class example of this type of landscape. There are ice-walled-lake plains (SB 15) among the irregular hummocks, kettles (SB 9), and disintegration ridges (SB 11). If you were to dig into these lake plains, you would find fine-grained lake sediment (sand, silt, and clay). You can pick out the ice-walled-lake plains because they often occupy high points in the landscape and they are commonly farmed.

The Grandfather Falls, Turtle Rock, and Averill–Kelly Creek Wilderness segments of the IAT are in a more central part of the lobe, behind the end moraines. In these segments you'll hike through lower-relief hummocky topography located near the southern edge of a group of northwest-southeast-trending drumlins.

On the Newwood and Camp 27 segments you'll approach the backside of the New Wood Moraine. Notice that the landscape steepens and is more rugged as you hike toward the outermost edge of the moraine. You'll see excellent examples of ice-walled-lake plains along the trail, as well as a small esker (SB 13), a feature not nearly as common in this area as along other parts of the IAT.

The Timberland Wilderness and Wood Lake segments are in the moraine as well and run more or less parallel to the former edge of the glacier. Hiking the IAT you'll cross several well-defined ice-walled-lake plains on the Timberland Wilderness Segment. In the Wood Lake Segment, you'll cross not only higher-relief hummocky topography but also a subglacial channel and a large ice-walled-lake plain in the Wood Lake Moraine.

## 90. Harrison Hills Segment
*First Lake Rd. to CTH J (15 miles)*

This segment of the IAT more or less parallels the front of the Harrison Moraine (figs. 216, 218, 219, 220). All of the terrain is high-relief hummocky topography (SB 11), atop which you'll have views of undeveloped kettle lakes (SB 9). The hummocky deposits are thick—probably more than 150 feet in this area. Much of the trail is wooded, so be prepared to have a limited view if you hike during a leafy season, but when the leaves are off, there are great views of hummocks and kettles (fig. 221). From a few places you can see the fairly flat outwash plain and pre-Wisconsin till surface to the east and south (fig. 222).

Boulders representing many different rock types litter the landscape in this area. All the sediment you see here was on glacier ice, which then collapsed as the buried ice melted out. Lookout Mountain (fig. 218) is the highest point in Lincoln County and is a great example of one of the hummocks that resulted. There are good views from the fire tower on top of Lookout Mountain if the tower is open. Notice the difference in landscape between the older glaciated surface to the southeast and the high-relief hummocky topography of the Harrison Moraine that surrounds you (fig. 222). For the most part, all of the hummocks contain till (SB 6) or supraglacial sediment that has been moved in mudflows and slides before coming to rest. As pointed out in the description of the Parrish Hills Segment, the eastern Wisconsin Valley Lobe was at its maximum position after the Langlade Lobe had retreated from the Parrish Moraine.

About a mile east of the CTH J trailhead (fig. 220), the trail descends from high-relief hummocky topography to outwash (SB 8) in the valley of the North Branch of the Prairie River. Numerous springs feed the flowing river along this segment. Low, flat terraces (SB 8) border both sides of the river north of CTH J.

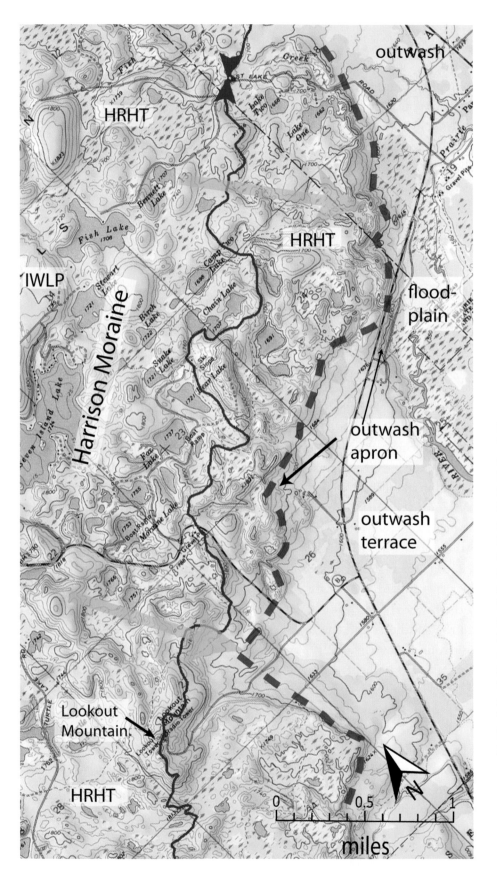

Figure 218.
Topography of the
eastern part of the
Harrison Hills
Segment. Dashed blue
line indicates the outer
edge of the Harrison
Moraine. Blue arrows
show ice-flow direction
(from northwest to
southeast). HRHT:
high-relief hummocky
topography (SB 11).
Note orientation of
compass arrow. (Map
is part of the Parrish
USGS Quadrangle
and was created with
TOPO! © 2011
National Geographic
Maps.)

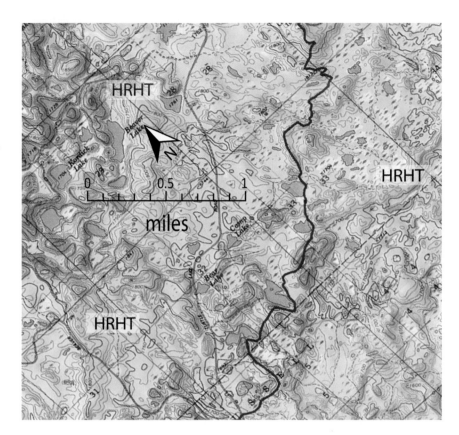

Figure 219. Topography of the central part of the Harrison Hills Segment. The outer edge of the Harrison Moraine is off the map to the right (southeast). Ice-flow direction was from left to right (northwest to southeast). HRHT: high-relief hummocky topography (SB 11). Note orientation of compass arrow. (Map is part of the Parrish, Harrison, and Bloomville USGS Quadrangles and was created with TOPO! © 2011 National Geographic Maps.)

## 91. Alta Junction Segment

*CTH J to CTH J (1.5 miles)*

At present the trail follows CTH J on outwash with a few shallow kettles. About 1 mile southeast of the trailhead and just east of CTH J is an ice-walled-lake plain (figs. 220, 223). Like many ice-walled-lake plains, it has a fairly flat top. Farmers use the plains for agriculture because the fine-grained, nutrient-rich, and stone-poor lake sediment makes good cropland.

## 92. Underdown Segment

*Copper Lake Ave. to Horn Lake Rd. (7 miles)*

Like the Harrison Segment, the Underdown Segment (figs. 216, 224) has world-class high-relief hummocky topography (SB 11). The trail passes spectacular hummocks separated by deep kettles. Although sphagnum peat now partly fills many of the kettles (fig. 225), some remain as lakes. The hummocks here are mostly composed of till and supraglacial sediment that formed a thick layer of rocks, sand, silt, and clay on the ice surface as the glacier started to retreat.

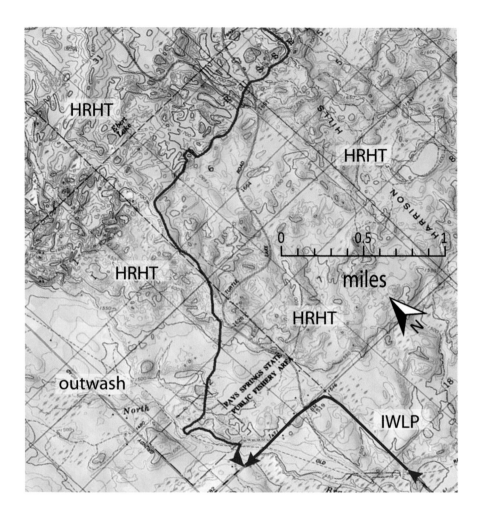

Figure 220. Topography of the western part of the Harrison Hills Segment (top) and the Alta Junction Segment (lower right). The outer edge of the Harrison Moraine is off the map to the right (southeast). Ice-flow direction was from left to right (northwest to southeast). HRHT: high-relief hummocky topography (SB 11); IWLP: ice-walled-lake plain (SB 15). Note orientation of compass arrow. (Map is part of the Parrish, Harrison, and Bloomville USGS Quadrangles and was created with TOPO! © 2011 National Geographic Maps.)

All this terrain is considered part of the Harrison Moraine of the Wisconsin Valley Lobe. The Underdown is a particularly good example of high-relief hummocky topography. For the most part, you won't have the long landscape views that you would find in other segments because of a thick forest cover, but the ups and downs of the trail will bring home the nature of high-relief hummocky topography!

## 93. Grandfather Falls Segment

*STH 107 at Camp New Wood County Park to STH 107 at*
*Grandfather Falls Hydro parking area (3.6 miles)*

This segment of the IAT has access at Camp New Wood County Park on STH 107 (fig. 226). From there you can go east over moderate-relief hummocky topography (SB 11), similar to that in the Turtle Rock Segment. From the parking area you can also cross the

Figure 221.  Hiking in high-relief hummocky topography on part of the Harrison Hills Segment during the spring 2009 Ice Age Alliance annual conference. (Photo by Vin Mickelson.)

highway and follow the trail for about 2 miles north along the river. Much of this is on outwash (SB 8), and part of it is low enough to be floodplain. Rocks in the bed of the stream have been water worn, and most are of granitic composition. The large outcrop in figure 227 is a Precambrian rock called tonalite, a kind of granite. These are among the oldest rocks crossed by the IAT.

## 94. Turtle Rock Segment

*CTH E to Burma Rd. (5 miles)*

The eastern trailhead of this segment of the IAT lies west of the Wisconsin River and well north of the Harrison Moraine (fig. 216) on rolling hills of Wisconsin Valley Lobe till

Figure 222.  Low oblique aerial view of the steep moraine front of the Harrison Moraine (foreground) and outwash along the Prairie River valley. Location is just southeast of area shown in figure 219.

(SB 5). Here the ice flowed from north-northwest as indicated by the drumlins (SB 14) north of the IAT (blue arrows on fig. 226). Unlike the landscape at the Harrison Hills and the Underdown segments, the ice here had relatively little debris on its surface—a few feet at most. As a result of its thin debris cover, the ice here probably retreated well before the buried ice in the Harrison Moraine, because the sun and warming atmosphere could melt the ice quickly.

From this point, the trail descends along a tributary to the dammed Wisconsin River and follows the flowage to Grandfather Falls Dam. From there, it follows the river below the dam for about a mile, where you will be able to see exposed Precambrian bedrock along the river edge. Figure 227 is a photo of a bedrock outcrop taken from the east side of the river (the Grandfather Falls Segment). The IAT is just behind the outcrop shown, and Turtle Rock itself is there (fig. 228). Where it leaves the river, the Turtle Rock Segment climbs steeply onto a hummocky till (SB 11) surface, then it follows the boundary between hummocky topography to the south and rolling till hills to the north to its western trailhead at

Figure 223. Low oblique aerial view of an ice-walled-lake plain (treeless area) close to the Alta Junction Segment. For location see IWLP in south (lower right) corner of figure 220. View looks northwest. CTH J is along far side of ice-walled-lake plain.

Burma Rd. Three-tenths of a mile east of this trailhead, the trail enters an area of low-relief hummocky topography. This formed in the same way as high-relief hummocky topography, but there was less debris on the ice surface to remain after meltout.

## 95. Averill–Kelly Creek Wilderness Segment
*Burma Rd. to CTH E (5.6 miles)*

On the Averill–Kelly Creek Wilderness Segment, you'll continue west through low-relief hummocks (SB 11) separated by peat bogs (figs. 216, 229). In some places the IAT follows an old railroad grade. Glacier ice flowed from north to south here, but there are no

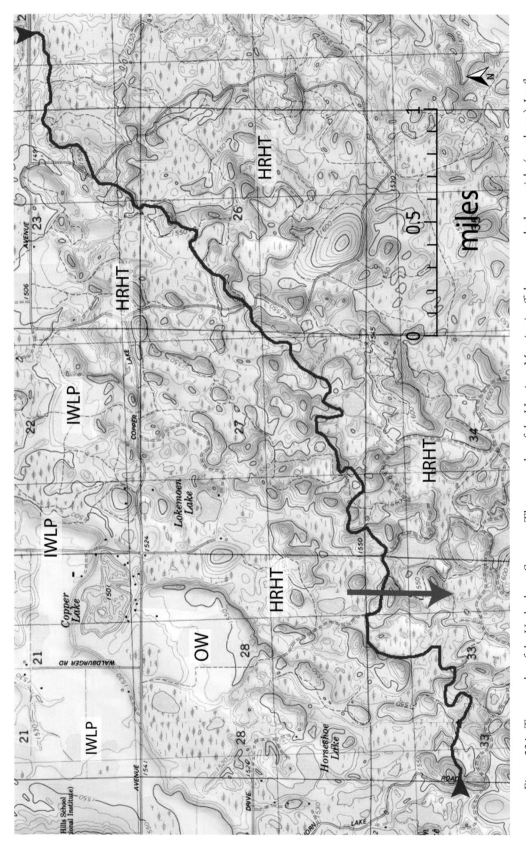

Figure 224. Topography of the Underdown Segment. The outer edge of the Harrison Moraine is off the map to the lower right (southeast). Ice-flow direction was from northwest to southeast. HRHT: high-relief hummocky topography (SB 11); IWLP: ice-walled-lake plain (SB 15); OW: outwash (SB 8). Pink arrow shows approximate location and camera direction of figure 225. (Map is part of the Bloomville and Irma USGS Quadrangles and was created with TOPO! © 2011 National Geographic Maps.)

Figure 225.  Low oblique aerial view of high-relief hummocky topography (SB 11) in the Underdown Segment. Ice Age Trail is in foreground. Location of photo is shown in figure 224.

indications of it in the landscape. Averill Creek increases its flow as it runs through spring-fed June Lake, about a mile north from where the creek joins the New Wood River. These waterways come together on a narrow outwash plain (SB 8), on which you'll walk to the segment's western trailhead.

## 96. Newwood Segment
*CTH E to primitive Conservation Ave. (6.6 miles)*

This segment begins on outwash close to the New Wood River (figs. 216, 230, 231). The trail goes straight for 0.4 miles on an old railroad bed used for logging. Then, as it heads north along the river for the next mile, it climbs onto low hummocks (SB 11) of till and

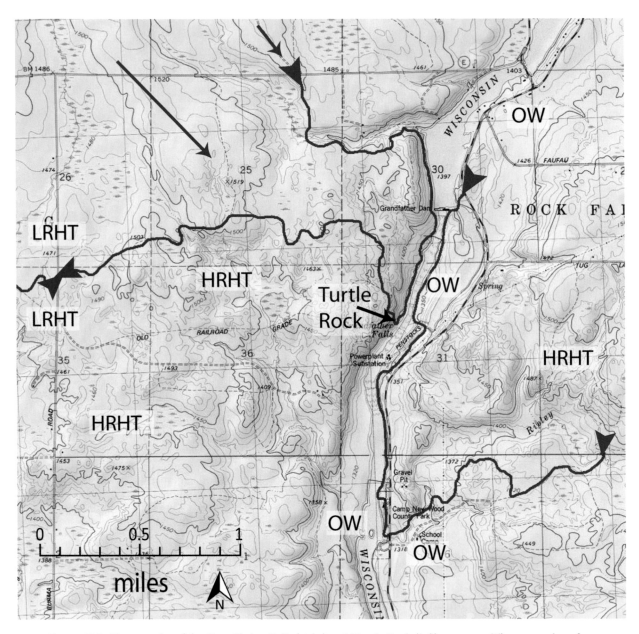

Figure 226. Topography of the Grandfather Falls (right) and Turtle Rock (left) segments. The outer edge of the Harrison Moraine is off the map to the south and east. Blue arrows are on poorly developed drumlins and show ice-flow direction. HRHT: high-relief hummocky topography (SB 11); IWLP: ice-walled-lake plain (SB 15); LRHT: low-relief hummocky topography (SB 11); OW: outwash (SB 8). (Map is part of the Grandfather Falls USGS Quadrangle and was created with TOPO! © 2011 National Geographic Maps.)

Figure 227.  View looking west across the Wisconsin River from the Grandfather Falls Segment to the Turtle Rock Segment (in the pines). (Photo by Vin Mickelson.)

back onto outwash (SB 8) several times. From there, the IAT rises about 30 feet into the low-relief hummocky topography (SB 11) of the Wood Lake Moraine, which it follows to the segment's western terminus. Much of the sediment here was on the ice surface as the ice beneath slowly melted. The debris layer on top of the ice was much thinner here than it was in the Harrison Hills or the Underdown segments, resulting in lower relief hummocky topography than in the Harrison Moraine.

The IAT winds through low-relief hummocky topography that formed when sediment on the glacier surface collapsed (figs. 216, 231). If you were to dig down below the surface, you would find that till (SB 5) and supraglacial sediment (SB 11) underlie the whole area. The ice here flowed toward the southwest. The trail crosses a small esker (SB 13; fig. 231)

Figure 228. Turtle Rock. Wisconsin River in the background. (Photo by Bruce Jaecks.)

and low ice-disintegration ridges before becoming the Camp 27 Segment at the rerouted Conservation Ave.

## 97. Camp 27 Segment

*Primitive Conservation Ave. to Tower Rd. (3 miles)*

The Camp 27 Segment (fig. 216) crosses the center of a small, low, ice-walled-lake plain (SB 15) just west of the rerouted Conservation Ave. (fig. 231). Ice-walled-lake plains often have a rim of coarser sediment around the edge, and the rim on this one is clearly seen on the north, west, and south sides of the feature (fig. 231). The segment continues through hummocky topography, where shallow kettles (SB 9) separate the hummocks. Just south of Camp 27 the trail crosses a low ridge that may be a small esker (SB 13). As you near the

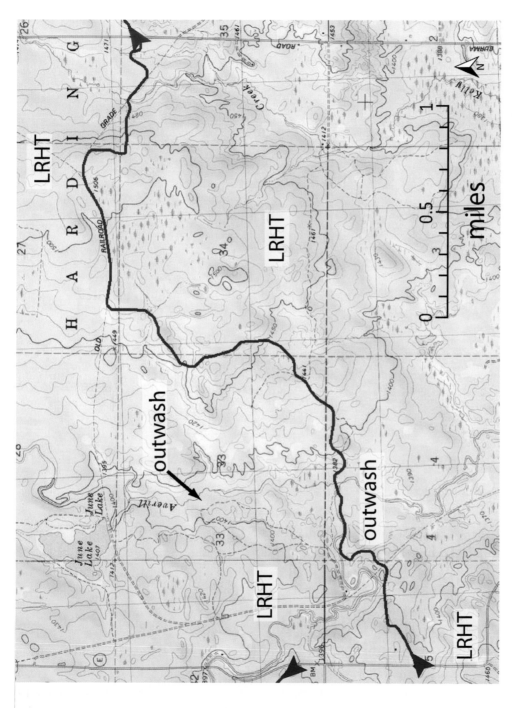

Figure 229. Topography of the Averill–Kelly Creek Wilderness Segment. The outer edge of the Harrison Moraine is off the map to the north. Ice-flow direction was from north to south, but there are no directional indicators. LRHT: low-relief hummocky topography (SB 11). (Map is part of the Natzke Camp USGS Quadrangle and was created with TOPO! © 2011 National Geographic Maps.)

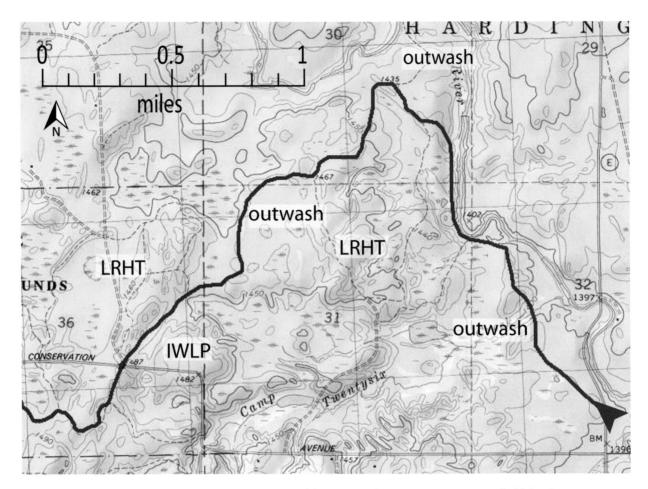

Figure 230.  Topographic map of the eastern part of the Newwood Segment. IWLP: ice-walled-lake plain (SB 15); LRHT: low-relief hummocky topography (SB 11). (Map is part of the Natzke Camp USGS Quadrangle and was created with TOPO! © 2011 National Geographic Maps.)

segment's western end on Tower Rd., the hummocks (SB 11) are higher and steeper as the trail approaches the outer edge of the Wood Lake Moraine. At its western trailhead, the outer edge of the Wood Lake Moraine is only 0.25 miles to the southwest (fig. 216).

## 98. Timberland Wilderness Segment

*Tower Rd. to Tower Rd. Wood Lake trailhead (3.7 miles)*

This segment of the IAT continues to traverse high-relief hummocky topography (SB 11) east of, and more or less parallel to, the Wood Lake Moraine (fig. 216). Note the small ice-walled-lake plain (SB 15) 0.7 miles from the trailhead; as you hike you'll skirt along its western edge (fig. 232). Also note the distinct rim along the northwest side of the lake plain,

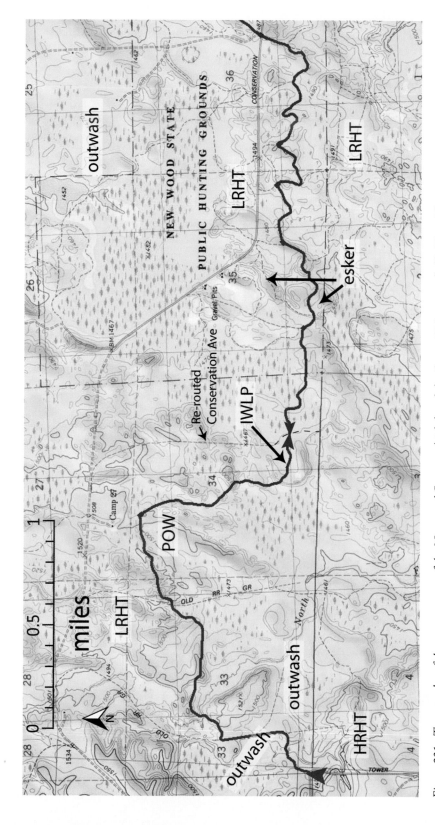

Figure 231. Topography of the western part of the Newwood Segment (right) and the Camp 27 Segment (left). Note: These segments now meet at an extension of Conservation Ave. that is shown in this figure. HRHT: high-relief hummocky topography (SB 11); IWLP: ice-walled-lake plain (SB 15); LRHT: low-relief hummocky topography (SB 11); POW: pitted outwash (SB 10). (Map is part of the Natzke Camp USGS Quadrangle and was created with TOPO! © 2011 National Geographic Maps.)

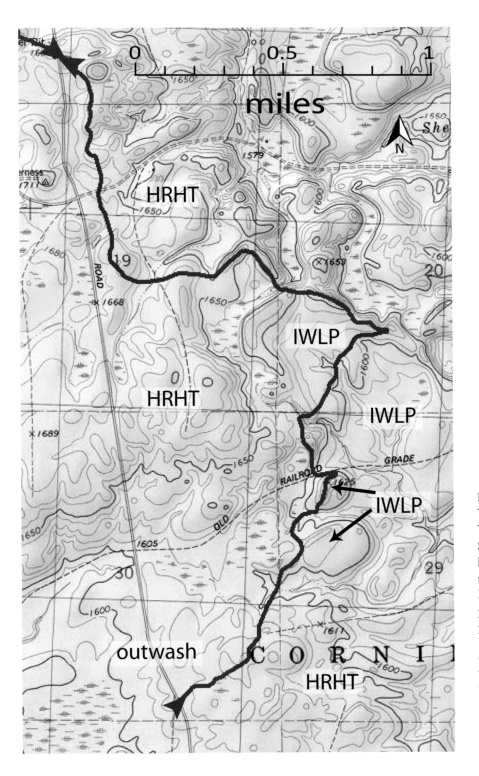

Figure 232. Topography of the Timberland Wilderness Segment. HRHT: high-relief hummocky topography (SB 11); IWLP: ice-walled-lake plain (SB 15). (Map is part of the Wood Lake USGS Quadrangle and was created with TOPO! © 2011 National Geographic Maps.)

just east of the trail. Two-tenths miles north of that point, the trail starts across a small ice-walled-lake plain. After crossing an old railroad grade, it climbs onto a larger plain that has a distinct raised rim on its northwest side. From there, the IAT passes though high-relief hummocky topography with several large erratics (SB 5) to its western terminus.

## 99. Wood Lake Segment

*Tower Rd. Wood Lake trailhead to STH 102 (12.1 miles)*

The Wood Lake Segment of the IAT exhibits spectacular high-relief hummocky topography (SB 11; figs. 233, 234). Kettles (SB 9) and erratics (SB 5) abound. This segment is in the outer part of the Wood Lake Moraine (figs. 216, 217). Springs feed Wood Lake, which partly fills a kettle between hummocks in the moraine. A large ice-walled-lake plain (SB 15) spans the land directly northwest of Wood Lake, but the trail skirts that plain as it remains in hummocky supraglacial sediment all the way to the end of the segment. Note the steep-sided valley along Gus Johnson Creek: water flowing toward the southwest and gushing from under the ice at the outer edge of the Wood Lake Moraine formed this subglacial channel (fig. 234). This is smaller than many of the tunnel channels seen along the IAT, but it has the basic characteristics of one (SB 17). You'll see that the bottom of the channel has depressions (kettles) in it that would likely have been filled with sand and gravel if this were an outwash, rather than a subglacial, channel. Also, the elevation of the channel bottom rises about 70 feet uphill toward the southwest, the direction the water flowed. Water can flow uphill when it is confined to a closed tunnel in the ice (SB 17) and pressure pushes it along. West of this channel the trail parallels the front of the Wood Lake Moraine through hummocky topography. Look for an excellent example of an ice-walled-lake plain (SB 15) with a high rim 0.25 miles north of the trail (fig. 234). At the western trailhead on STH 102, the IAT is at the outer edge of the Wood Lake Moraine. Low areas to the south are underlain by outwash (SB 8) and high areas by older till (SB 5).

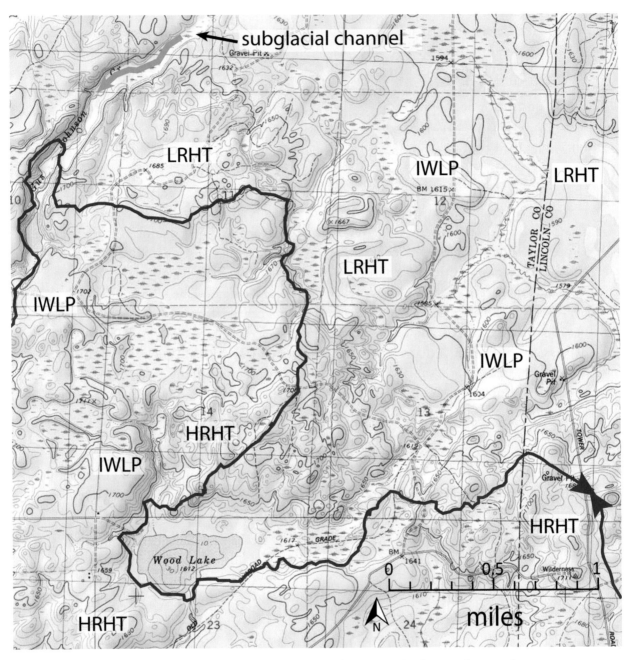

Figure 233. Topography of the eastern part of the Wood Lake Segment. Ice flow was from northeast to southwest. HRHT: high-relief hummocky topography (SB 11); IWLP: ice-walled-lake plain (SB 15); LRHT: low-relief hummocky topography (SB 11). Green arrow shows direction of water flow in subglacial channel. (Map is part of the Wood Lake USGS Quadrangle and was created with TOPO! © 2011 National Geographic Maps.)

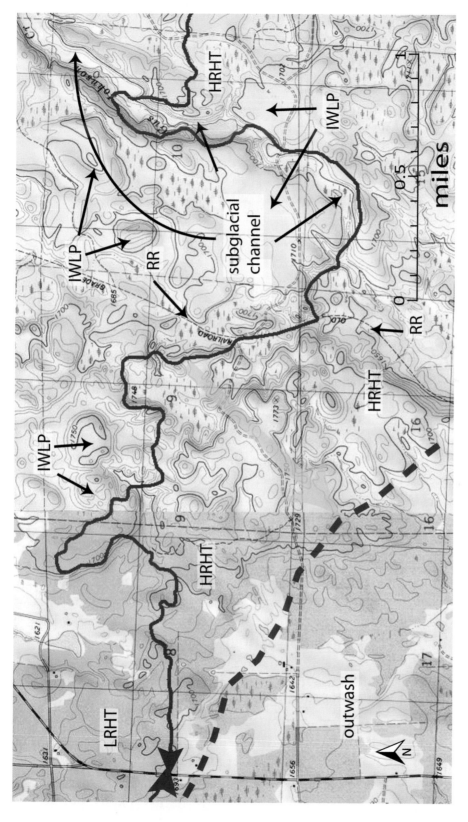

Figure 234. Topography of the western part of the Wood Lake Segment. Dashed blue line shows outer edge of the Wood Lake Moraine. Blue arrow shows ice-flow direction. HRHT: high-relief hummocky topography (SB 11); IWLP: ice-walled-lake plain (SB 15); LRHT: low-relief hummocky topography (SB 11); RR: railroad. (Map is part of the Wood Lake and Rib Lake USGS Quadrangles, and was created with TOPO! © 2011 National Geographic Maps.)

# Chippewa Lobe
# Ice Age Trail Segments

THE CHIPPEWA LOBE OF THE LATE WISCONSIN GLACIER developed textbook examples of high-relief hummocky topography (SB 11) and particularly large ice-walled-lake plains (SB 15). Ice flowed generally southwestward out of the Lake Superior basin into northwestern Wisconsin (figs. 1, 3, 235). Earlier glaciations had extended farther into Chippewa, Clark, and Marathon counties. The publications listed at the back of the book will give you more information about these deposits.

The Chippewa Lobe overrode Precambrian metamorphic rocks (SB 21; fig. 236). In fact, some of the oldest rocks in Wisconsin are exposed in the Chippewa River valley a short distance outside the edge of the late Wisconsin Glaciation. Precambrian quartzite is exposed in some places in the Blue Hills (Southern Blue Hills and Hemlock Creek segments; fig. 236) and is a common rock type in the till along the west side of the Chippewa Lobe. This quartzite is about the same age as the Baraboo quartzite at Devil's Lake State Park: about 1.6 billion years old! Rocks that the glacier carried into this area include a range of metamorphic and igneous types. If you look carefully in areas of exposed gravel, you may see banded iron formation, a metamorphic rock with alternating gray and reddish bands of mostly iron and silica (fig. 5).

Ice lobes—or even parts of lobes—advance and retreat at different rates and reach their maximum extent at slightly different times. The result is that the end moraines of the lobes crosscut one another: we can use their overlapping patterns, along with the vertical sequence of deposits that the glacier left behind, to interpret the deposits' relative age. The earliest late Wisconsin advance probably took place between 30,000 and 25,000 cal. years ago and extended to the areas marked "Stanley" and "early Chippewa" in figure 237. We can tell this because the Perkinstown and late Chippewa advance sediments partly cover the Stanley and

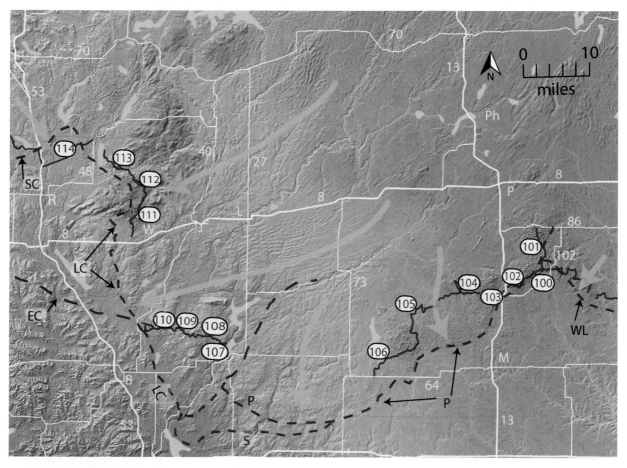

Figure 235. Shaded relief of the Chippewa Lobe IAT segments (red): (100) Rib Lake, (101) Timm's Hill, (102) East Lake, (103) Pine Line, (104) Mondeaux Esker, (105) Jerry Lake, (106) Lake Eleven, (107) Chippewa River, (108) Firth Lake, (109) Harwood Lakes, (110) Chippewa Moraine, (111) Southern Blue Hills, (112) Northern Blue Hills, (113) Hemlock Creek, (114) Tuscobia. Yellow lines and numbers indicate major highways. Cities and villages shown (yellow): (B) Bloomer, (C) Cornell, (L) Ladysmith, (M) Medford, (P) Perkinstown, (Ph) Phillips, (R) Rice Lake, (W) Weyerhauser. Blue arrows show ice-flow direction. Dashed blue lines indicate the edge of the late Chippewa (LC), Perkinstown (P), and Stanley (S) advances of the Chippewa Lobe, the Wood Lake (WL) advance of the Wisconsin Valley Lobe, and the St. Croix (SC) advance of the Superior Lobe. See also figure 237. The other dashed blue line is the approximate outer edge of the early Chippewa (EC) advance, but much of this evidence has been removed by later outwash streams. (Base map constructed from USGS National Elevation Dataset and modified by WGNHS.)

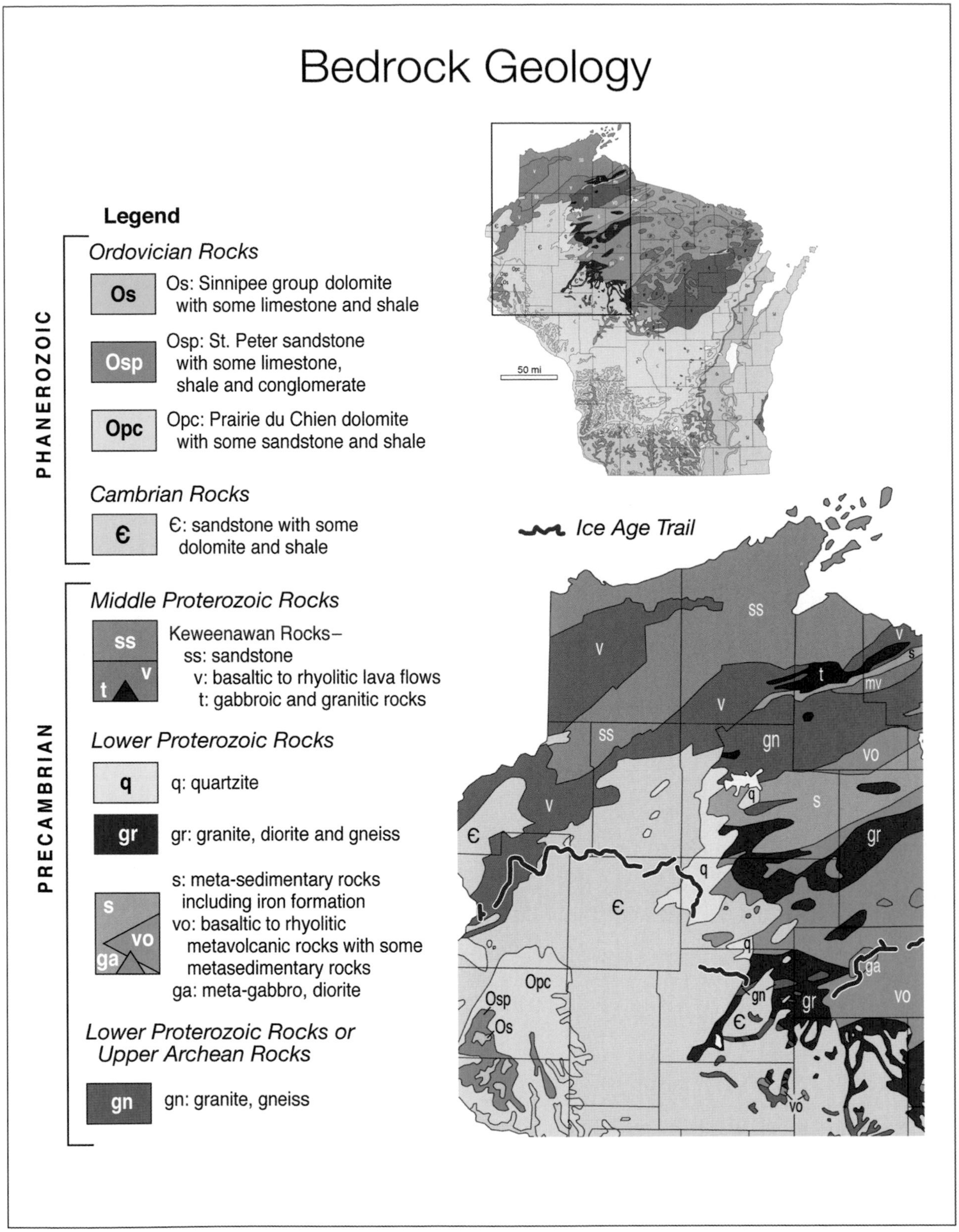

# Bedrock Geology

## Legend

### PHANEROZOIC

#### Ordovician Rocks

**Os** — Os: Sinnipee group dolomite with some limestone and shale

**Osp** — Osp: St. Peter sandstone with some limestone, shale and conglomerate

**Opc** — Opc: Prairie du Chien dolomite with some sandstone and shale

#### Cambrian Rocks

**Є** — Є: sandstone with some dolomite and shale

### PRECAMBRIAN

#### Middle Proterozoic Rocks

**ss / t / v** — Keweenawan Rocks–
ss: sandstone
v: basaltic to rhyolitic lava flows
t: gabbroic and granitic rocks

#### Lower Proterozoic Rocks

**q** — q: quartzite

**gr** — gr: granite, diorite and gneiss

**s / vo / ga** — s: meta-sedimentary rocks including iron formation
vo: basaltic to rhyolitic metavolcanic rocks with some metasedimentary rocks
ga: meta-gabbro, diorite

#### Lower Proterozoic Rocks or Upper Archean Rocks

**gn** — gn: granite, gneiss

50 mi

~~ Ice Age Trail

Figure 236. Bedrock geology of part of northwestern Wisconsin with Ice Age Trail superimposed. Note the Barron quartzite, which forms the Blue Hills, in yellow. (Modified from WGNHS; drafted by Mary Diman.)

early Chippewa deposits. The late Chippewa advance truncates the line marking the outer edge of the Perkinstown advance, which gives us further clues that the Perkinstown advance must be older. At the eastern edge of the Chippewa Lobe, the Perkinstown Moraine truncates the Wood Lake Moraine of the Wisconsin Valley Lobe and therefore is younger than the Wood Lake advance.

Unfortunately, there is no way to date most of these advances except relatively. The radiocarbon dating method requires finding organic material that the ice overrode and therefore preserved. So far, we haven't found any sediment like this. A recently developed technique, called cosmogenic dating, uses isotopes of beryllium or aluminum to date deposits (SB 18). It suggests that boulders at the surface in the area that the late Chippewa advance covered have been there about 22,000 cal. years (Ullman et al. 2011). All of these results are difficult to interpret in areas where debris-covered ice may have lasted for thousands of years before the final meltout. The advances were probably closely spaced in time, which, geologically speaking, means hundreds to a few thousand years apart.

The climate was very cold as the ice advanced, resulting in probably 5,000 to 10,000 years of relative ice-margin stability. Permafrost similar to what occurs on the North Slope of Alaska today covered the landscape for at least tens of miles away from the front of the

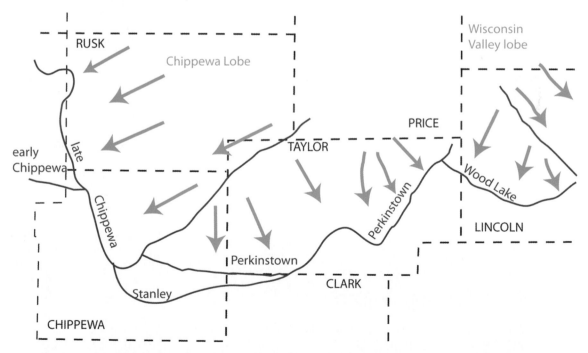

Figure 237. Maximum positions of ice advance in the Chippewa Lobe and western Wisconsin Valley Lobe. Blue arrows show ice-flow direction. (Modified from Syverson 2007.)

glacier. The ice surface probably only melted during the brief Ice Age summers. Much of unglaciated Wisconsin had just sparse tundra vegetation that eked out a living in the harsh climate. Periods of intense freezing and thawing crumbled the quartzite into many talus blocks, forming a sea of rocks called felsenmeer. There is a good example of this talus sea at the Blue Hills Felsenmeer State Natural Area (fig. 238).

Several authors have suggested that after the main glacier edge retreated from this area, debris-covered ice still covered this landscape. This lingering ice may have been a remnant of a glacier surge (SB 2) that caused the late Chippewa advance. A sudden surge may have produced the high-relief hummocky topography and abundant ice-walled-lake plains that you see along the southwest side of the lobe. We know that when modern glaciers experience the rapid, out-of-equilibrium advance of an ice surge, large amounts of sediment accumulate on the ice surface. This may explain why the supraglacial sediment (SB 11)—which then

Figure 238.  Talus on both sides of valley in the Blue Hills Felsenmeer State Natural Area. Note person for scale.

became the high-relief topography after the ice melted out from underneath—was so exten-sive and thick in this area. A frozen bed zone near the ice margin (SB 3, SB 11) probably made the supraglacial sediment thicker than it otherwise would have been, too, because the frozen bed zone acted as a wall that faster-moving ice from behind pushed on top of. In addi-tion, Syverson (2007) suggests that prior glacial action influenced the Chippewa advance: when the first Perkinstown advance occurred, it left debris-covered ice that then deflected the late Chippewa advance to the west (see ice-flow arrows in figures 235 and 237).

## 100. Rib Lake Segment
*STH 102 to CTH D (4.5 miles)*

On the part of the Rib Lake Segment between STH 102 and CTH C (figs. 235, 239), you'll traverse outwash (SB 8) and go by old gravel pits where you'll see some of the outwash. Braided streams flowed to the south from the Wood Lake Moraine across this surface. The connector road to the rest of the segment goes north and then west. Where the connector turns to the west on Rustic Rd. No. 1, the trail enters the area covered by the Chippewa Lobe when it sat at the Perkinstown Moraine. This is not a moraine here, however; it is pitted out-wash (SB 10). The Timm's Hill National Trail strikes out to the north from a trail junction just west of CTH C. About a half mile west of CTH C, you'll enter a landscape of low-relief hummocky topography (SB 11) with lake-filled kettles (SB 9; fig. 239). Sediment that settled in ice-walled-lake plains (SB 15) underlies just beneath your feet where the trail leaves Harper Dr. The ridge that the trail follows to CTH D appears to be a rim of one of these lake plains.

## 101. Timm's Hill National Trail
*Rib Lake Segment of the IAT to Timm's Hill (10 miles)*

The Timm's Hill National Trail is not part of the Ice Age Trail but instead is a spur trail that extends north from the Rib Lake Segment of the IAT (figs. 235, 239, 240). As the trail heads northward from its southern end, it crosses a low-relief pitted-outwash plain (SB 10) and collapsed lake sediment. All the lakes here are kettles (SB 9). The trail traverses a large ice-walled-lake plain (SB 15) as it crosses CTH C for the third time. North of this intersection, the hummock relief (SB 11) increases, and many small kettle lakes and peat bogs fill the ket-tles between high hummocks. Look for several ice-walled-lake plains south of Timm's Hill.

Timm's Hill is the highest point in Wisconsin. At 1,951 feet above sea level, it just hap-pens to be the *highest* hummock in an area of already high landscape. The hill is probably composed of supraglacial sediment and till (SB 5) rather than solid rock. Because other hum-mocks surround the state's tallest point, the landmark does not stand out from a distance as does Rib Mountain (about 1,940 feet) near Wausau. Still, you'll appreciate all 1,951 of those feet after you scale the tower on top of Timm's Hill for a good view of the surrounding

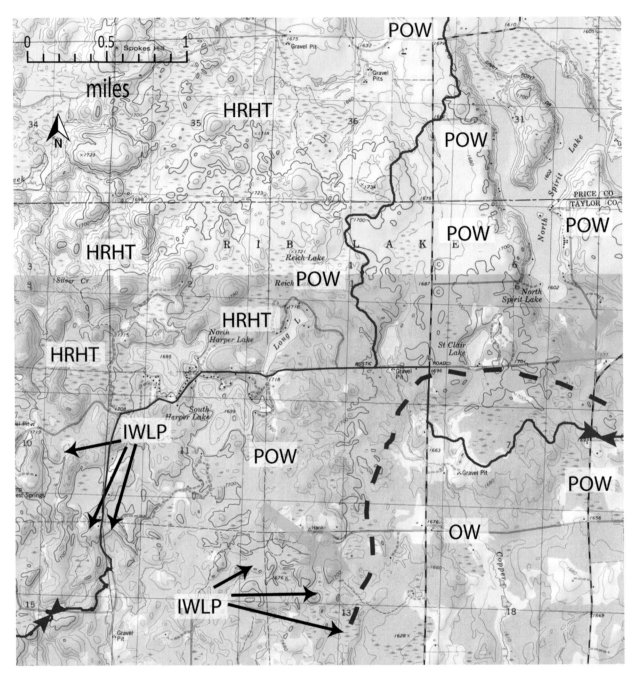

Figure 239. Topography of the Rib Lake Segment (east to west) and the southern part of the Timm's Hill National Trail (north to south). Dashed blue line indicates the outer edge of late Wisconsin glacial deposits (Perkinstown Moraine on the west, Wood Lake Moraine on the east). Blue arrows show ice-flow direction. HRHT: high-relief hummocky topography (SB 11); IWLP: ice-walled-lake plain (SB 15); OW: outwash (SB 8); POW: pitted outwash (SB 10). (Map is part of the Rib Lake and Timm's Hill USGS Quadrangles and was created with TOPO! © 2011 National Geographic Maps.)

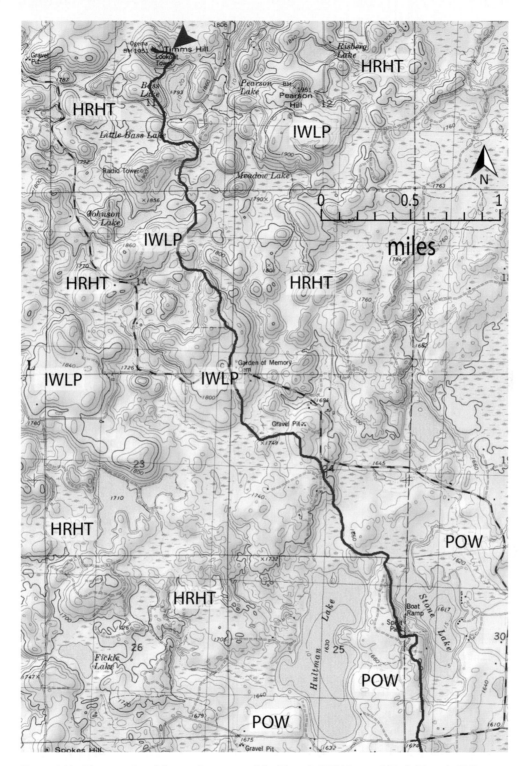

Figure 240. Topography of the northern part of the Timm's Hill National Trail. Timm's Hill is at the northern trailhead (red arrow). All the lakes are kettles (SB 9). HRHT: high-relief hummocky topography (SB 11); IWLP: ice-walled-lake plain (SB 15); POW: pitted outwash (SB 10). (Map is part of the Timm's Hill USGS Quadrangle and was created with TOPO! © 2011 National Geographic Maps.)

Figure 241.  View looking southward from the Timm's Hill tower. All the hills in the distance are glacial hummocks (SB 11) or ice-walled-lake plains (SB 15). (Photo by Vin Mickelson.)

hummocks and kettles (figs. 241, 242). Note the ice-walled-lake plain southeast of Timm's Hill in figure 240 and the one northwest of the tower in the upper left of figure 242.

## 102. East Lake Segment

*CTH D to STH 13 wayside (6.5 miles)*

This segment of the IAT is within the Perkinstown Moraine (SB 6), which here stretches northeast to southwest (figs. 235, 243). Ice flowed from northwest to southeast. The village of Rib Lake, only 2 miles southeast of the IAT (fig. 243), is on an outwash plain (SB 8) just southeast of the outer edge of the moraine.

On the East Lake Segment you'll hike exclusively on high-relief hummocky moraine (SB 11). Some of the hummocks are about 100 feet high; the trail ascends the most vertical feet

Figure 242. Oblique aerial view of Timm's Hill (tower on top of hill) looking to the north. Bass Lake is in the foreground and Timm's Lake is in the background. An ice-walled-lake plain (SB 15) is in the upper left.

on the hummocks in the central part of the segment. All the lakes here are kettles (SB 9). Low ridges along and across STH 13 are the gravel rim remnants of ice-walled-lake plains (SB 15). The ridge under and just east of the wayside park on STH 13 is probably one of these coarse-grained, gravelly rims.

## 103. Pine Line Segment
*STH 13 to Fisher Creek Rd. at Fawn Ave. (1 mile)*

The mile-long Pine Line Segment (fig. 243) follows the south edge of collapsed ice-walled-lake sediment (SB 15). From there it crosses into a shallow kettle (SB 9) filled with peat in a pitted-outwash plain (SB 10).

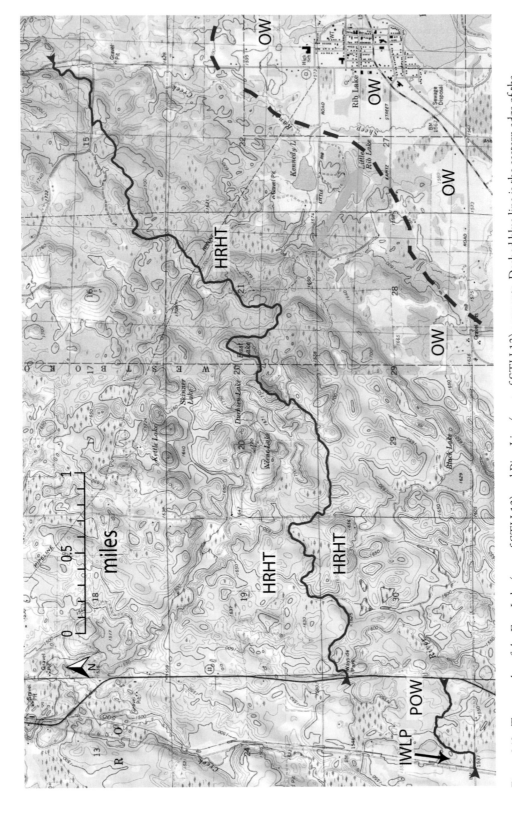

Figure 243. Topography of the East Lake (east of STH 13) and Pine Line (west of STH 13) segments. Dashed blue line is the eastern edge of the Perkinstown Moraine. Blue arrow shows ice-flow direction. HRHT: high-relief hummocky topography (SB 11); IWLP: ice-walled-lake plain (SB 15); OW: outwash (SB 8); POW: pitted outwash (SB 10). (Map is part of the Rib Lake and Westboro USGS Quadrangles and was created with TOPO! © 2011 National Geographic Maps.)

## 104. Mondeaux Esker Segment

*Shady Dr. to CTH E (13 miles)*

For 0.4 miles west of the Shady Dr. trailhead, this segment of the IAT crosses the high-relief hummocky topography (SB 11) of the Perkinstown Moraine (SB 6), most of which lies to the south and east (figs. 235, 244). The trail then continues on rolling topography underlain by till (SB 5) before dropping into the tunnel channel (SB 17) of the Mondeaux Flowage.

The most striking glacial features here are the Mondeaux Flowage itself (fig. 245) and a large esker (SB 13) that occupies parts of that valley. Like many tunnel channels, this one partly filled with debris-rich ice after it stopped carrying water. We know that because hummocky sand and gravel released from the melting debris-rich ice has partially filled the channel in places. When the tunnel was active, water flowed toward the south-southeast and filled the valley, at least to the extent that the lake fills the valley now. Picture the gently flowing water you see now as it was then: a torrential subglacial river that roared for a few days to weeks then suddenly stopped. As the flow of water decreased, the weight of the overlying ice pushed debris-rich glacial ice into the valley to fill the open space that had been occupied by water.

Later, during deglaciation, when the ice had warmed to the melting point (SB 3), water again flowed along this path, but it created a much smaller tunnel through the ice. This flow lasted longer—several years to decades—as the subglacial river deposited sand and gravel in the tunnel. The deposits this river left are the esker you see today. As is typical throughout Wisconsin, the tunnel channel is substantially wider than the esker you see within it. Note that water in this valley now flows to the north, and that the dam is at the north end of the lake. When glacial meltwater occupied both the tunnel channel and the later esker tunnel, it flowed toward the south.

Spearhead Point Campground and West Point Campground both sit on broad, lower parts of the esker. Where the IAT turns abruptly west and passes a picnic area on the west side of the lake, there is a steep, narrow part of the esker a few hundred yards to the south that shows up well in figure 245.

Just west of the Mondeaux Flowage, the trail crosses an elongate depression filled with peat (fig. 244). This depression is likely the path of an old tunnel channel, one that carried water until the younger tunnel channel, which we see holding Mondeaux Flowage today, formed.

Your entire hike on this IAT segment is behind the Perkinstown Moraine. West of Mondeaux Flowage the IAT crosses rolling topography; you'll see some localized hummocky topography, but most of the ground beneath your feet is till that was dropped from the base of the glacier.

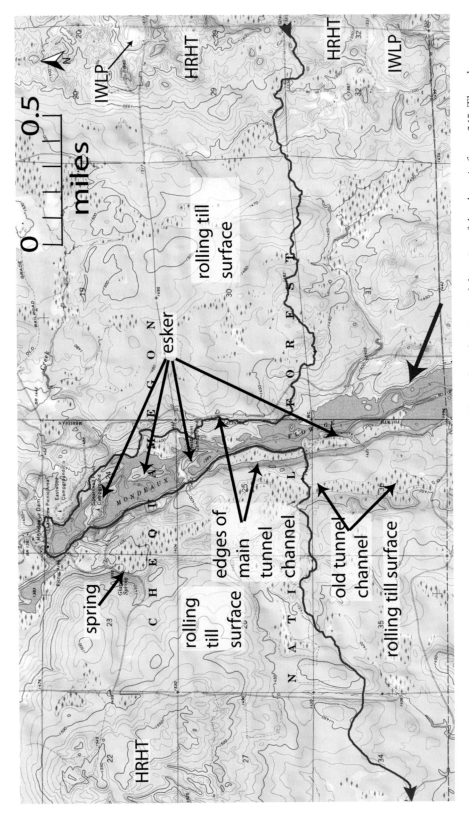

Figure 244. Topography of the Mondeaux Esker Segment. Blue arrow at base of map shows location and direction of the photo in figure 245. The valley that contains Mondeaux Flowage is a tunnel channel (SB 17) and contains an esker (SB 13). HRHT: high-relief hummocky topography (SB 11); IWLP: ice-walled-lake plain (SB 15). (Map is part of the Westboro and Mondeaux Dam USGS Quadrangles and was created with TOPO! © 2011 National Geographic Maps.)

Figure 245. Low oblique aerial view of part of Mondeaux Flowage looking toward the northwest. The esker runs down the middle of the lake. Location of photo is shown on figure 234.

## 105. Jerry Lake Segment

*CTH E to Sailor Creek Rd. (Forest Rd. 571) (15 miles)*

Between the CTH E trailhead and the North Fork of the Yellow River, the trail crosses rolling topography underlain by till (SB 5; fig. 246). You are behind the Perkinstown Moraine (SB 6; fig. 235), where there was little supraglacial sediment (SB 11) on the glacier surface. From there, the trail descends onto outwash (SB 8) along the North Fork of the Yellow River before climbing into low-relief hummocky topography (SB 11) with several water-filled kettles (SB 9). If you could look under the surface you would see a combination of lake sediment, outwash sand, and debris-flow sediment. As in many of the landscapes in northern Wisconsin, the hummocks here formed when the irregular layer of sand and silt on top of the glacier—the supraglacial sediment—settled down to the ground as the ice slowly

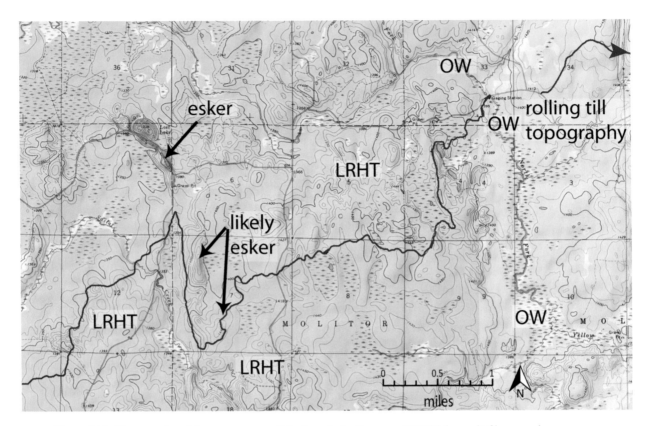

Figure 246. Topography of the eastern part of the Jerry Lake Segment. LRHT: low-relief hummocky topography (SB 11); OW: outwash (SB 8). (Map is part of the Mondeaux Dam and Jump River Fire Tower USGS Quadrangles and was created with TOPO! © 2011 National Geographic Maps.)

melted away. The supraglacial sediment layer was thinner here than it was where high-relief hummocky topography formed a few miles to the south, so you won't have as many high hummocks to climb.

About a mile west of Lake 19 Rd., the rise is a gravel ridge that is probably an esker (SB 13; fig. 246). It appears to be a continuation of a better-defined esker about a mile north of where the IAT crosses Sailor Creek Road. The latter is called the Lost Lake Esker, which is part of Lost Lake State Natural Area. The road itself follows the top of this sharp-crested ridge, which rises more than 50 feet above Lost Lake. Although this esker is not directly connected to a large esker several miles farther north, it seems likely that they were part of the same drainage system. In many cases, eskers that form with no ice beneath the tunnel are sharp-crested. Ones that formed in tunnels with even a small amount of ice beneath the tunnel usually are broader and less distinct because sediment tends to slide off the underlying ice in the late stages of melting.

About a mile and a half north of Jerry Lake, the trail enters the high-relief hummocky topography (fig. 247) of the Perkinstown Moraine (SB 6). Much of what underlies this topography is supraglacial sediment, ranging from boulder-sized material to fine-grained lake sediment deposited in ice-walled lakes. Jerry Lake fills a deep kettle.

## 106. Lake Eleven Segment
*Sailor Lake Rd. (Forest Rd. 571) to STH 64 (14 miles)*

When you hike this segment, you will be exclusively on the Perkinstown Moraine (fig. 235), which is the major moraine (SB 6) built during the last glaciation along the eastern side of the Chippewa Lobe. Most of the trail in the Lake Eleven Segment is in high-relief hummocky topography (SB 11; figs. 248, 249). Contrast figures 247 and 248. These are adjacent segments, but they have very different relief. Because the forest floor here is lush, you won't have very many good views of the soil below the surface. Like other hummocky topography around the edge of the Chippewa Lobe, however, you can be assured that supraglacial sediment dominates its composition (SB 11). Many of the flat-topped hills you'll see on this segment are ice-walled-lake plains (SB 15). The highest relief—and the steepest hike—along the segment is in the eastern part, where abundant kettle lakes (SB 9) fill deep cavities (fig. 248). Lake Eleven and all other lakes that can be seen from the trail are good examples of these kettles. Several large ice-walled-lake plains lie less than a mile east of the trail (upper right corner of figure 248). Figure 250 shows a view of an ice-walled-lake plain located about 3 miles east of where the IAT crosses CTH M (east of fig. 248).

Although it is not on the IAT, there is a good example nearby of an outwash fan with a prominent ice-contact face on its eastern edge (fig. 248). It is crossed by Butternut Hill Rd. The fan developed when stagnant, debris-covered ice sat in the vicinity of Kathryn Lake and Foss Lake near Perkinstown. Meltwater flowed off the ice surface toward the west, perhaps into an ice-walled lake, carrying gravel and building an alluvial fan against the stagnant ice. Later, the ice melted out, leaving the steep ice-contact face and the relatively uncollapsed outwash fan surface. This landform is also called an outwash head (SB 8, SB 10).

Close to the western end of this segment, the IAT skirts several ice-walled-lake plains that are small: only about 20 feet high (fig. 249). All the depressions seen here are kettles, and many are peat-filled.

From here to the west, you will walk on IAT segments that are in or behind the Chippewa Moraine instead of the Perkinstown Moraine, which you have been traversing (fig. 235).

## 107. Chippewa River Segment
*CTH Z to CTH CC (1.8 miles)*

The Chippewa River Segment is the easternmost segment of the IAT in the Chippewa Moraine, which formed on the west side of the Chippewa Lobe (figs. 235, 237). On the

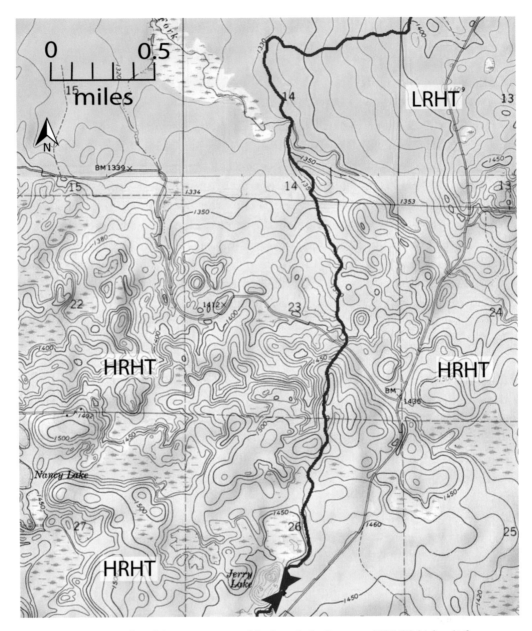

Figure 247. Topography of the western part of the Jerry Lake Segment. HRHT: high-relief hummocky topography (SB 11); LRHT: low-relief hummocky topography (SB 11). (Map is part of the Jump River Fire Tower and Perkinstown USGS Quadrangles and was created with TOPO! © 2011 National Geographic Maps.)

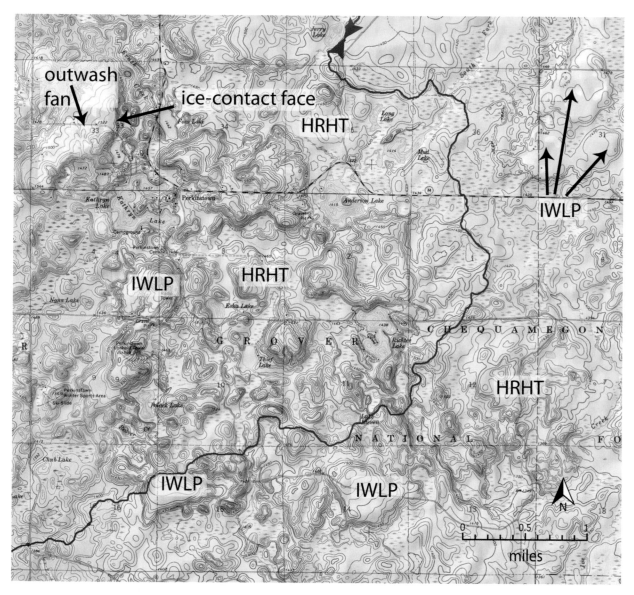

Figure 248. Topography of the eastern part of the Lake Eleven Segment. The entire area has high-relief hummocky topography (HRHT) (SB 11). IWLP: ice-walled-lake plain (SB 15). (Map is part of the Perkinstown and Lublin NW USGS Quadrangles and was created with TOPO! © 2011 National Geographic Maps.)

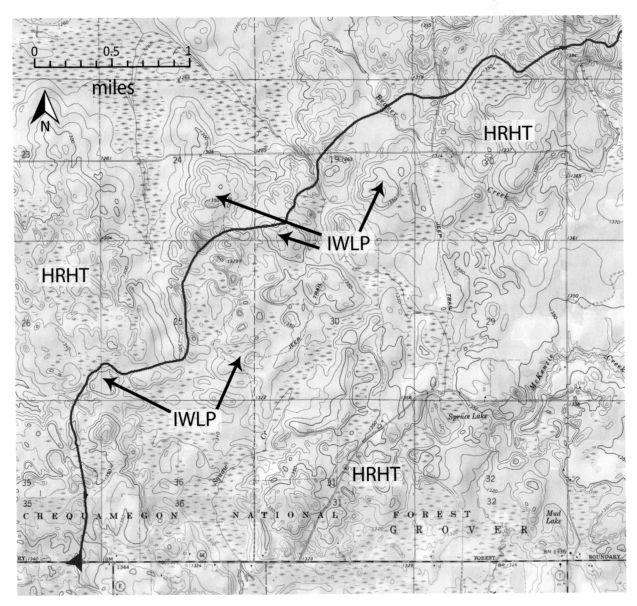

Figure 249. Topography of the western part of the Lake Eleven Segment. The entire area has high-relief hummocky topography (HRHT) (SB 11). IWLP: ice-walled-lake plain (SB 15). (Map is part of the Lublin NW and Lublin USGS Quadrangles and was created with TOPO! © 2011 National Geographic Maps.)

Figure 250.  Low oblique aerial view of ice-walled-lake plain (IWLP) (SB 15) in high-relief hummocky topography (SB 11). Note raised rim with lighter soil cover and low center with darker (moister) color. This ice-walled-lake plain is about 3 miles west of the Ice Age Trail on the north side of CTH M.

eastern end of this segment, the trail is on the low-relief hummocky topography (SB 11) of the Chippewa Moraine (fig. 251). The outer edge of this hummocky complex is more than 10 miles to the southwest (fig. 235). Ice flowed from northeast to southwest. Perch Lake is in a kettle (SB 9), but the ridge that CTH CC occupies is human-made.

## 108. Firth Lake Segment
### CTH CC to 245th Ave. (Moonridge Tr.) (6 miles)

One of the easternmost distinctive landforms you'll see on the Firth Lake Segment is a low, mostly collapsed, ice-walled-lake plain (SB 15) just west of CTH CC (fig. 251). As you continue west from here, the trail crosses a low-relief hummocky surface (SB 11) underlain by supraglacial sediment that extends to the east edge of the marsh around Firth Lake. From

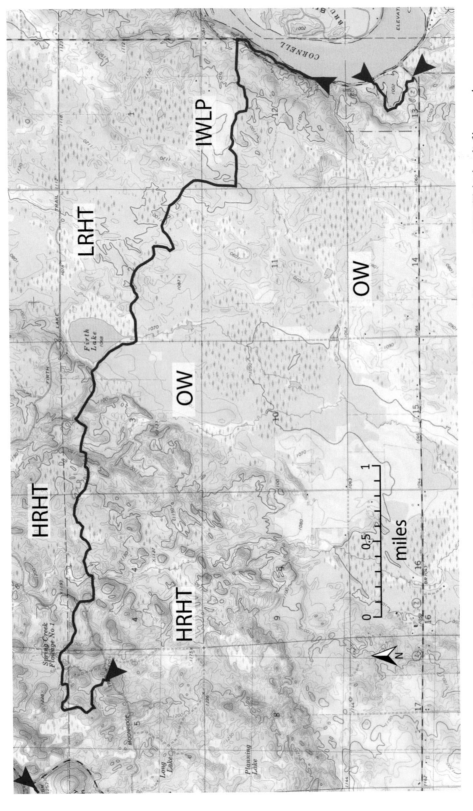

Figure 251. Topography of the Chippewa River (lower right) and Firth Lake (lower right to upper left) segments. HRHT: high-relief hummocky topography (SB 11); IWLP: ice-walled-lake plain (SB 15); LRHT: low-relief hummocky topography (SB 11); OW: outwash (SB 8). (Map is part of the Cornell and Bob Lake USGS Quadrangles and was created with TOPO! © 2011 National Geographic Maps.)

there the IAT passes onto an outwash plain (SB 8). The braided stream that deposited this outwash flowed from north to south. Firth Lake is a shallow kettle (SB 9) partly filled with peat. Walk to the boardwalk at its edge for a good view of a typical northern wetland and a lake slowly filling with organic sediment. Peat will eventually take over the rest of the open water, and all of Firth Lake will be a peat bog with a small stream flowing through it. Immediately west of Firth Lake, the trail climbs into high-relief hummocky topography underlain by supraglacial sediment. All the depressions you'll see along the rest of your hike are kettles.

## 109. Harwood Lakes Segment

*CTH E to 167th St. (Plummer Lake Rd.) (5 miles)*

Your entire hike on the Harwood Lakes Segment will be in the Chippewa Moraine (fig. 235). Here the high-relief hummocky topography (SB 11) is dramatic (fig. 252). Near its eastern trailhead, the IAT skirts the base of Baldy Mountain, a 130-foot-tall ice-walled-lake plain (SB 15). When the lake walls collapsed, they left the conical hill with the very small flat surface at the top that you see today. From Baldy Mountain westward, you'll climb up and drop down into many—at least twenty—other hummocks and kettles (SB 9). As you hike toward the western end of this segment at Plummer Lake, consider that each hummock formed from settling supraglacial sediment.

About a half mile east of the Plummer Lake Rd. trailhead, the trail goes along the ice-contact face of an ice-walled-lake plain. When the lake still lapped its shores, the water surface was over 100 feet above the present level of the trail on the lake's north side. To hold the water in, the ice had to be more than 100 feet thick where the trail is now, and topping that was likely 50 to 100 feet of supraglacial sediment.

## 110. Chippewa Moraine Unit of the Ice Age National Scientific Reserve and the Chippewa Moraine Segment

*167th St. (Plummer Lake Rd.) to 267th Ave. (Oak Ln.) (7.8 miles)*

The Chippewa Moraine exhibits impressive high-relief hummocky topography (SB 11) and some of the best developed ice-walled-lake plains (SB 15) anywhere in the United States (figs. 235, 253). The moraine is more than 5 miles wide here and has numerous deep kettles (SB 9) and high hummocks. Many of the kettles you see are small lakes or peat bogs.

Glacier ice advanced into this area from the northeast. The outermost edge of the late Wisconsin advance cuts across the southwest corner of the reserve (fig. 253). Outwash (SB 8) shown in the southwest corner of the map is outside the terminal moraine (SB 6). Look for a short esker (SB 13) along the loop trail northeast of the outwash. The relief of the present topography (kettle bottom to hummock top) is an expression of the thickness of supraglacial sediment. The sediment may have been particularly thick here because of an

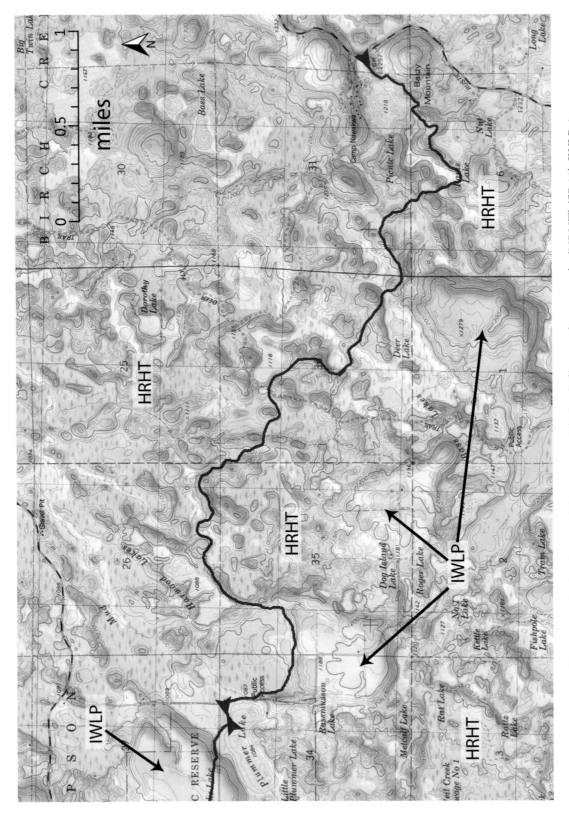

Figure 252. Topography of the Harwood Lakes Segment. The entire area is high-relief hummocky topography (HRHT) (SB 11). IWLP: ice-walled-lake plain (SB 15). (Map is part of the Bob Lake USGS Quadrangle and was created with TOPO! © 2011 National Geographic Maps.)

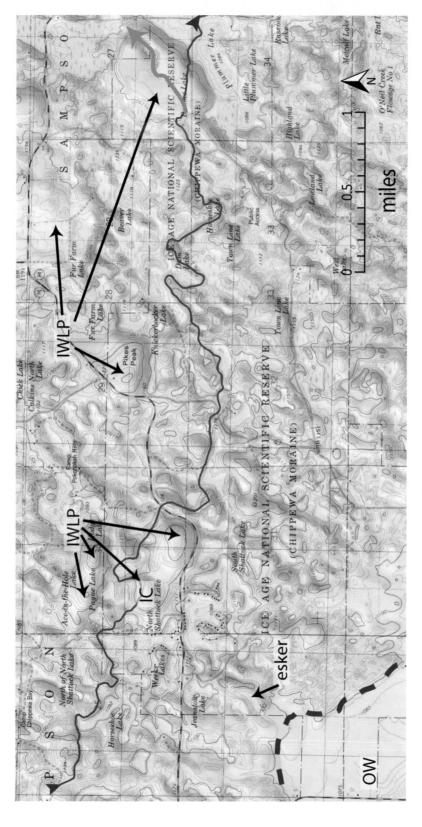

Figure 253. Topography of the Chippewa Moraine Segment. The entire area is high-relief hummocky topography (HRHT) (SB 11) except for the outwash (OW) (SB 8). Green arrow near upper right is meltwater channel described in text. Dashed blue line is outer edge of the Chippewa Moraine. IWLP: ice-walled-lake plain (SB 15); IC: interpretive center for the Chippewa Moraine Unit of the Ice Age National Scientific Reserve. (Map is part of the Bob Lake and Marsh Miller Lake USGS Quadrangles and was created with TOPO! © 2011 National Geographic Maps.)

extensive frozen-bed zone near the ice margin (SB 3). Alternatively, it may be that high land to the northeast near Flambeau Ridge, combined with a frozen bed, slowed the flow of the ice near the margin. Ice from behind pushed up and onto the slower-moving ice. As it did so, debris-rich ice and frozen sediment moved upward into the glacier as well. As the ice began to melt, all the debris that was released by melting piled up on the glacier's surface, getting thicker and thicker, sliding and flowing from high spots to low spots until all the buried ice had melted (fig. 38). This slow melting process may have lasted several thousand years after the active glacier margin had retreated to Lake Superior.

The IAT crosses a large ice-walled-lake plain just north of Plummer Lake, near the east end of the segment. A now-dry channel that cuts across the top of the ice-walled-lake plain (green arrow in fig. 253) must have been eroded just after the lake drained but before the surrounding ice had melted out. The only source of water to cut the channel would have been melting glacier ice at its head (southwest end).

You'll have an excellent view of this hummocky topography from the Chippewa Moraine Interpretive Center off CTH M, where you can purchase a hiking field-trip guide that describes in more detail the features found on other trails around the interpretive center and along other parts of the IAT (fig. 253).

## 111. Southern Blue Hills Segment
*Bass Lake Rd. (Old 14 Rd.) to Yuker Rd. (6.3 miles)*

At the trailhead on Bass Lake Rd. you will be standing on pitted outwash (SB 10) and looking out at a landscape of low hummocks (SB 11), but from there you'll quickly climb into higher-relief hummocky topography (fig. 254). These high hummocks are part of the Chippewa Moraine (fig. 235). After skirting the north side of North Lake, the trail climbs into the Blue Hills, which are erosion-resistant Precambrian quartzite that glaciers partially covered with stoney till. You'll also see angular, tan to gray quartzite boulders, many of which are locally derived, along the route. The extent of this Barron quartzite is shown in figure 236. (See the introduction to this book and the Devil's Lake description for more details on the formation of quartzites in Wisconsin.)

Ice covered all of the Blue Hills during the early Chippewa advance. It appears, however, that the late Chippewa advance was not as thick as its predecessor: the later glacier flowed partly into the Blue Hills but didn't cover the hill tops at their western end. The outermost edge of this advance was 1 to 2 miles west of the area shown in figure 254, so this segment is on deposits of the late Chippewa advance. Good evidence for the intense permafrost conditions that once existed here is talus—rock pieces that crumbled during extreme freeze-thaw events—piled at the base of steep rock walls in what is now the Blue Hills Felsenmeer State Natural Area, which is a few miles west of here (fig. 238).

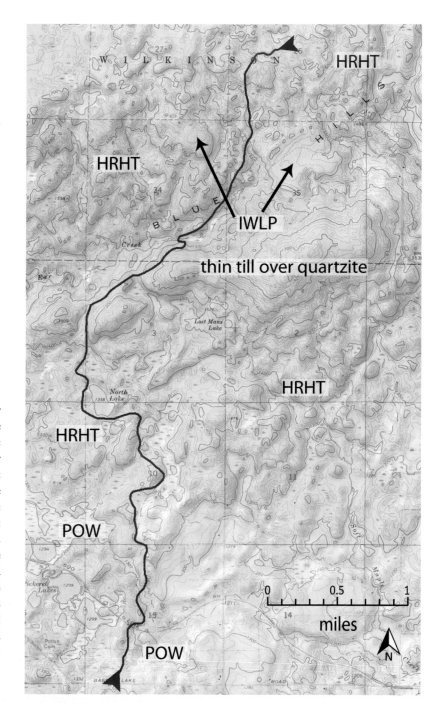

Figure 254. Topography of the Southern Blue Hills Segment. HRHT: high-relief hummocky topography (SB 11); IWLP: ice-walled-lake plain (SB 15); OW: outwash (SB 8); POW: pitted outwash (SB 10). (Map is part of the Weyerhauser and Bucks Lake USGS Quadrangles and was created with TOPO! © 2011 National Geographic Maps.)

About one mile north of North Lake the trail descends from the quartzite hills with thin till back into high-relief hummocky topography. Here you'll see several ice-walled-lake plains (SB 15), but not as many as you did in the Chippewa Moraine Segment. The IAT remains in this hummocky topography of the Chippewa Moraine to the trailhead on Yuker Rd.

## 112. Northern Blue Hills Segment
*CTH F southern trail crossing to CTH F at the Murphy Flowage*
*Recreation Day Use Area (8.5 miles)*

In the Northern Blue Hills Segment (fig. 235), the IAT follows the edge of Devil's Creek, then crosses it and climbs onto a till (SB 5) and supraglacial sediment-covered upland with low-relief hummocks (SB 11; fig. 255). The broad valley west of the trailhead appears to have been a lake that the glacier dammed as it was retreating. Another ice-dammed lake probably formed in debris-covered ice 1.5 miles north of the trailhead. When you pass the parking area at the north end of Stout Rd., you will cross the divide between streams that flow eastward to the Chippewa River and those that flow westward to the Red Cedar River. From there, you'll hike in low-relief hummocky topography to the trail's intersection with Bucks Lake Road, then cross onto an outwash terrace pitted with kettles (SB 9, SB 10). Most of the hilltops that you can see rising on the horizon all around you have Precambrian Barron Quartzite close to their surface, with just a thin layer of till or supraglacial sediment overtop.

## 113. Hemlock Creek Segment
*CTH F at Murphy Flowage Recreation Day Use Area to Finohorn Rd. (5.5 miles)*

The eastern trailhead of the Hemlock Creek Segment (fig. 235) sits on pitted outwash (SB 10; fig. 256). From the trailhead you'll drop into a deep, complex kettle through which Hemlock Creek flows until it empties into Hemlock Lake. A short piece of the trail is along the crest of a small esker that splits into several small gravel ridges. Coming out of the kettle, the IAT is on pitted outwash to Bolgers Rd., west of which you'll have a quick climb into Precambrian Barron quartzite hills. From the hills the trail descends onto a pitted-outwash plain and then across smooth outwash to the western trailhead. Figure 256 shows the edge of the late Chippewa advance. The ice margin wrapped around and only partly covered the Blue Hills. An ice-marginal stream flowed toward the southwest, between the glacier and the Blue Hills.

For more information on the tunnel channel (green arrow) in figure 256, in the position of the present Red Cedar Lake, read (and hike to) the next segment, Tuscobia.

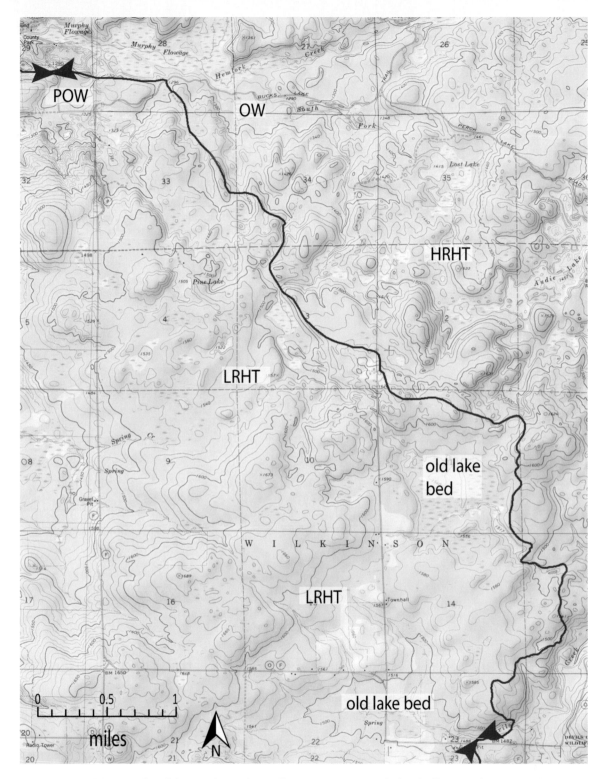

Figure 255. Topography of the Northern Blue Hills Segment. HRHT: high-relief hummocky topography (SB 11); LRHT: low-relief hummocky topography (SB 11); OW: outwash (SB 8); POW: pitted outwash (SB 10). (Map is part of the Bucks Lake and Mikana USGS Quadrangles and was created with TOPO! © 2011 National Geographic Maps.)

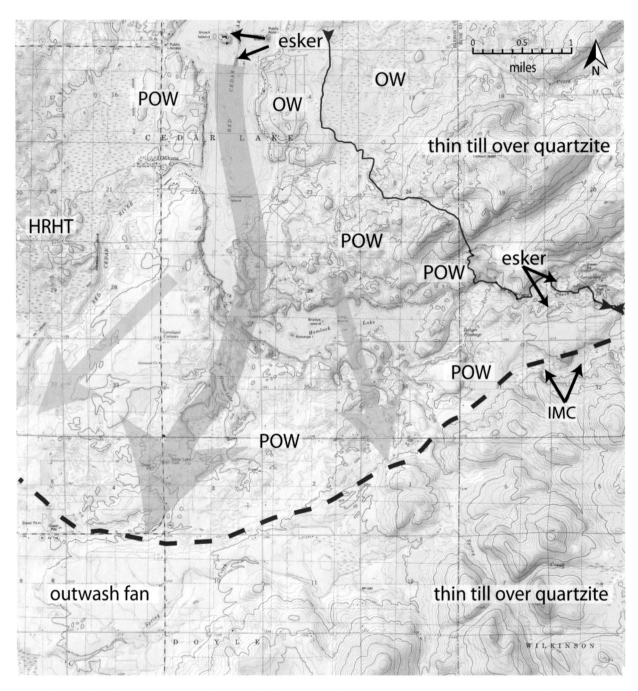

Figure 256. Topography of the Hemlock Creek Segment. Green arrow shows path of tunnel channel. Dashed blue line is outer edge of late Chippewa advance. Blue arrows show ice-flow direction. The esker (SB 13) is discussed in the Tuscobia Segment description. HRHT: high-relief hummocky topography (SB 11); IMC: ice-marginal channel (SB 7); OW: outwash (SB 8); POW: pitted outwash (SB 10). (Map is part of the Mikana USGS Quadrangle and was created with TOPO! © 2011 National Geographic Maps.)

## 114. Tuscobia Segment

*Featherstone Rd. to CTH SS (11.1 miles)*

The Tuscobia Segment (fig. 235) of the IAT follows the Tuscobia State Trail westward across late-Chippewa-advance deposits, a tunnel channel (SB 17), and outwash (SB 8) from both the Chippewa Lobe and the Superior Lobe (figs. 257, 258). Look for Balsam Lake and Red Cedar Lake, both of which occupy a tunnel channel that formed during the late Chippewa advance. Water in this tunnel channel flowed toward the south and exited from beneath the ice just south of Red Cedar Lake (fig. 256). Notice the distinct change in topography where the tunnel channel exited: the pitted outwash (SB 10) dotted with kettles

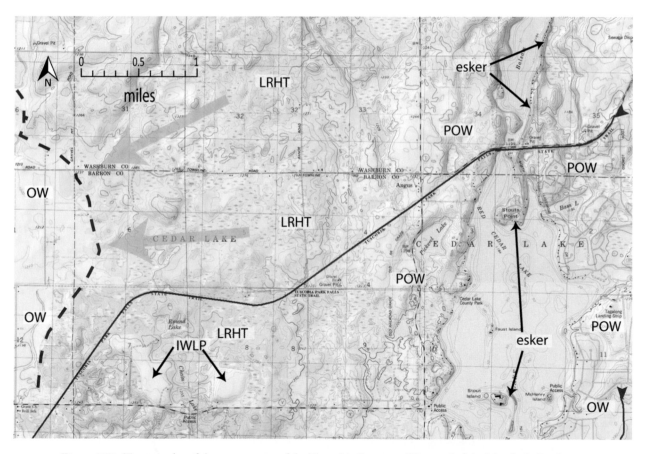

Figure 257. Topography of the eastern part of the Tuscobia Segment. West end of the Hemlock Creek Segment is visible in the lower right. Dashed blue line shows outer edge of the late Chippewa advance. Blue arrows show ice-flow direction. IWLP: ice-walled-lake plain (SB 15); LRHT: low-relief hummocky topography (SB 11); OW: outwash (SB 8); POW: pitted outwash (SB 10). (Map is part of the Birchwood, Mikana, and Rice Lake North USGS Quadrangles and was created with TOPO! © 2011 National Geographic Maps.)

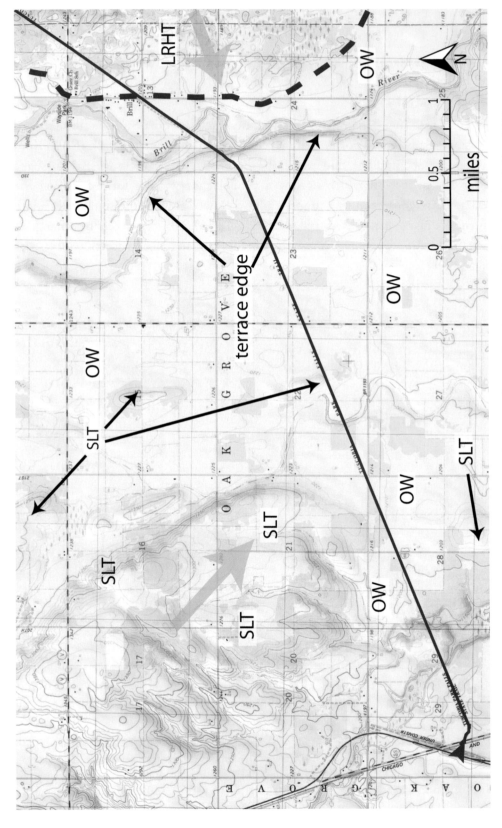

Figure 258. Topography of the western part of the Tuscobia Segment. Blue arrows show ice-flow direction of the Superior Lobe (left) and Chippewa Lobe (right). Dashed blue line indicates outer edge of late Chippewa advance. LRHT: low-relief hummocky topography (SB 11); OW: outwash (SB 8); SLT: low hills of Superior Lobe till (SB 5). (Map is part of the Rice Lake North and Haugen USGS Quadrangle and was created with TOPO! © 2011 National Geographic Maps.)

(SB 9) where the ice sat shifts to smooth outwash that formed as a fan in front of the ice edge. The tunnel channel formed when the bed of the glacier was still frozen (SB 3) and a sudden, short-lived torrent of water gouged a tunnel under the ice.

There are deep kettles just north and south of the trail as you proceed west from the trailhead. Just east of where the trail and STH 48 cross the tunnel channel at Red Cedar Lake Narrows, you'll hike through Honeymoon Esker. This esker (SB 13) formed after the tunnel channel, when the bed of the glacier had warmed to the melting point and as the ice was melting away. The esker goes below the surface of Red Cedar Lake to the south and shows up as small islands here and there (figs. 256, 257) above the water's surface. North of the trail, the esker borders the east edge of Balsam Lake. Balsam Lake Road runs along the esker crest. Water in the tunnel flowed to the south as the sand and gravel settled to form the esker. You can see a good cross section of the esker where the trail and highway cut through it.

West of the tunnel channel, the IAT crosses pitted outwash that braided streams (SB 8) flowing to the south deposited as ice of the late Chippewa phase melted back. Another very deep kettle is just north of the trail less than a half mile west of the tunnel channel, and Pickerel Lake occupies a kettle as well (fig. 257). From there, the trail passes through low-relief hummocky topography—this is actually the moraine (SB 6) that the late Chippewa advance built. At the village of Brill you'll leave the moraine and walk over a low outwash terrace before crossing the floodplain of the Brill River. From there, it is a short climb onto an outwash terrace that is about 30 feet higher than the floodplain (fig. 258). Braided streams flowing south from both the Chippewa and Superior lobes deposited the high outwash surface here. As the ice retreated to the north, the braids of the stream merged to a single channel that then eroded its bed, downcutting to the present stream level.

Near the west end of this segment, look for low hills of Superior Lobe till (SB 5) partly buried by the outwash (fig. 235). This till is late Wisconsin age, but the Emerald advance of the Superior lobe deposited it here. About 1 mile east of the CTH SS trailhead (fig. 258), you'll again hike on outwash deposited by streams flowing toward the south, away from the Superior Lobe.

# Superior Lobe
# Ice Age Trail Segments

Betweeen 25,000 and 30,000 cal. years ago, the Superior Lobe of the Laurentide Ice Sheet advanced into northwestern Wisconsin out of the Lake Superior basin. It advanced across a landscape over which, in turn, several other glaciers had slid and scraped in the past. The deposits from these older glaciers sit at the surface south of the late Wisconsin glacial deposits that the IAT segments cross (fig. 259; Baker et al. 1983). You will see till that the Superior Lobe deposited: it is reddish brown because of iron staining, similar to the till in other lobes that came out of the Lake Superior basin. There are substantial differences, however, between the rock types that make up the till from each successive glacier lobe. To explain why, we need to look at much earlier geologic history that shaped the rocks the glaciers later carved.

Late in the Precambrian, about 1,100 million years ago, long northeast-southwest-trending fractures and faults developed as tectonic forces began to pull apart the North American continent (fig. 260). This tearing of the earth's crust produced valleys similar to the rift valleys in East Africa today. The rift in this area of Wisconsin is called the Keweenawan, or Midcontinent, Rift, and it extends from the Lake Superior basin toward the southwest as far as Kansas (fig. 261). For the most part, it is covered south of Minnesota by younger sedimentary rocks.

As the rifting occurred, volcanic eruptions of lava flowed into the new rift valleys. The valleys flooded time and again with hot lava, producing overlapping layers of basalt (SB 21) across the entire area. Streams of water flowing down the steep slopes on the sides of the rift valleys carried gravel and sand that accumulated on top of the basalt flows in many places. The resulting distribution of Keweenawan rocks is shown in figure 236. You'll

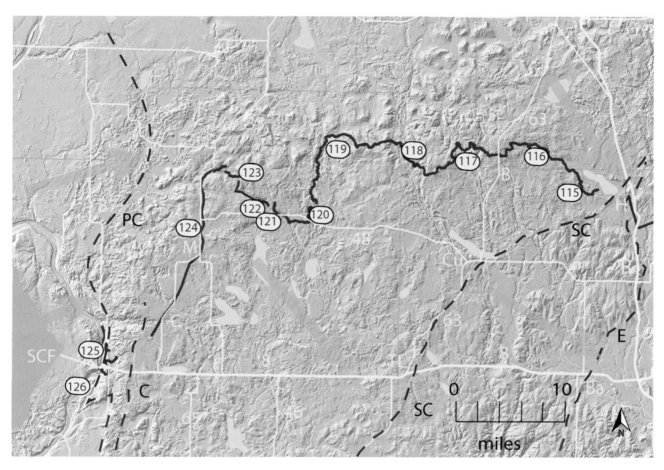

Figure 259. Shaded relief of the Superior Lobe IAT segments (red): (115) Bear Lake, (116) Grassy Lake, (117) Timberland Hills Area, (118) Sand Creek, (119) McKenzie Creek, (120) Pine Lake, (121) Straight River, (122) Straight Lake, (123) Trade River, (124) Gandy Dancer State Trail, (125) St. Croix Falls. The Interstate State Park Unit of the Ice Age National Scientific Reserve (126) is located at the western terminus of the Ice Age Trail. Yellow lines and numbers indicate major highways. Cities and villages shown (yellow): (B) Baronette, (Ba) Barron, (C) Centuria, (Cu) Cumberland, (F) Frederic, (M) Milltown, (R) Rice Lake, (SCF) St. Croix Falls, (TL) Turtle Lake. Dashed blue line SC is outer edge of the St. Croix advance of the Superior Lobe. Dashed blue line E is approximate outermost extent of the Emerald Phase of the Superior Lobe. Dashed blue line C is approximate outermost extent of the Centuria Phase of the Superior Lobe. Dashed blue line (PC) is the approximate outermost extent of the Pine City Phase of the Grantsburg Sublobe. Blue arrows show ice-flow direction. (Base map constructed from USGS National Elevation Dataset and modified by WGNHS.)

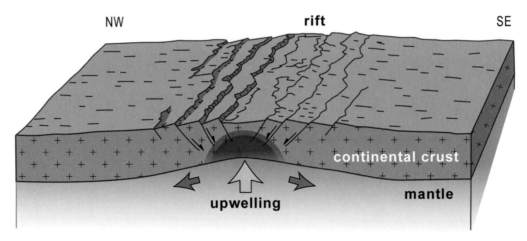

Figure 260. Early development of the Keweenawan Midcontinent Rift in late Precambrian time. This was followed by partial filling of the rift by flood basalt (SB 21) and locally derived sand and gravel, and ultimately by compression of the earth's crust at this location. (Modified from LaBerge 1994; drafted by Mary Diman.)

see the exposed basalt flows on the IAT near its western terminus at Interstate State Park, especially along the walls of the gorge.

In places, air bubbles were trapped in the lava as it cooled and hardened. Groundwater rich in dissolved calcium, silica, iron, and other elements later flowed through these openings, forming crystals that partly or completely filled the bubbles. Common crystal fillings you can see today include the minerals calcite, quartz, and feldspar.

Many hobbyists in northwestern Wisconsin and Minnesota collect Lake Superior agates. These brightly colored stones formed when concentric bands of very small crystals of the mineral quartz ($SiO_2$), variously stained with iron, accumulated in voids in the basalt. Weathering and erosion for tens of millions of years released these weather-resistant pieces of agate, and now you're likely to find these relatively common stones in the till or sand and gravel of the Superior Lobe.

Glacial deposits overlie Cambrian sandstone beneath trail segments east of the Straight River Segment, but basalt boulders that the glacier carried from the northeast are present all along the trail. From there to the western terminus of the IAT, the trail overlies Keweenawan basalt.

The earliest late Wisconsin advance of the Superior Lobe is called the Emerald advance (fig. 259). The glacier deposited a thin, discontinuous till layer without building an end moraine (SB 6). The St. Croix advance, which came along slightly later, built the broad, hummocky end moraine that the Bear Lake and Grassy Lake segments traverse (fig. 259). Although it contains high-relief hummocky topography (SB 11), the relief over which you'll

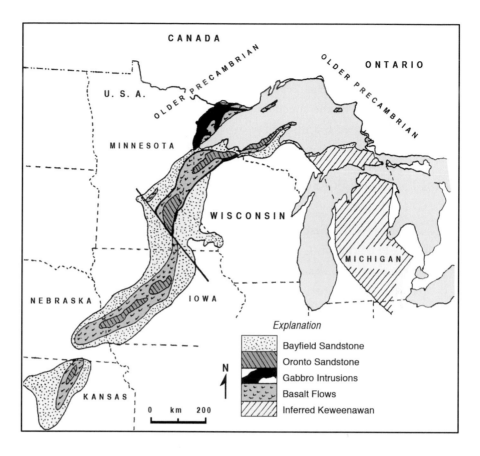

Figure 261. Extent of the Midcontinent Rift. The rift continues beneath Lake Superior and is responsible for the location of the Lake Superior basin. (Modified from LaBerge 1994; drafted by Mary Diman.)

climb is not as great here as it is in the Chippewa Moraine or farther east in the Harrison Moraine. You'll come upon some ice-walled-lake plains (SB 15) in and behind the moraine, but they are also not as tall or abundant as they are farther east along the IAT.

Instead of following the St. Croix Moraine toward the southwest, the IAT goes more or less directly west for about 20 miles before turning to the south (fig. 259). Relief along this stretch is generally fairly low, with some areas of gently rolling hills built from the gradual deposition of till (SB 5) at the base of the glacier. The Superior Lobe IAT segments cross several spectacular tunnel channels (SB 17). The Timberland Hills Area Segment goes through the bottom of the junction between two tunnel channels (SB 17). The Sand Creek Segment follows hummocky topography in the floor of a tunnel channel, and most spectacular of all is the Straight Lake tunnel channel, which the Pine Lake, Straight River, and Straight Lake segments all follow (fig. 259). These segments also follow a very long esker (SB 13).

Just northwest of the IAT you'll come to a high-relief upland dominated by basalt hills rather than glacial hummocks (fig. 259). There are several low outcrops of the basalt close to the trail. Look for their striations, which indicate that the ice flowed toward the southeast.

For the most part, the trail is on pitted outwash (SB 10) throughout much of its length between the Trade River Segment and the St. Croix Falls Segment. Interstate State Park, at the western terminus of the Ice Age Trail, is a unit of the Ice Age National Scientific Reserve. The trail follows the crest of a tall esker right in the city of St. Croix Falls in addition to providing views of the spectacular basalt-lined gorge. Here you'll also see deep potholes that were cut beneath a huge flow of water from Lake Superior about 16,000 cal. years ago, as well as striations scratched into some of the bedrock surfaces.

## 115. Bear Lake Segment
*CTH VV (28th Ave.) to 30th Ave. (4.2 miles)*

The Bear Lake Segment (fig. 259) is entirely in high- and low-relief hummocky terrain (SB 11) of the St. Croix Moraine (fig. 262). The outermost edge of the moraine is about one and a half miles southeast of the trailhead on CTH VV. All the lakes fill glacial kettles (SB 9), as do most of the wetlands. This topography is typical of much of the St. Croix Moraine. The relief of the hummocks here is not as great as it is in parts of the Wisconsin Valley Lobe or the Chippewa Lobe. The depression that Bear Lake occupies is part of a tunnel channel (SB 17) formed when ice sat at the St. Croix Moraine. Note that the lake is more than 80 feet deep directly north of the CTH VV trailhead, whereas other parts of the channel are much shallower. This difference in water depth resulted from an uneven buildup of sediment in the tunnel channel. Water from the channel deposited coarse sand and gravel in front of the ice margin. The city of Haugen is located on this sand and gravel, where you'll also find large gravel pits in the outwash fan (SB 8) on the south side of town on both sides of CTH V.

## 116. Grassy Lake Segment
*30th Ave. to Pershing Rd. (7.2 miles)*

This segment of the IAT is all in high-relief hummocky topography (SB 11; figs. 259, 263) of the St. Croix Moraine (SB 6). Based on the depth of the valley, much of the water in the tunnel channel (SB 17) beneath apparently came down the present Bear Creek valley (fig. 262), and not down the valley of Boyer Creek. Boyer Creek valley, which is northeast of the western part of the Grassy Lake Segment (fig. 263), is an underfit stream (SB 8), indicating that at times in the past more water came down the valley than you see now in the flow of the present Boyer Creek. Note, however, the kettles (SB 9) in the valley bottom, which indicate that ice lay buried beneath the sediment of the stream bottom as meltwater flowed overtop. The broad St. Croix Moraine here may represent several minor readvances. The somewhat higher-relief ridge that you'll hike across about a half mile east of Leaman Lake (fig. 263) may represent one of these minor readvance positions.

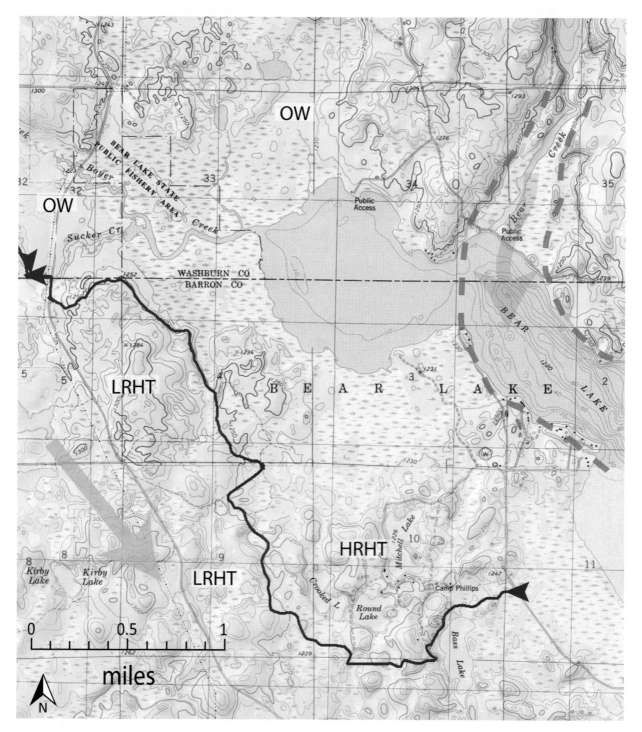

Figure 262. Topography of the Bear Lake Segment. Blue arrow shows ice-flow direction. Dashed green lines show edges of tunnel channel (SB 17). Green arrow shows water-flow direction in channel. HRHT: high-relief hummocky topography (SB 11); LRHT: low-relief hummocky topography (SB 11); OW: outwash (SB 8). (Map is part of the Haugen USGS Quadrangle and was created with TOPO! © 2011 National Geographic Maps.)

A connector road between this segment and the Timberland Hills Area Segment is called Brick Yard Rd. (fig. 263). The low, nearly circular hill south of this road is probably an ice-walled-lake plain (SB 15). We have not been able to determine the location of the brickyard, but the clay may have been mined for bricks from the north edge of the lake plain. Miller Creek valley (fig. 263), which is about a mile south of the western trailhead, is a small tunnel channel.

## 117. Timberland Hills Area Segment
*Leach Lake Rd. to Lake 32 Rd. (9.7 miles)*

The Timberland Hills Area Segment (figs. 259, 264, 265) is all in high-relief hummocky topography (SB 11) of the St. Croix Moraine (SB 6). The eastern trailhead of this segment (red arrow in fig. 264) is on the north edge of an ice-walled-lake plain (SB 15), most of which has collapsed. The high spot on Leach Lake Rd. 0.3 miles southeast of the trailhead appears to be on a rim of this feature. The Leach Lake Rd. trailhead is at the south end of a low ridge that is probably an ice-disintegration ridge (SB 11) rather than an esker (SB 13). The ridge formed as supraglacial sediment slid off buried ice. The trail drops onto the floor of a deep tunnel channel (SB 17) north of Offers Lake. This spot is also where two tunnel channels come together: one from the north that now contains Leach Lake, and the other from the west, which the IAT follows. After passing an unnamed pond, you'll climb onto a ridge where you'll remain for about a half mile. The ridge is probably an esker, but it could also be an ice-disintegration ridge that formed as buried ice melted out. The trail climbs to the north out of the tunnel channel onto what appears to be a high sand-and-gravel surface. It stays on this for only a short distance before dropping steeply into the tunnel channel again.

From the point where you climb out of the tunnel channel to the end of this segment, the IAT is on low- to moderate-relief hummocky topography. Several ridges along this stretch appear to be ice-disintegration ridges as opposed to eskers (fig. 265).

## 118. Sand Creek Segment
*Lake 32 Rd. to CTH E (15th St.) (5.5 miles)*

On this segment of the IAT, you'll follow a tunnel channel (SB 17) for most of the segment's length (figs. 259, 266). This is the same tunnel channel that is in the southwest corner of figure 265. In this area, the landscape is high-relief hummocky topography (SB 11) containing sand and gravel, unlike the typical topography of this type in northern Wisconsin that is comprised of till-like supraglacial sediment. As the flow of water in the tunnel at the base of the ice began to decrease, water and wet sediment froze onto the bottom of the glacier. When this debris-rich ice melted, it produced the hummocky topography. Look for remnants of what may be a small esker (SB 13) that formed during the final meltout of the

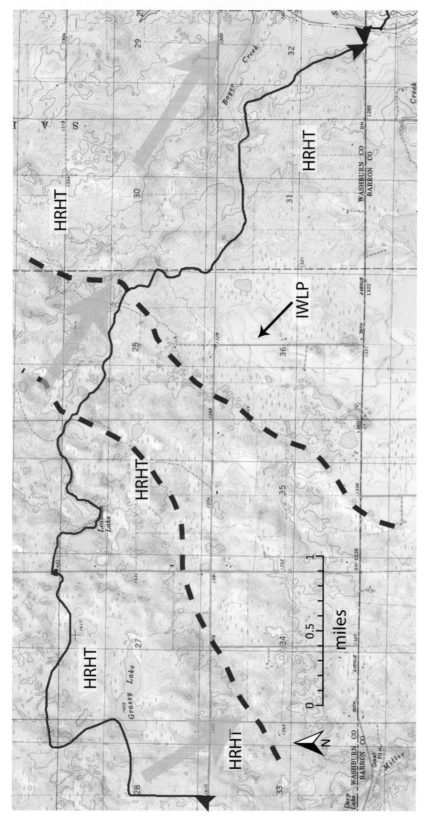

Figure 263. Topography of the Grassy Lake Segment. Blue arrows show ice-flow direction. Dashed blue lines show minor positions of ice-margin stability. HRHT: high-relief hummocky topography (SB 11); IWLP: ice-walled-lake plain (SB 15). (Map is part of the Sarona and Shell Lake USGS Quadrangles and was created with TOPO! © 2011 National Geographic Maps.)

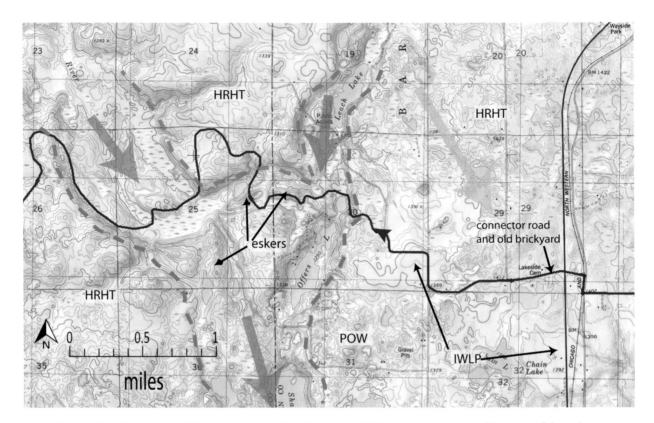

Figure 264. Topography of the eastern part of the Timberland Hills Area Segment. Red line east of the red arrow is a connector road. Blue arrow shows ice-flow direction. Dashed green lines show edges of tunnel channels. Green arrows show water-flow direction in tunnel channels. HRHT: high-relief hummocky topography (SB 11); IWLP: ice-walled-lake plain (SB 15); POW: pitted outwash (SB 10). (Map is part of the Timberland USGS Quadrangle and was created with TOPO! © 2011 National Geographic Maps.)

ice to the east of the trail. All the depressions in this area are kettles (SB 9). Water in the tunnel channel and in the much younger esker tunnel flowed toward the southeast, in the same direction as the ice flow.

## 119. McKenzie Creek Segment
*CTH E (15th St.) to CTH O (270th Ave.) (14 miles)*

The McKenzie Creek Segment of the IAT is behind the St. Croix Moraine (figs. 259, 267). Along much of the trail, you'll hike over a rolling till surface that formed at the base of the ice with relatively little supraglacial sediment on top. In some places along the trail, streams carried sand and gravel on top of the ice and deposited it in a thicker layer. This sand and gravel produced the area's high-relief hummocky topography (SB 11) as the ice underneath

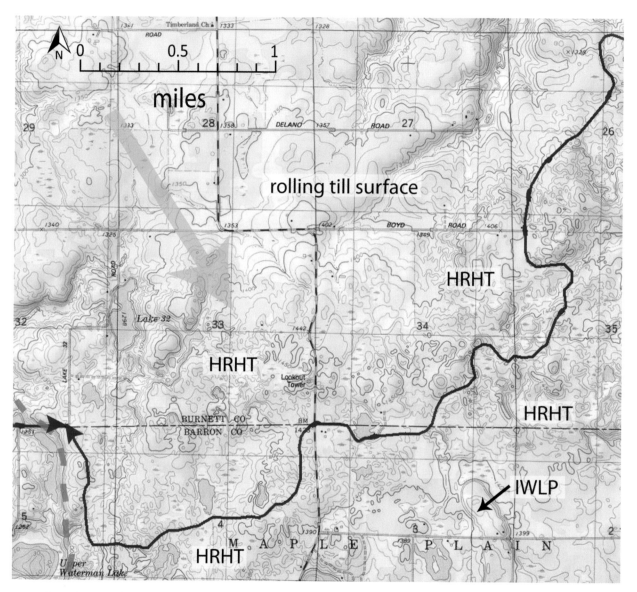

Figure 265.   Topography of the western part of the Timberland Hills Segment. Blue arrow shows ice-flow direction. Dashed green line shows part of the edge of the tunnel channel (SB 17) shown on figure 266. HRHT: high-relief hummocky topography (SB 11); IWLP: ice-walled-lake plain (SB 15). (Map is part of the Timberland USGS Quadrangle and was created with TOPO! © 2011 National Geographic Maps.)

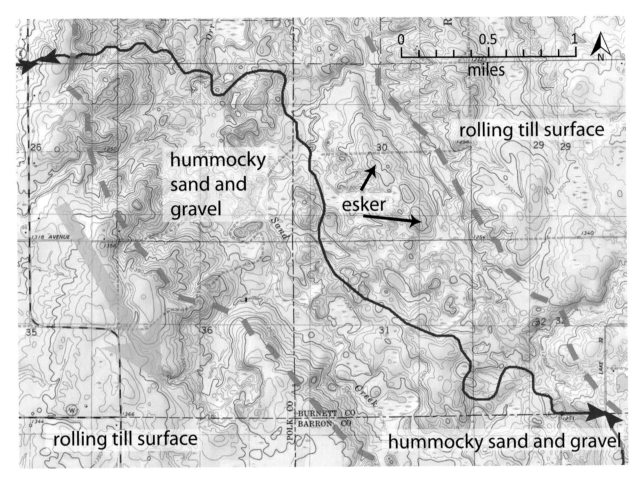

Figure 266. Topography of the Sand Creek Segment. Blue arrow shows ice-flow direction. Dashed green lines show edges of a tunnel channel (SB 17). Water-flow direction was toward the southeast. (Map is part of the Timberland and Indian Creek USGS Quadrangles and was created with TOPO! © 2011 National Geographic Maps.)

melted (fig. 38). On the western side of figure 267, you can see more high-relief hummocky topography; a broad river flowing on top of or just under the ice surface probably carried sand and gravel toward the southeast, depositing the foundation for this rolling landscape. This hummocky sand and gravel continues southward on figure 268. Notice that it is bounded on both sides: on the east by somewhat higher ice-walled-lake plain (SB 15) sediment, and on the west by pitted outwash (SB 10). The hummocky sand and gravel may show the path of a tunnel channel (SB 17), but if it does, most of the evidence of that subglacial channel was destroyed during later meltout and collapse of supraglacial sediment. All lakes in this segment are in kettles (SB 9).

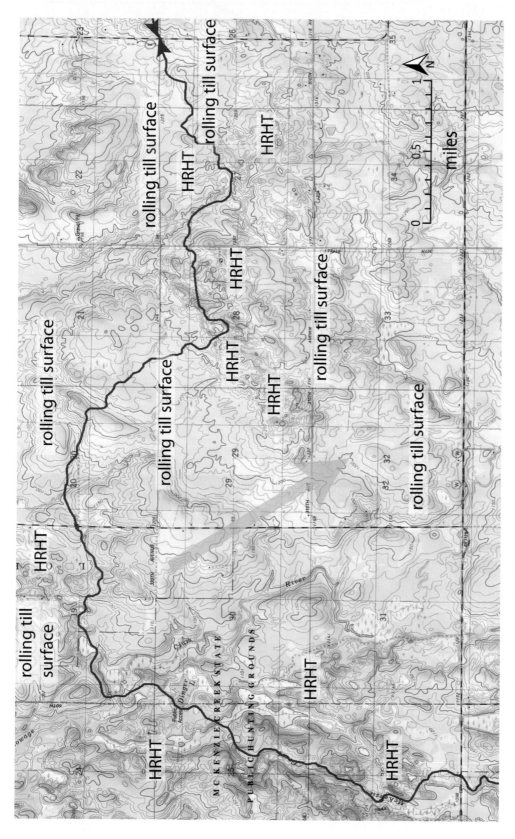

Figure 267. Topography of the eastern part of the McKenzie Creek Segment. Blue arrow shows ice-flow direction. HRHT: high-relief hummocky topography (SB 11). (Map is part of the Indian Creek and Clam Falls USGS Quadrangles and was created with TOPO! © 2011 National Geographic Maps.)

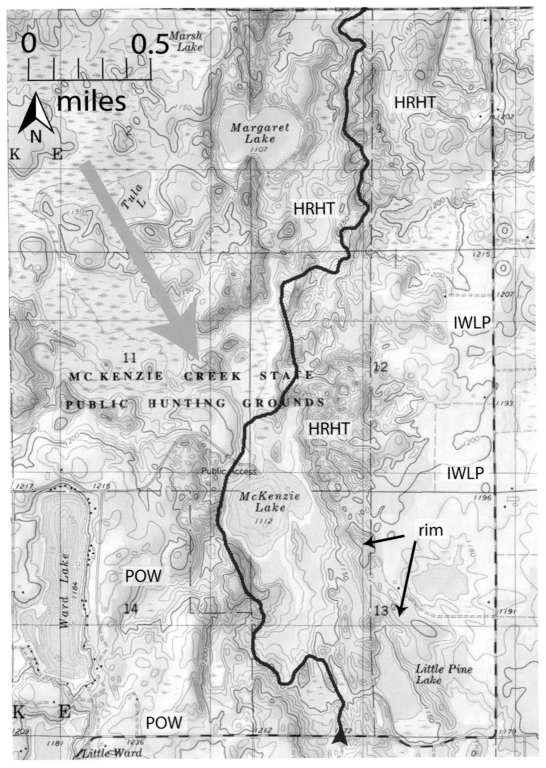

Figure 268. Topography of the western part of the McKenzie Creek Segment. Blue arrow shows ice-flow direction. HRHT: high-relief hummocky topography (SB 11); IWLP: ice-walled-lake plain (SB 15). Position of the rim of an IWLP is shown. (Map is part of the Clam Falls and Big Round Lake USGS Quadrangles and was created with TOPO! © 2011 National Geographic Maps.)

## 120. Pine Lake Segment

*70th St. to Round Lake Rd. (3 miles)*

The eastern trailhead of the Pine Lake Segment (figs. 259, 269) is on a high pitted-outwash surface (SB 10). It descends rapidly into kettles (SB 9) between high-relief hummocks (SB 11). All the steep slopes along the trail are ice-contact slopes produced when debris-covered ice melted. West of STH 48 the trail runs on pitted outwash with a few shallow kettles but overall low relief. This and the adjacent segments lie well behind the St. Croix Moraine (fig. 259).

## 121. Straight River Segment

*Round Lake Rd. to 270th Ave. (3.2 miles)*

At the eastern trailhead of the Straight River Segment (fig. 269), you'll be standing on pitted outwash (SB 10) with shallow kettles (SB 9). About a half mile west of there the trail descends into one of the best-preserved tunnel channels (SB 17) anywhere along the IAT. Water flowed in the tunnel channel while this area was completely ice covered and when the glacier edge was at the St. Croix Moraine (SB 6; fig. 259). The glacier bed was frozen (SB 3), and water under high pressure flowed through this channel to the ice margin. You'll have to take this part of the story on faith: much of the evidence for the outer part of the tunnel channel has been destroyed by the collapse of thick supraglacial sediment in the St. Croix Moraine.

After the ice warmed and water could coexist indefinitely with ice at the bed of the glacier, water collected in a much smaller tunnel and also flowed toward the southeast. Sediment accumulated in this tunnel, forming the low ridge of an esker (SB 13) that runs along the floor of the tunnel channel. Also, look for the somewhat winding ridge just north of STH 48 (center of fig. 269); this appears to be a tributary esker to the main one in the tunnel channel. By the time the pitted outwash beneath your feet was deposited, most of the ice had melted, as evidenced by the few relatively shallow kettles interrupting the surface. The tributary esker must have been ice free by this time and was partly buried by glacial outwash. We know this because had the esker been beneath and surrounded by ice, there would be deep kettles on either side of it. Conversely, we know there had to be ice filling the Straight Lake channel when the pitted outwash was deposited, because if that were not the case, outwash would have completely filled the channel.

The long Straight Lake tunnel channel extends northwestward. You'll follow it as you hike the Straight River and Straight Lake segments of the IAT. The trail follows the crest of the esker that sits in the bottom of the Straight Lake tunnel channel west of the STH 48 crossing. The IAT follows the crest of this esker for about a half mile before dropping off the east side. The trail then climbs out of the tunnel channel via a tributary channel onto pitted

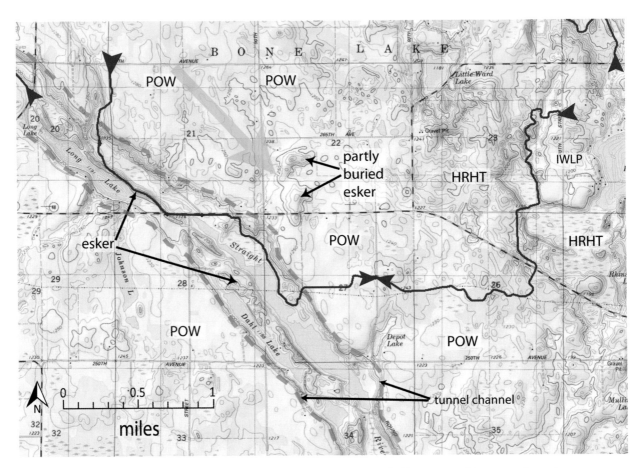

Figure 269.  Topography of the Pine Lake (right) and Straight River (left) segments. Blue arrow shows ice-flow direction. Dashed green line is edge of tunnel channel (SB 17). Water flowed toward the southeast in the tunnel. HRHT: high-relief hummocky topography (SB 11); IWLP: ice-walled-lake plain (SB 15); POW: pitted outwash (SB 10). (Map is part of the Big Round Lake USGS Quadrangle and was created with TOPO! © 2011 National Geographic Maps.)

outwash. This channel is not just a post-glacial channel but also was a tributary to the tunnel channel—or more likely to the esker tunnel—when it was water filled, as well.

## 122. Straight Lake Segment
*CTH I (100th St.) to 280th Ave. (3.7 miles)*

As you hike on the Straight Lake Segment (fig. 270), you'll follow the south side of the Straight River tunnel channel (SB 17) from its eastern trailhead for about one and a half miles. The trail then drops into the tunnel channel and follows the crest of a small esker.

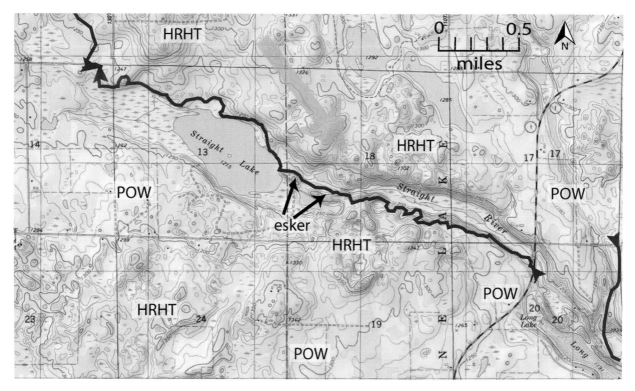

Figure 270. Topography of the Straight Lake Segment. Blue arrow shows ice-flow direction. HRHT: high-relief hummocky topography (SB 11); POW: pitted outwash (SB 10). (Map is part of the Luck USGS Quadrangle and was created with TOPO! © 2011 National Geographic Maps.)

Straight Lake, a kettle (SB 9) now partly filled with organic sediment, sits in the bottom of the channel. The steep sides of the tunnel channel bound the lake on its southwest and northeast sides, and the high land rising on both sides of the tunnel channel is pitted outwash (SB 10) that must have been deposited while there was still glacier ice occupying the position of the tunnel channel. The flow of water in the tunnel channel was toward the southeast, the same direction that the glacier and the water in the esker tunnel flowed (fig. 270). (For a discussion of the age relationships of landforms in this area, see the description of the Straight River Segment.)

## 123. Trade River Segment
*280th Ave. to 150th St. (4.3 miles)*

This segment of the IAT will take you well behind the St. Croix Moraine (fig. 259). Between the eastern trailhead and 140th St., the trail crosses gently rolling till (SB 5) that you may notice is distinctly different from the high-relief hummock topography (SB 11) just to the east (fig. 271). Continuing west, the trail crosses low rolling topography. Basalt bedrock (SB 21) is close to the ground surface. There is a very large basalt boulder at B in figure 271. This is not an erratic, because basalt is the local bedrock. The high hills to the northwest and northeast are made up of very thin till over basalt bedrock. Ice flowed toward the southeast in this area. Johnson (1998) reported striations (SB 4) on basalt outcrops along CTH W about a half mile north of the western part of this segment that tell us the ice-flow direction, but these have since weathered away.

## 124. Gandy Dancer State Trail Segment
*150th St. to 160th Ave. (15.1 miles)*

At the eastern terminus of the long Gandy Dancer State Trail Segment, the general orientation of the IAT changes toward the south and southwest (figs. 259, 271). The hike is easygoing not only because of the naturally occurring low relief, but also because the trail follows an old railroad grade shared with the Gandy Dancer State Trail. For the first mile from the eastern trailhead, you'll walk on pitted-outwash (SB 10) sand and gravel, and from there to just north of Luck, you'll traverse rolling till topography (fig. 271). The trail changes elevation by less than 50 feet.

After crossing nearly flat outwash near Luck, you'll climb onto a pitted-outwash plain (fig. 272). Note the shallow kettles (SB 9) that dot the landscape. Four-tenths of a mile north of Milltown, you will pass close to the edge of a deep, long kettle trending toward the south. This kettle marks the location of a former subglacial tunnel (probably a small tunnel channel) that never filled with sediment. From this depression to the trail's western trailhead south of Centuria, you'll hike on pitted outwash. Figure 273 shows several more parts of former subglacial tunnels.

At the western trailhead on 160th Ave., the trail is on pitted outwash. Look for the natural depression just southwest and west of the intersection of 160th Ave. and 200th St. This is another former subglacial tunnel, but it is also the head of the Big Rock Creek. This creek flows westward to the St. Croix River, which the St. Croix Falls Segment of the IAT crosses at the creek's lower end. This drainage developed after deglaciation as the St. Croix River Valley eroded. When the pitted-outwash plain you are on now was deposited, the channel through which the St. Croix River flows probably hadn't yet been carved. As water cut the St. Croix River channel, fast-flowing tributary streams had steep gradients and downcut rapidly through the sand and gravel to a point where they encountered more erosion-resistant till or bedrock. As the downcutting slowed, the water was still erosive, so it widened the creek bottom. Erosion intensified through a process called spring sapping. Rainfall that percolates from the surface flows through the sand and gravel layer to the water table. Groundwater sitting atop the dense till or rock layer then moves laterally

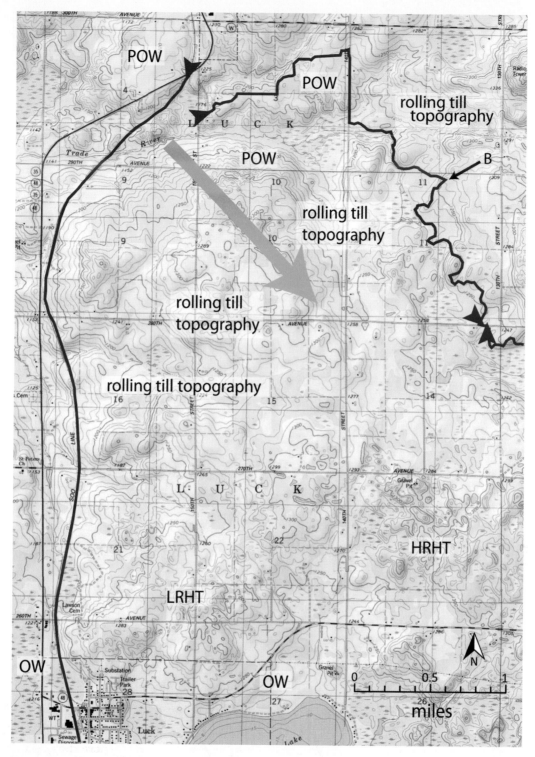

Figure 271.  Topography of the Trade River Segment (upper right) and the eastern part of the Gandy Dancer State Trail Segment (upper left to lower left). Blue arrow shows ice-flow direction. B: large basalt boulder; HRHT: high-relief hummocky topography (SB 11); LRHT: low-relief hummocky topography (SB 11); OW: outwash (SB 8); POW: pitted outwash (SB 10). (Map is part of the Frederic and Luck USGS Quadrangles and was created with TOPO! © 2011 National Geographic Maps.)

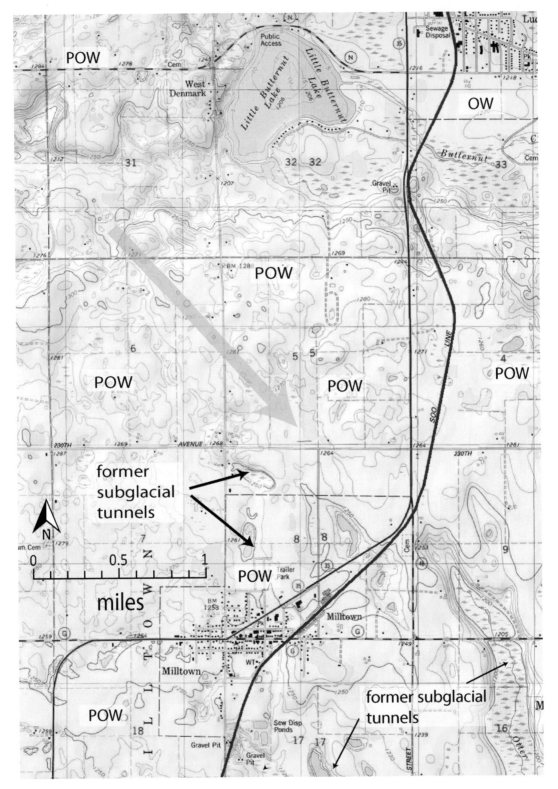

Figure 272. Topography of the central part of the Gandy Dancer State Trail Segment. Blue arrow shows ice-flow direction. OW: outwash (SB 8); POW: pitted outwash (SB 10). (Map is part of the Luck and Milltown USGS Quadrangles and was created with TOPO! © 2011 National Geographic Maps.)

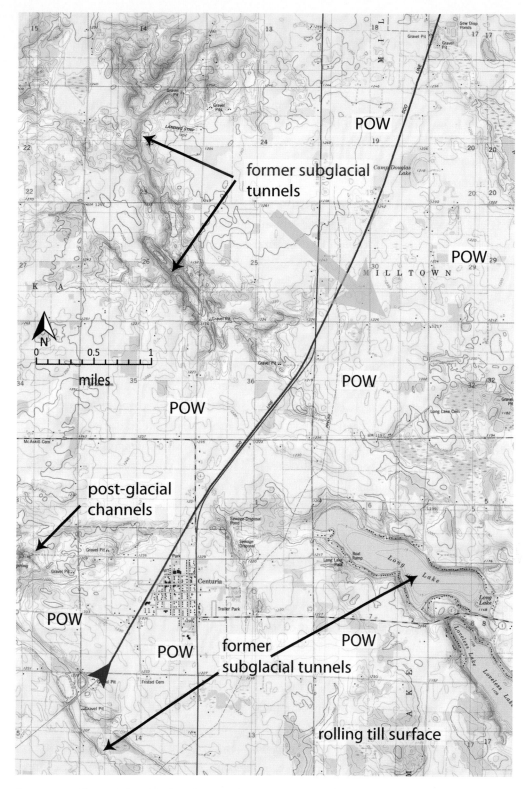

Figure 273. Topography of the western part of the Gandy Dancer State Trail Segment. Blue arrow shows ice-flow direction. POW: pitted outwash (SB 10). (Map is part of the Milltown and Centuria USGS Quadrangles and was created with TOPO! © 2011 National Geographic Maps.)

to the channel edge, where it emerges as springs. All of this water carries sand into the creek, making the channel bottom even broader.

## 125. St. Croix Falls Segment
*River Rd. to Interstate State Park (7 miles)*

The River Road trailhead of the St. Croix Falls Segment is on the floodplain of the St. Croix River (figs. 259, 274). This is the westernmost segment of the IAT, and the western terminus of the trail is at Potholes Trail in Interstate State Park to the south. From several places along this trail you'll have excellent views of the St. Croix River Gorge. The geologic history of the St. Croix River valley is discussed in more detail in the description of Interstate State Park.

About 3 miles from the eastern trailhead, the trail climbs onto a small remnant of a St. Croix River terrace (SB 8) before descending to cross Big Rock Creek. Across the St. Croix River you can see a larger terrace (SB 8) at about the same elevation. As mentioned in the discussion of the Gandy Dancer State Trail Segment, rapid downcutting of the St. Croix River caused the deep incision of streams like Big Rock Creek through sand and gravel. When the stream bed had been carved down to more resistant material, such as till or bedrock, the channel's bottom widened, producing steep-walled, wide channels. That erosion extended upstream in post-glacial time. In Lions Park you'll rise onto a narrow, low terrace, then climb steeply out of the valley. When leaves are off the trees, you'll enjoy a view of the St. Croix River from the high part of the trail before it descends into the Mindy Creek valley (fig. 274). The trail descends from a bedrock ridge with a thin cover of till, sand, and gravel. The angular, sharp-edged basalt boulders you'll see along the trail and in the channel (fig. 275) indicate that they are locally derived and that the glacier or meltwater did not carry them very far, if at all. If they had been transported any significant distance, you would see rounded edges from their tumbling and grinding journey across the landscape.

The low ridge with houses on it just east of North Day Rd. marks the outer edge of an advance of the Grantsburg Sublobe of the Des Moines Lobe from the west (fig. 274). This advance took place while the Superior Lobe was still in the St. Croix River basin, after the ice had retreated from the area near St. Croix Falls, but probably not before all the buried ice had melted out from beneath Superior Lobe deposits. (See the description of Interstate State Park for more discussion of the Grantsburg Sublobe.)

As you walk through Florence Baker Riegel Memorial Park (fig. 276), look for exposed basalt bedrock along the trail. You won't see any definite striations because weathering of the rock surface has made them difficult to distinguish from cracks in the rock.

The trail goes on to an esker (SB 13; fig. 276) in back of the hospital and runs along the esker's crest to a point just north of USH 8/STH 35. You'll hike along the backbone

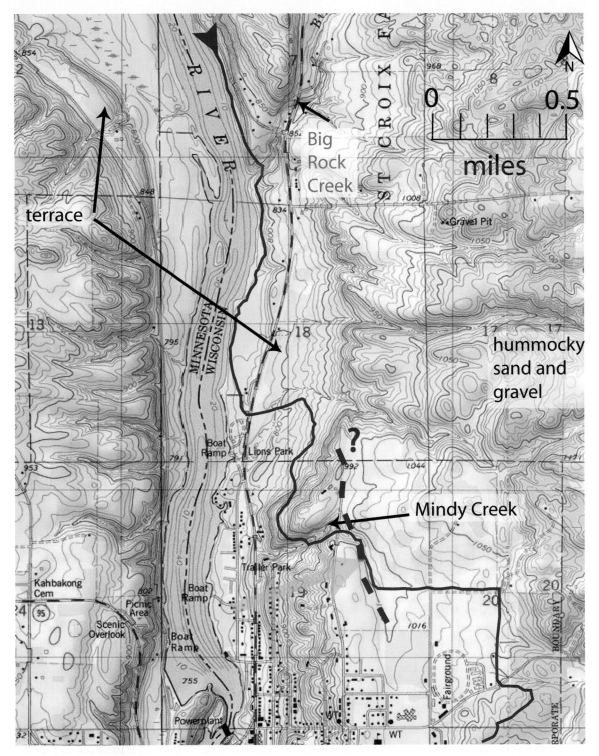

Figure 274. Topography of the eastern part of the St. Croix Falls Segment. Several terraces (SB 8) are present on both sides of the St. Croix River. Only major terraces are labeled. Dashed blue line is approximate outer (eastern) extent of the Pine City advance of the Grantsburg Sublobe. It cannot be traced with certainty north of the line shown. Blue arrow shows ice-flow direction of that advance. (Map is part of the St. Croix Dalles USGS Quadrangle and was created with TOPO! © 2011 National Geographic Maps.)

Figure 275. View along the St. Croix Segment near Mindy Creek. Note the abundant angular basalt (SB 21) boulders that indicate that bedrock is close to the surface and that they have not been transported far by ice or water.

of this high esker just north and east of the park. This esker formed beneath the Superior Lobe when ice flowed toward the southeast. As you hike south of USH 8, you'll parallel the esker for a short distance before climbing onto a till- and gravel-covered bedrock surface.

The western terminus of the Ice Age Trail is at the Potholes Trail, a loop trail with excellent examples of potholes. Not to be confused with kettle (SB 9) holes, these features form when rocks carried by fast-flowing water cut into bedrock. Potholes are quite common where rivers flow on rock, but the potholes you'll see in the rest of the state are much smaller (a foot or two deep) compared to the ones in Interstate State Park. (See, for example, the shallow potholes along the Eau Claire River by the west side of the Green Bay Lobe or the potholes

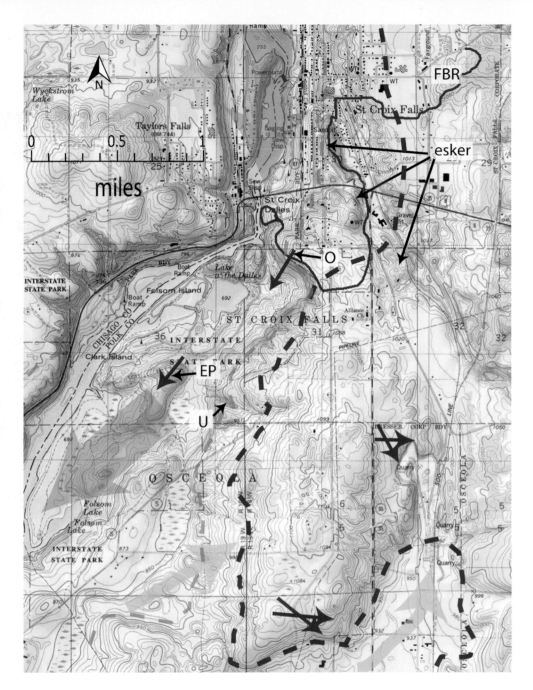

Figure 276. Topography of the western part of the St. Croix Falls Segment and of Interstate State Park. The western terminus of the Ice Age National Scenic Trail is the loop trail (Potholes Trail) just east of the St. Croix River. Dark blue arrows show striations, thus ice-flow directions of the Superior Lobe. Dashed blue line is approximate outer edge of the Pine City advance of the Grantsburg Sublobe. Light blue arrows show ice-flow direction of only the Pine City advance. Dashed green lines show minimum extent of flood water at the time of the major flood down the St. Croix River valley discussed in the text. Green arrows show meltwater-flow direction. O: Observation Rock; EP: Eagle Peak; FBR: Florence Baker Riegel Memorial Park in St. Croix Falls; U: location where the unconformity between basalt and overlying conglomerate can be seen. (Map is part of the St. Croix Dalles USGS Quadrangle and was created with TOPO! © 2011 National Geographic Maps.)

along the Wisconsin River near Grandfather Falls.) You can find much older potholes high on the bluffs at Devil's Lake. The potholes on the Minnesota side of the St. Croix River are even deeper and wider than the ones on the Wisconsin side and are worth a visit as well.

## 126. Interstate State Park Unit of the Ice Age National Scientific Reserve

Please note that this is a state park. Just as you've been a good steward of the geologic resources on the rest of the trail, you should not hammer on or collect anything in the park!

The complex landscape of Interstate State Park results from glaciers and floods of water flowing over and around hills of fractured basalt (fig. 276). The St. Croix River gorge, which separates Minnesota and Wisconsin, was cut by meltwater very late during retreat of the Superior Lobe. Various geologic features on the upland adjacent to the gorge help tell a complex story.

Interstate State Park features thick layers of basalt bedrock. This rock was extruded from the Midcontinent Rift onto the ground surface from volcanoes about 1,100 million years ago (figs. 260, 261). This is the same Keweenawan basalt that underlies about half of the Superior Lobe IAT segments. Cordua (1989) mapped eight individual basalt flows, some of which you'll see exposed in the area. The tilted flows create a series of "steps" that run more or less north to south in the park. One of these long steps forms the summit of Eagle Peak (fig. 276). The top of the flow tilts gently to the west, whereas the steeper eastern slope of the hill has been eroded across the layer of basalt. The light-colored minerals you see in the small holes in the basalt are quartz, feldspar, and several other minerals. There are great views of the basalt flows along the walls of the gorge from the Potholes Trail.

In the eastern part of the park, you can spot a good example of an unconformity between the basalt and the overlying Cambrian conglomerate. As discussed in the introduction to the Superior Lobe IAT segments, the basalt formed when flows of liquid lava came to the earth's surface in a rift: a place where the earth's crust was being pulled apart. After the tectonic forces changed, the rifting stopped and the area was compressed as the North American plate closed the gap. The basalt flows were tilted and uplifted.

Weathering and erosion took place for several hundred million years after that, leading to a low-relief, gently rolling plain across which a shallow sea sat in the late Cambrian. For the most part in Wisconsin, sand settled on the sea's floor of weathered Precambrian rocks, forming evenly layered sandstone. In a few places, however, sea cliffs stood at the edge of the sea, and boulders fell into the wave zone, where the surf tumbled, broke, and ground them smooth, resulting in a conglomerate (SB 21) rock layer at the base of the much thicker overlying sandstone. You can glimpse this conglomerate near the west end of the Skyline Trail (U on fig. 276). You can see a similar conglomerate between Cambrian sandstone and underlying Precambrian rock at Devil's Lake State Park and in Parfrey's Glen.

Glaciation is the next part of the geologic story preserved at Interstate State Park. The park is located more than 20 miles northwest of the outer edge of the St. Croix advance (fig. 259), so thick ice covered it during the Emerald and St. Croix phases of the last glaciation. The best record of this late Wisconsin ice cover is preserved on some rock surfaces as striations (SB 4) and shallow, wider grooves etched into the rock as the bottom of the glacier slid overtop. Johnson (2000) interpreted the striations as having been formed by the Superior Lobe during the last glaciation, although they exhibit several flow directions. Striations oriented northeast to southwest apparently indicate the direction of the oldest flow, when the Superior Lobe was moving toward the southwest during either the Emerald or St. Croix phases. You'll see striations with this orientation at the tops of Observation Rock and Eagle Peak (fig. 276). The next youngest striations indicate flow toward the southeast, the main flow direction of the Superior Lobe in this area. The youngest set of striations, according to Johnson (2000), are gouged directly east-west and indicate flow toward the east (fig. 276). These may have been cut when the ice was at the Centuria ice-margin position (fig. 259). Note that there is no way to date striations, so this is somewhat speculative.

After the Superior Lobe retreated north of the St. Croix Falls area, another lobe of ice entered the picture. The Des Moines Lobe advanced southward into southern Minnesota and Iowa, about to the location of Des Moines. A small sublobe of that ice advanced toward the northwest, just entering what is now Wisconsin. This is called the Pine City advance of the Grantsburg Sublobe (fig. 1), and it deposited fine-grained, brown till that is quite different from the reddish-brown, sandy till deposited by the Superior Lobe (figs. 259, 274, 276). It did not make significant landforms in this area, but the thin till can be mapped (Johnson 2000).

At some time during retreat of the glacier, meltwater began to cut the St. Croix River Dalles. More recent glaciers and stream erosion have erased evidence of exactly when this took place, but it was at least after the ice retreated from the Centuria ice margin and perhaps after its retreat from the Pine City ice margin. The Pine City advance dammed a lake called glacial Lake Grantsburg in the St. Croix River valley north of here, and it is possible that lake drainage flowing beneath and along the edge of the glacier began cutting the present valley as the Pine City ice started to retreat. It seems unlikely, however, that drainage of glacial Lake Grantsburg caused the major erosion of the gorge and formation of the potholes.

Instead, the major downcutting and pothole formation must have been the result of a huge flood that took place after the ice margin retreated farther north, out of the St. Croix River basin and into the Lake Superior basin. The flood water at Interstate State Park must have reached at least an elevation of about 900 feet, because at Eagle Peak, which is about that high, the rock has been eroded by the river. This means that rushing water filled the entire area between the green dashed lines in figure 276! Lake O' the Dalles is probably a plunge pool, a depression cut by water coming over the hills to the north and crashing down onto the rock below. Where did the water come from, and why the huge flood?

From the time the retreating ice first exposed the Lake Superior basin, the water in the basin was higher than it presently is. It could not flow to the east, as it does today, because glacier ice covered these lower outlets (fig. 277). The higher lake that resulted is called glacial Lake Duluth. This body of water first drained though the Moose Lake outlet, but as soon as the glacier edge retreated far enough for water to escape through the slightly lower Bois Brule–St. Croix outlet, that valley captured the flow. Torrents of water flushed downstream, flooding this valley as the level of Lake Duluth fell. We don't know how much the lake level

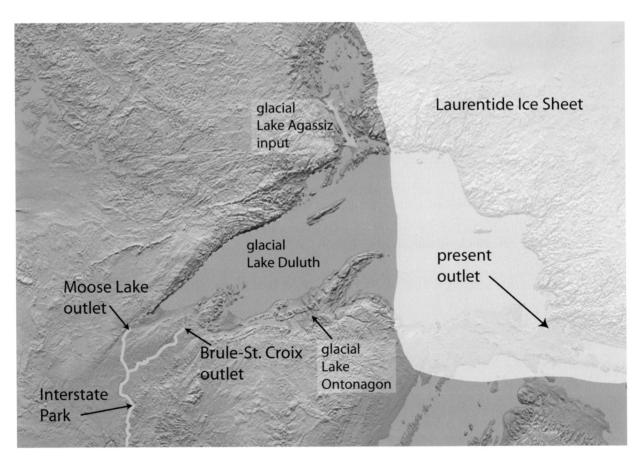

Figure 277. Interstate State Park, the Moose Lake outlet of glacial Lake Duluth, and the Brule–St. Croix outlet of glacial Lake Duluth. Large amounts of water entered the lake (blue arrows) from glacial Lake Ontonagon and glacial Lake Agassiz (northwest of area shown on map). Dashed blue line shows approximate extent of glacial Lake Duluth. At that time, probably shortly after 16,000 cal. years ago, the lake was high enough to drain into the St. Croix River drainage via those routes because the present lower outlet was still covered by the Laurentide Ice Sheet. (Base map from NASA website: http://photojournal.jpl.nasa.gov/catalog/PIA03377.)

fell, but it was perhaps as much as 30 feet in a very short time. Was this the flood that cut the Dalles and created the potholes? Perhaps, but there are other possibilities. For a time, a large lake called glacial Lake Ontonagon was present along the ice margin in the Upper Peninsula of Michigan, and it could have drained rapidly into Lake Superior, causing a big flood down the St. Croix River valley. At times, water certainly drained into Lake Superior from giant ice-marginal lakes to the west as well. The largest of these was glacial Lake Agassiz. Complicating this story is the fact that we can't date the megaflood that rushed down the St. Croix valley, because the water swept most deposits away. We do know that major downcutting downstream in the Mississippi River valley took place just over 16,000 cal. years ago (Knox, pers. comm. 2010); a giant flood down the St. Croix River valley could have initiated the valley cutting.

Besides the mystery of the flood, the most unusual features to ponder at Interstate State Park are the abundant large potholes (fig. 278). The ones you see are very big compared to other Ice Age potholes around the world: here they are up to 16 feet deep and 6 feet across. You can find even larger potholes on the Minnesota side of the river. Next time you're on the banks of a babbling, clear river, look for the process that forms these potholes. It starts in places where you can see tumbling rocks caught beneath eddies of spinning water at the bottom of fast-flowing water (fig. 279). Over time, the spinning rocks are worn round, and the rock bottom of the river itself is eroded into a smooth bowl. Large potholes may have formed at Interstate State Park because the basalt here is fractured and so probably allowed rushing water and fluctuating water pressure to pry apart pieces of rock that then fell into the river. Although the flood carried some of these away, many remained in the river, spinning and rolling on the bottom, smoothing their own rough edges and grinding potholes into the surrounding basalt. These smooth, spherical rocks are called grinders (fig. 280).

The western terminus of the Ice Age National Scenic Trail is the Potholes Trail at the edge of the gorge, overlooking the scenic St. Croix River (fig. 281). We hope you have enjoyed this book and your journey back into geologic time!

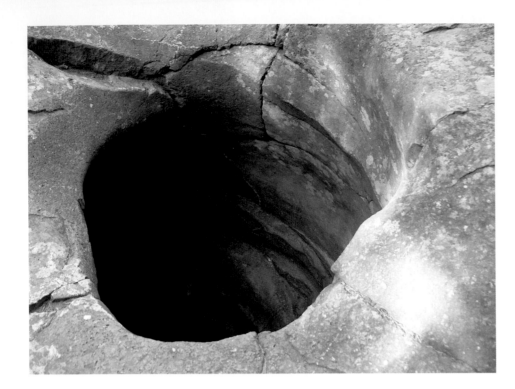

Figure 278. View into a pothole along the Potholes Trail. Note the grooves on the inside of the pothole and compare with figure 279. Pothole is about 3 feet across. (Photo by Vin Mickelson.)

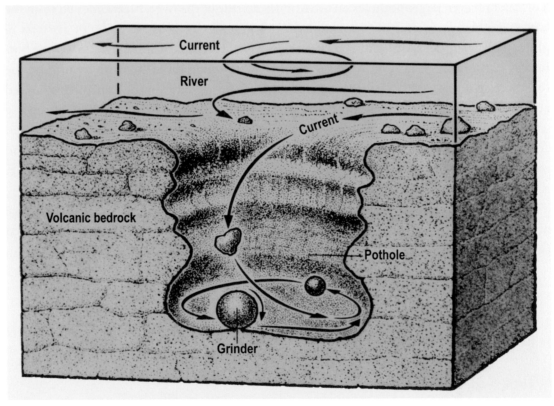

Figure 279. Formation of a pothole beneath rapidly flowing water. Grinders are rocks, mostly basalt, that have been worn to a spherical shape by spinning in the pothole. (Drawing courtesy of National Park Service; drafted by Mary Diman.)

Figure 280. Grinder, a basalt boulder, found at the bottom of a pothole at Interstate State Park. See figure 279 and text for the proposed origin of a grinder. This grinder is about 15 inches across.

Figure 281. View of the lower St. Croix Dalles from the western terminus of the Ice Age National Scenic Trail. Photo was taken on the Potholes Trail looking toward the southwest across the St. Croix River. Note the fractured basalt (SB 21) on the walls of the channel. (Photo by Vin Mickelson.)

# FURTHER READINGS

Many ideas presented in this book merit further exploration by those interested in geologic processes or Wisconsin's landscape. The list of references that follows, grouped into divisions of the book, will guide you to more sources of information available in bookstores, libraries, or online.

## INTRODUCTION

Black, R. F. 1974. *Geology of Ice Age National Scientific Reserve of Wisconsin*. National Park Service Scientific Monograph Series 2. Washington, DC: National Park Service.

Dott, R. H., Jr., and J. W. Attig. 2004. *Roadside Geology of Wisconsin*. Missoula, MT: Mountain Press Publishing.

Farrand, W. R., D. M. Mickelson, W. R. Cowan, and J. E. Goebel. 1984. *Quaternary Geologic Map of the Lake Superior 4″ × 6″ Quadrangle, United States and Canada*. U.S. Geological Survey Map I-1420 (NL-16). Reston, VA: U.S. Geological Survey.

Goebel, J. E., D. M. Mickelson, W. R. Farrand, Lee Clayton, J. C. Knox, Adam Cahow, H. C. Hobbs, and M. S. Walton. 1983. *Quaternary Geologic Map of the Minneapolis 4″ × 6″ Quadrangle, United States*. U.S. Geological Survey Map I-1420 (NL-15). Reston, VA: U.S. Geological Survey.

Hallberg, G. R., J. A. Lineback, D. M. Mickelson, J. C. Knox, J. E. Gobel, H. C. Hobbs, J. W. Whitfield, R. A. Ward, J. D. Boellsdorf, J. B. Swinehart, and J. H. Dreeszen. 1991. *Quaternary Geologic Map of the Des Moines 4″ × 6″ Quadrangle, United States*. U.S. Geological Survey Map I-1420 (NK-15). Reston, VA: U.S. Geological Survey.

Knox, J. C., L. Clayton, and D. M. Mickelson, eds. 1982. *Quaternary History of the Driftless Area*. Field Trip Guide Book 5. Madison: University of Wisconsin–Extension, Geological and Natural History Survey.

LaBastille, A. 1977. "On the Trail of Wisconsin's Ice Age." *National Geographic* 152 (2): 182–205.

LaBerge, G. L. 1994. *Geology of the Lake Superior Region*. Phoenix, AZ: Geoscience Press.

Lineback, J. A., N. K. Bleuer, D. M. Mickelson, W. R. Farrand, and R. P. Goldthwait. 1983. *Quaternary Geologic Map of the Chicago 4″ × 6″ Quadrangle, United States*. U.S. Geological Survey Map I-1420 (NK-16). Reston, VA: U.S. Geological Survey.

Martin, L. 1965. *Physical Geography of Wisconsin*. Madison: University of Wisconsin Press.

Mickelson, D. M. 1997. "Wisconsin's Glacial Landscapes." In *Wisconsin Land and Life*, edited by R. C. Ostergren and T. R. Vale, 35–48. Madison: University of Wisconsin Press.

Paull, R. K., and R. A. Paull. 1977. *Geology of Wisconsin and the Upper Peninsula of Michigan*. Dubuque, IA: Kendall/Hunt.

Reuss, H. S. 1990. *On the Trail of the Ice Age: A Guide to Wisconsin's Ice Age National Reserve and Trail for Hikers, Bikers, and Motorists*. Sheboygan, WI: Ice Age Park and Trail Foundation.

Schultz, G. M. 2009. *Wisconsin's Foundations*. 2nd ed. Madison: University of Wisconsin Press.

Scotese, C. R. 1987. "Development of the Circum-Pacific Panthalassic Ocean during the Early Paleozoic." In *Circum-Pacific Orogenic Belts and the Evolution of the Pacific Ocean Basin*, edited by J. W. Monger, 49–57. Geodynamics Series 18. Washington, DC: American Geophysical Union.

Smith, B., E. Sherman, and A. Hanson. 2008. *Along Wisconsin's Ice Age Trail*. Madison: University of Wisconsin Press.

Syverson, K. M., and P. M. Colgan. 2004. "The Quaternary of Wisconsin: A Review of Stratigraphy and Glaciation History." In *Quaternary Glaciations: Extent and Chronology*, pt. 2, *North America*, edited by J. Ehlers and P. L. Gibbard, 295–311. Amsterdam: Elsevier.

## Science Briefs

Attig, J. W., D. M. Mickelson, and L. Clayton. 1989. "Late Wisconsin Landform Distribution and Glacier-bed Conditions in Wisconsin." *Sedimentary Geology* 62:399–405.

Benn, D. I., and D. J. A. Evans. 1998. *Glaciers and Glaciation*. New York: John Wiley and Sons.

Bennett, M. R., and N. F. Glasser. 2009. *Glacial Geology: Ice Sheets and Landforms*. Oxford: John Wiley and Sons.

Birkby, R. 2005. *Lightly on the Land: The Student Conservation Association Trail Building and Maintenance Manual*. 2nd ed. Seattle, WA: Mountaineers Books.

Clayton, L. 2001. *Pleistocene Geology of Waukesha County, Wisconsin*. Wisconsin Geological and Natural History Survey, Bulletin 99. Madison: University of Wisconsin–Extension.

Clayton, L., J. W. Attig, and D. M. Mickelson. 1999. "Tunnel Channels Formed in Wisconsin during the Last Glaciation." In *Glacial Processes Past and Present*, edited by D. M. Mickelson and J. A. Attig, 69–82. Geological Society of America Special Paper 337. Boulder, CO: Geological Society of America.

———. 2001. "Effects of Late Pleistocene on the Landscape of Wisconsin." *Boreas* 30:173–88.

Cutler, P. M., P. M. Colgan, and D. M. Mickelson. 2002. "Sedimentalogic Evidence for Outburst Floods from the Laurentide Ice Sheet Margin in Wisconsin, USA: Implications for Tunnel-channel Formation." *Quaternary International* 90:23–40.

Ehlers, J., and P. L. Gibbard. 2004. *Quaternary Glaciations: Extent and Chronology*. Pt. 2, *North America*. Amsterdam: Elsevier.

Evans, D. J., ed. 2003. *Glacial Landsystems*. London: Arnold.

Gillespie, A. R., S. C. Porter, and A. F. Atwater, eds. 2004. *Quaternary Period in the United States*. Amsterdam: Elsevier.

Hansel, A. K., and D. M. Mickelson. 1988. "A Reevaluation of the Timing and Causes of High Lake Phases in the Lake Michigan Basin." *Quaternary Research* 29:113–28.

Hooyer, T. S., ed. 2007. *Late-glacial History of East-central Wisconsin: Guide Book for the 53rd Midwest Friends of the Pleistocene Field Conference, May 18–20, 2007, Oshkosh, Wisconsin*. Wisconsin Geological and Natural History Survey, Open-file Report 2007-01. Madison: University of Wisconsin–Extension.

Libby, W. F. 1952. *Radiocarbon Dating*. Chicago: University of Chicago Press.

Mickelson, D. M. 2007. *Landscapes of Dane County*. Wisconsin Geological and Natural History Survey Educational Series 43. Madison: University of Wisconsin–Extension.

Syverson, K. M. 2007. *Pleistocene Geology of Chippewa County, Wisconsin*. Wisconsin Geological and Natural History Survey, Bulletin 103. Madison: University of Wisconsin–Extension.

## Northeast Ice Age Trail Segments

Acomb, L. J., D. M. Mickelson, and E. B. Evenson. 1982. "Till Stratigraphy and Late Glacial Events in the Lake Michigan Lobe of Eastern Wisconsin." *Geological Society of America Bulletin* 93:289–96.

Black, R. F. 1970. *Glacial Geology of Two Creeks Forest Bed, Valderan Type Locality, and Northern Kettle Moraine State Forest*. Wisconsin Geological and Natural History Survey, Information Circular 13. Madison: University of Wisconsin–Extension.

Dott, E. R., and D. M. Mickelson. 1995. "Lake Michigan Water Levels and the Development of Holocene Beach-ridge Complexes at Two Rivers, Wisconsin: Stratigraphic, Geomorphic, and Radiocarbon Evidence." *Geological Society of America Bulletin* 107:286–96.

Dutch, S. I. 1980. "Structure and Landform Evolution in the Green Bay, Wisconsin Area." In *Geology of Eastern and Northeastern Wisconsin: 44th Annual Tri-State Geological Field Conference*, edited by R. D. Stieglitz, 119–34. Green Bay: Earth Science Discipline, University of Wisconsin.

Garry, C. E., R. W. Baker, D. P. Schwert, and A. F.Schneider. 1990. "Environmental Analyses of a Twocreekan-aged Beetle (Coleoptera) Assemblage from Kewaunee, Wisconsin." In *Late Quaternary History of the Lake Michigan Basin*, edited by A. F Schneider and G. S. Fraser, 57–66. Geological Society of America Special Paper 251. Boulder, CO: Geological Society of America.

Goldthwait, J. W. 1907. *The Abandoned Shore-lines of Eastern Wisconsin*. Wisconsin Geological and Natural History Survey, Bulletin 17. Madison: University of Wisconsin–Extension.

Hooyer, T. S. 2007. *Late-glacial History of East-central Wisconsin: Guide Book for the 53rd Midwest Friends of the Pleistocene Field Conference, May 18–20, 2007, Oshkosh, Wisconsin*. Wisconsin Geological and Natural History Survey, Open-file Report 2007-01. Madison: University of Wisconsin–Extension.

Hooyer, T. S., and W. N. Mode. 2008. *Pleistocene Geology of Winnebago County, Wisconsin*. Wisconsin Geological and Natural History Survey, Bulletin 105. Madison: University of Wisconsin–Extension.

Kaiser, K. F. 1994. "Two Creeks Interstate Dated through Dendrochronology and AMS." *Quaternary Research* 42:288–98.

Kowalke, O. L., and E. F. Kowalke. 1938. "Topography of Abandoned Beach Ridges at Ellison Bay, Door County, Wisconsin." *Transactions of the Wisconsin Academy of Sciences, Arts, and Letters* 31:547–53.

Leavitt, S. W, I. P. Panyushkina, T. Lange, A. Wiedenhoeft, L. Cheng, R. D. Hunter, J. Hughes, F. Pranschke, A. F. Schneider, J. Moran, and R. Stieglitz. 2006. "Climate in the Great Lakes Region between 14,000 and 4000 Years Ago from Isotopic Composition of Conifer Wood." *Radiocarbon* 48 (2): 205–17.

Libby, W. F. 1952. *Radiocarbon Dating.* Chicago: University of Chicago Press.

McCartney, M. C., and D. M. Mickelson. 1982. "Late Woodfordian and Greatlakean History of the Green Bay Lobe, Wisconsin." *Geological Society of America Bulletin* 91:297–302.

Mickelson, D. M., and B. J. Socha. Forthcoming. *Quaternary Geology of Manitowoc and Calumet Counties.* Wisconsin Geological and Natural History Survey Bulletin. Madison: University of Wisconsin–Extension.

Morgan, A. V., and A. Morgan. 1979. "The Fossil Coleoptera of the Two Creeks Forest Bed, Wisconsin." *Quaternary Research* 12:226–40.

Palmquist, J. C., ed. 1989. *Wisconsin's Door Peninsula: A Natural History.* Appleton, WI: Perin Press.

Schneider, A. F. 1993. *Pleistocene Geomorphology and Stratigraphy of the Door Peninsula, Wisconsin (Road Log): 40th Midwest Friends of the Pleistocene Field Trip Guidebook.* Kenosha: College of Science and Technology, University of Wisconsin–Parkside.

Schweger, C. E. 1966. "Pollen Analysis of Iola Bog and Paleoecology of the Two Creeks Interval." M.S. thesis, University of Wisconsin–Madison.

Socha, B. J., P. M. Colgan, and D. M. Mickelson. 1999. "Ice-surface Profiles and Bed Conditions of the Green Bay Lobe from 13,000 to 11,000 $^{14}$C-years B.P." In *Glacial Processes Past and Present,* edited by J. W. Attig and D. M. Mickelson, 151–58. Geological Society of America Special Paper 337. Boulder, CO: Geological Society of America.

Stieglitz, R. D., and W. E. Schuster. 1993. "Glaciation and Karst Features of the Door Peninsula, Wisconsin." In *Pleistocene Geomorphology and Stratigraphy of the Door Peninsula, Wisconsin (Road Log): 40th Midwest Friends of the Pleistocene Field Trip Guidebook,* edited by A. F Schneider, 47–52. Kenosha: College of Science and Technology, University of Wisconsin–Parkside.

Thwaites, F. T., and K. Bertrand. 1957. "Pleistocene Geology of the Door Peninsula, Wisconsin." *Geological Society of America Bulletin* 68:831–80.

Wilson, L. R. 1932. "The Two Creeks Forest Bed, Manitowoc County, Wisconsin." *Transactions of the Wisconsin Academy of Sciences, Arts and Letters* 27:31–46.

———. 1936. "Further Studies of the Two Creeks Forest Bed, Manitowoc County, Wisconsin." *Torrey Botanical Club Bulletin* 63:317–25.

## Northern Kettle Moraine Ice Age Trail Segments

Attig, J. W. 1986. "Glacial Geology of the Kettle Moraine." *Wisconsin Natural Resources* 10 (5): 17–20.

Black, R. F. 1974. *Geology of Ice Age National Scientific Reserve of Wisconsin.* National Park Service Scientific Monograph Series 2. Washington: National Park Service.

Carlson, A. E., D. M. Mickelson, S. M. Principato, and D. M. Chapel. 2004. "Genesis of the Northern Kettle Moraine, Wisconsin." *Geomorphology* 67:365–74.

Colgan, P. M. 1999. "Reconstruction of the Green Bay Lobe, Wisconsin, United States, from 26,000 to 13,000 Radiocarbon Years B.P." In *Glacial Processes Past and Present,* edited by D. M. Mickelson and J. A. Attig, 137–50. Geological Society of America Special Paper 337. Boulder, CO: Geological Society of America.

Mode, W. M. 1989. "Glacial Geology of East-central Wisconsin." In *Wisconsin's Door Peninsula: A Natural History*, edited by J. C. Palmquist, 66–81. Appleton, WI: Perin Press.

Winguth, C., D. M. Mickelson, P. M. Colgan, and B. Laabs. 2004. "Modeling the Deglaciation of the Green Bay Lobe of the Southern Laurentide Ice Sheet." *Boreas* 33 (1): 34–47.

MIDDLE KETTLE MORAINE ICE AGE TRAIL SEGMENTS

Alden, W. C. 1918. *The Quaternary Geology of Southeastern Wisconsin, with a Chapter on the Older Rock Formations*. United States Geological Survey Professional Paper 106. Washington, DC: Government Printing Office.

Attig, J. W. 1986. "Glacial Geology of the Kettle Moraine." *Wisconsin Natural Resources* 10 (5): 17–20.

Battista, J. R. 1990. "Quaternary Geology of the Horicon Marsh Area." M.S. thesis, University of Wisconsin–Madison.

Mickelson, D. M., and K. M. Syverson. 1997. *Quaternary Geology of Ozaukee and Washington Counties, Wisconsin*. Wisconsin Geological and Natural History Survey, Bulletin 91. Madison: University of Wisconsin–Extension.

Syverson, K. M. 1988. "The Glacial Geology of the Kettle Interlobate Moraine Region, Washington Co., Wisconsin." M.S. thesis, University of Wisconsin–Madison.

SOUTHERN KETTLE MORAINE ICE AGE TRAIL SEGMENTS

Alden, W. C. 1918. *The Quaternary Geology of Southeastern Wisconsin, with a Chapter on the Older Rock Formations*. United States Geological Survey Professional Paper 106. Washington, DC: Government Printing Office.

Attig, J. W. 1986. "Glacial Geology of the Kettle Moraine." *Wisconsin Natural Resources* 10 (5): 17–20.

Chamberlin, T. C. 1878. "On the Extent and Significance of the Wisconsin Kettle Moraine." *Transactions of the Wisconsin Academy of Sciences, Arts and Letters* 4:201–34.

Clayton, L. 2001. *Pleistocene Geology of Waukesha County*. Wisconsin Geological and Natural History Survey, Bulletin 99. Madison: University of Wisconsin–Extension.

Stoelting, P. K. 1978. "The Concept of an Esker, Esker Form, and Esker Form System in Eastern Wisconsin." Ph.D. thesis, University of Wisconsin–Milwaukee.

SOUTHERN GREEN BAY LOBE ICE AGE TRAIL SEGMENTS

Alden, W.C. 1905. *Drumlins of Southeastern Wisconsin*. United States Geological Survey, Bulletin 273. Washington, DC: Government Printing Office.

———. 1918. *The Quaternary Geology of Southeastern Wisconsin, with a Chapter on the Older Rock Formations*. United States Geological Survey Professional Paper 106. Washington, DC: Government Printing Office.

Anderson, R. C. 2005. *Geomorphic History of the Rock River, South-central Wisconsin and Northwestern Illinois*. Illinois State Geological Survey, Circular 565. Champaign: Illinois Department of Natural Resources.

Attig, J. W., L. Clayton, K. I. Lange, and L. J. Maher. 1990. *Ice Age Geology of Devil's Lake State Park*. Wisconsin Geological and Natural History Survey Educational Series 35. Madison: University of Wisconsin–Extension.

Attig, J. W., D. M. Mickelson, and L. Clayton. 1989. "Late Wisconsin Landform Distribution and Glacier-Bed Conditions in Wisconsin." *Sedimentary Geology* 62:399–405.

Black, R. F. 1974. *Geology of Ice Age National Scientific Reserve of Wisconsin.* National Park Service Scientific Monograph Series 2. Washington, DC: National Park Service.

Bleuer, N. K. 1970. *Glacial Stratigraphy of South-central Wisconsin.* Wisconsin Geological and Natural History Survey, Information Circular 15. pp. J1–J35.

Clayton, J. A., and J. C. Knox. 2008. "Catastrophic Flooding from Glacial Lake Wisconsin." *Geomorphology* 93 (3–4): 384–97.

Clayton, L., and J. W. Attig, 1989. *Glacial Lake Wisconsin.* Geological Society of America Memoir 173. Boulder, CO: Geological Society of America.

———. 1990. *Geology of Sauk County, Wisconsin.* Wisconsin Geological and Natural History Survey, Information Circular 67. Madison: University of Wisconsin–Extension.

———. 1997. *Pleistocene Geology of Dane County, Wisconsin.* Wisconsin Natural History and Geological Survey, Bulletin 95. Madison: University of Wisconsin–Extension.

Colgan, P. M. 1999. "Reconstruction of the Green Bay Lobe, Wisconsin, United States, from 26,000 to 13,000 Radiocarbon Years B.P." In *Glacial Processes Past and Present,* edited by D. M. Mickelson and J. A. Attig, 137–50. Geological Society of America Special Paper 337. Boulder, CO: Geological Society of America.

Colgan, P. M., and D. M. Mickelson. 1997. "Genesis of Streamlined Landforms and Flow History of the Green Bay Lobe, Wisconsin, USA." *Sedimentary Geology* 111:7–25.

Dalziel, I. W. D., and R. H. Dott Jr. 1970. *Geology of the Baraboo District, Wisconsin.* Wisconsin Geological and Natural History Survey, Information Circular 14. Madison: University of Wisconsin–Extension.

Dott, R. H., Jr. 1999. "Van Hise Rock in the Baraboo Hills." *Wisconsin Academy Review* 45 (3): 35–36.

Lange, K. I. 1989. *Ancient Rocks and Vanished Glaciers: A Natural History of Devil's Lake State Park, Wisconsin.* Baraboo: Wisconsin Department of Natural Resources.

Maher, L. J., Jr. 1982. "The Palynology of Devil's Lake, Sauk County, Wisconsin." In *Quaternary History of the Driftless Area,* edited by J. C. Knox, L. Clayton, and D. M. Mickelson, 119–35. Wisconsin Geological and Natural History Survey Field Trip Guide Book 5. Madison: University of Wisconsin–Extension.

Mickelson, D. M. 1983. *A Guide to Glacial Landscapes of Dane County, Wisconsin.* Wisconsin Geological and Natural History Survey Field Trip Guide Book 6. Madison: University of Wisconsin–Extension.

———. 2007. *Landscapes of Dane County.* Wisconsin Geological and Natural History Survey Educational Series 43. Madison: University of Wisconsin–Extension.

Miller, J. 2000. "Glacial Stratigraphy and Chronology of Central Southern Wisconsin, West of the Rock River." M.S. thesis, University of Wisconsin–Madison.

Muir, J. 1915. *Travels in Alaska.* New York: Houghton Mifflin.

Thwaites, F. T. 1958. "Land Forms of the Baraboo District, Wisconsin." *Transactions of the Wisconsin Academy of Sciences, Arts and Letters* 47:137–59.

## Western Green Bay Lobe Ice Age Trail Segments

Attig, J. W., D. M. Mickelson, and L. Clayton. 1989. "Late Wisconsin Landform Distribution and Glacier-bed Conditions in Wisconsin." *Sedimentary Geology* 62:399–405.

Black, R. F. 1974. *Geology of Ice Age National Scientific Reserve of Wisconsin.* National Park Service Scientific Monograph Series 2. Washington, DC: National Park Service.

Clayton, J. A., and Knox, J. C. 2008. "Catastrophic Flooding from Glacial Lake Wisconsin." *Geomorphology* 93:384–97.

Clayton, L. 1986. *Pleistocene Geology of Portage County, Wisconsin.* Wisconsin Geological and Natural History Survey, Information Circular 56. Madison: University of Wisconsin–Extension.

———. 1987. "Pleistocene Geology of Adams County, Wisconsin." Wisconsin Geological and Natural History Survey, Report of Investigations 59. Madison: University of Wisconsin–Extension.

Clayton, L., and J. W. Attig. 1987. *Glacial Lake Wisconsin.* Geological Society of America Memoir 173. Boulder, CO: Geological Society of America.

Clayton, L., J. W. Attig, and D. M. Mickelson. 1999. "Tunnel Channels Formed in Wisconsin during the Last Glaciation." In *Glacial Processes Past and Present*, edited by D. M. Mickelson and J. W. Attig, 69–82. Geological Society of America Special Paper 337. Boulder, CO: Geological Society of America.

Colgan, P. M. 1999. "Reconstruction of the Green Bay Lobe, Wisconsin, United States, from 26,000 to 13,000 Radiocarbon Years B.P." In *Glacial Processes Past and Present*, edited by D. M. Mickelson and J. W. Attig, 137–50. Geological Society of America Special Paper 337. Boulder, CO: Geological Society of America.

Hooyer, T. S. 2007. *Late-glacial History of East-central Wisconsin: Guide Book for the 53rd Midwest Friends of the Pleistocene Field Conference, May 18–20, 2007, Oshkosh, Wisconsin.* Wisconsin Geological and Natural History Survey, Open-file Report 2007-01. Madison: University of Wisconsin–Extension.

Portage County Ice Age Trail Chapter. 2002. "Field Trips and History to See and Understand our Central Wisconsin Landforms." Portage County Ice Age Trail Chapter.

Thwaites, F. T. 1943. "Pleistocene of Part of Northeastern Wisconsin." *Geological Society of America Bulletin* 54:87–144.

## Northern Green Bay Lobe and Langlade Lobe Ice Age Trail Segments

Attig, J. W., N. R. Ham, and D. M. Mickelson. 1998. *Environments and Processes along the Margin of the Laurentide Ice Sheet in North-central Wisconsin: Guidebook for the 44th Midwest Friends of the Pleistocene Field Conference.* Wisconsin Geological and Natural History Survey, Open-file Report 1998-01. Madison: University of Wisconsin–Extension.

Attig, J. W., D. M. Mickelson, and L. Clayton. 1989. Late Wisconsin Landform Distribution and Glacier-bed Conditions in Wisconsin." *Sedimentary Geology* 62:399–405.

Attig, J. W., and M. A. Muldoon. 1989. *Pleistocene Geology of Marathon County, Wisconsin.* Wisconsin Geological and Natural History Survey, Information Circular 65. Madison: University of Wisconsin–Extension.

Ham, N. R., and J. W. Attig. 1997. *Pleistocene Geology of Lincoln County, Wisconsin.* Wisconsin Geological and Natural History Survey, Bulletin 93. Madison: University of Wisconsin–Extension.

Mickelson, D. M. 1986. *Glacial and Related Deposits of Langlade County.* Wisconsin Geological and Natural History Survey, Information Circular 52. Madison: University of Wisconsin–Extension.

Mode, W. N., and J. W. Attig. 1988. "Pleistocene Geology of the Marathon County Area of Central Wisconsin." *Geoscience Wisconsin* 12:25–44.

Nelson, A. R., and D. M. Mickelson. 1977. "Landform Distribution and the Genesis in the Langlade and Green Bay Glacial Lobes, North Central Wisconsin." *Transactions of the Wisconsin Academy of Sciences, Arts and Letters* 65:41–57.

Thwaites, F. T. 1943. "Pleistocene of Part of Northeastern Wisconsin." *Geological Society of America Bulletin* 54:87–144.

## Wisconsin Valley Lobe Ice Age Trail Segments

Attig, J. W. 1993. *Pleistocene Geology of Taylor County, Wisconsin.* Wisconsin Geological and Natural History Survey, Bulletin 90. Madison: University of Wisconsin–Extension.

Attig, J. W., N. R. Ham, and D. M. Mickelson, 1998. *Environments and Processes along the Margin of the Laurentide Ice Sheet in North-central Wisconsin: Guidebook for the 44th Midwest Friends of the Pleistocene Field Conference.* Wisconsin Geological and Natural History Survey, Open-file Report 1998-01. Madison: University of Wisconsin–Extension.

Attig, J. W., D. M. Mickelson, and L. Clayton. 1989. "Late Wisconsin Landform Distribution and Glacier-bed Conditions in Wisconsin." *Sedimentary Geology* 62:399–405.

Clayton, L., J. W. Attig, N. R. Ham, M. D. Johnson, C. J. Patterson, and K. M. Syverson. 2008. "Ice-walled-lake Plains in the Upper Midwest and Their Relationship to Glacial Hummocks." *Geomorphology* 97:237–48.

Ham, N. R., and J. W. Attig. 1997. *Pleistocene Geology of Lincoln County, Wisconsin.* Wisconsin Geological and Natural History Survey, Bulletin 93. Madison: University of Wisconsin–Extension.

Nelson, A. R., and D. M. Mickelson. 1977. "Landform Distribution and the Genesis in the Langlade and Green Bay Glacial Lobes, North Central Wisconsin." *Transactions of the Wisconsin Academy of Sciences, Arts, and Letters* 65:41–57.

## Chippewa Lobe Ice Age Trail Segments

Attig, J. W. 1993. *Pleistocene Geology of Taylor County, Wisconsin.* Wisconsin Geological and Natural History Survey, Bulletin 90. Madison: University of Wisconsin–Extension.

Black, R. F. 1974. *Geology of Ice Age National Scientific Reserve of Wisconsin.* National Park Service Scientific Monograph Series 2. Washington, DC: National Park Service.

Cahow, A. C. 1976. "Glacial Geomorphology of the Southwest Segment of the Chippewa Lobe Moraine Complex, Wisconsin." Ph.D. diss., Michigan State University–East Lansing.

Clayton, L., J. W. Attig, N. R. Ham, M. D. Johnson, C. E. Jennings, and K. M. Syverson. 2008. "Ice-walled-lake Plains: Implications for the Origin of Hummocky Glacial Topography in Middle North America." *Geomorphology* 97:237–48.

Clayton, L., J. W. Attig, and D. M. Mickelson. 2001. "Effects of Late Pleistocene Permafrost on the Landscape of Wisconsin." *Boreas* 30:173–88.

Holmes, M. A., and K. M. Syverson. 1997. "Permafrost History of Eau Claire and Chippewa Counties, Wisconsin." *The Compass* 73 (3): 91–96.

Johnson, M. D. 1986. *Pleistocene Geology of Barron County*. Wisconsin Geological and Natural History Survey, Information Circular 55. Madison: University of Wisconsin–Extension.

Syverson, K. M. 1998. "Glacial Geology of the Chippewa Moraine Ice Age National Scientific Reserve Unit, Chippewa County, Wisconsin." In *Geology of Western Wisconsin: Guidebook for 61st Annual Tri State Geological Field Conference and University of Wisconsin System Geological Field Conference*, edited by K. M. Syverson and K. G. Havholm, 57–65. Eau Claire: Dept. of Geology, University of Wisconsin–Eau Claire.

Syverson, K. M. 2007. *Pleistocene Geology of Chippewa County, Wisconsin*. Wisconsin Geological and Natural History Survey, Bulletin 103. Madison: University of Wisconsin–Extension.

Syverson, K. M., R. W. Baker, S. Kostka, and M. D. Johnson. 2005. "Pre-Wisconsinan and Wisconsinan Glacial Stratigraphy, History, and Landscape Evolution, Western Wisconsin." In *Field Trip Guidebook for Selected Geology in Minnesota and Wisconsin*, edited by L. Robinson, 238–78. Minnesota Geological Survey Guidebook 21. St. Paul: Minnesota Geological Survey.

Syverson, K. M., P. Bement, H. Bohl, T. Hogue, K. Johnson, M. Koepp-Yarrington, G. Michael, C. Schmidt, S. Toshner, and E. Wieland. 1995. "Hiking Field Trip Guide for Glacial Landforms in the Chippewa Moraine Ice Age National Scientific Reserve, Wisconsin." Unpublished field guide available through the Wisconsin Geological and Natural History Survey and the Chippewa Moraine Visitors' Center.

Ullman, D. J., Carlson, A. E., Syverson, K. M., and Caffee, M. W. 2011. "Enhancing the Deglaciation Chronology of Wisconsin Using In-Situ Cosmogenic Radionuclides: EOS Transactions, Fall AGU, PP21A-1330." Poster presentation, UW–Madison Geoscience Graduate Symposium, April 27–28.

## Superior Lobe Ice Age Trail Segments

Baker, R. W., J. F. Diehl, T. W. Simpson, L. W. Zelazny, and S. Beske-Diehl. 1983. "Pre-Wisconsinan Glacial Stratigraphy, Chronology, and Paleomagnetics of West-central Wisconsin." *Geological Society of America Bulletin* 94:1442–9.

Black, R. F. 1974. *Geology of Ice Age National Scientific Reserve of Wisconsin*. National Park Service Scientific Monograph Series 2. Washington, DC: National Park Service.

Clayton, L. 1984. *Pleistocene Geology of the Superior Region, Wisconsin*. Wisconsin Geological and Natural History Survey, Information Circular 46. Madison: University of Wisconsin–Extension.

Cordua, W. S. 1989. "A Summary of the Bedrock Geology of the Dresser-St. Croix Falls Area, Polk County, Wisconsin and Chisago County, Minnesota." In *Paleogeography and Structure of the St. Croix River Valley: 53rd Annual Tri-State Geological Field Conference, River Falls, Wisconsin, October 13, 14, 15, 1989*, edited by I. S. Williams, 1–8. River Falls: Dept. of Plant and Earth Science, University of Wisconsin.

Johnson, M. D. 1986. *Pleistocene Geology of Barron County*. Wisconsin Geological and Natural History Survey, Information Circular 55. Madison: University of Wisconsin–Extension.

———. 1986. *Pleistocene Geology of Polk County*. Wisconsin Geological and Natural History Survey, Bulletin 92. Madison: University of Wisconsin–Extension.

LaBerge, G. 1994. *Geology of the Lake Superior Region*. Phoenix, AZ: Geoscience Press.

Mudrey, M. G., Jr., G. L. LaBerge, P. E. Meyers, and W. S. Cordua. 1987. *Bedrock Geology of Wisconsin, Northwest Sheet.* Wisconsin Geological and Natural History Survey, Regional Map Series 87-11. Madison: University of Wisconsin–Extension.

Ojakangas, R. W. 2009. *Roadside Geology of Minnesota.* Missoula, MT: Mountain Press.

Ojakangas, R. W., and C. L. Matsch. 1982. *Minnesota's Geology.* Minneapolis: University of Minnesota Press.

Syverson, K. M., R. W. Baker, S. Kostka, and M. D. Johnson. 2005. "Pre-Wisconsinan and Wisconsinan Glacial Stratigraphy, History, and Landscape Evolution, Western Wisconsin." In *Field Trip Guidebook for Selected Geology in Minnesota and Wisconsin,* edited by L. Robinson, 238–78. Minnesota Geological Survey Guidebook 21. St. Paul: Minnesota Geological Survey.

Williams, I. S. 1989. *Paleogeography and Structure of the St. Croix River Valley: 53rd Annual Tri-State Geological Field Conference, River Falls, Wisconsin, October 13, 14, 15, 1989.* River Falls: Dept. of Plant and Earth Science, University of Wisconsin.

# INDEX